Proceedings of the
Steklov Institute of Mathematics
Volume 198 Issue 1 of 4 1994

Proceedings of an All-Union School on the Theory of Functions (Miass, July 1989)

Edited by
S. B. Stechkin

A Translation of

ТРУДЫ
МАТЕМАТИЧЕСКОГО ИНСТИТУТА
имени В. А. СТЕКЛОВА

ТОМ 198 1992

ABSTRACT: This collection consists of papers presented at the All-Union School on the Theory of Functions, which was held in Miass in July 1989. The papers are devoted to topical trends in the theory of functions and the theory of approximation of functions: extremal properties of functions, approximation and interpolation of functions by trigonometric polynomials and splines, widths of function classes, best approximation of operators, cubature formulas, and classical problems of analytic number theory.

1991 *Mathematics Subject Classification.* Primary 11L05, 11P05, 30A05, 33C45, 41-xx, 42A10, 42A20, 46E35.

РОССИЙСКАЯ АКАДЕМИЯ НАУК

ТРУДЫ

МАТЕМАТИЧЕСКОГО ИНСТИТУТА
имени В. А. СТЕКЛОВА

CXCVIII

ТРУДЫ ВСЕСОЮЗНОЙ ШКОЛЫ ПО ТЕОРИИ ФУНКЦИЙ

(Миасс, июль 1989 г.)

Под редакцией
С. Б. СТЕЧКИНА

Ответственный редактор Трудов МИРАН
академик Е. Ф. МИЩЕНКО
Заместитель ответственного редактора Трудов МИРАН
доктор физико-математических наук Е. А. ВОЛКОВ

Рецензенты
доктора физико-математических наук
В. Н. ТЕМЛЯКОВ, В. М. ТИХОМИРОВ

Москва «Наука» 1992

Correspondence between *Trudy Mat. Inst. Steklov* and *Proc. Steklov Inst. Math.*

Russian imprint	Vol.	English imprint	Russian imprint	Vol.	English Imprint	Issue	Russian imprint	Vol.	English Imprint	Issue	Russian imprint	Vol.	English Imprint	Issue
1966	74	1967	1971	109	1974		1979	144	1980	2	1988	179	1989	2
1965	75	1967	1970	110	1972		1980	145	1981	1	1987	180	1989	3
1965	76	1967	1970	111	1972		1978	146	1980	3	1988	181	1989	4
1965	77	1967	1971	112	1973		1980	147	1981	2	1988	182	1990	1
1965	78	1967	1970	113	1972		1978	148	1980	4	1990	183	1991	4
1965	79	1966	1970	114	1974		1978	149	1981	3	1990	184	1991	2
1965	80	1968	1971	115	1974		1979	150	1981	4	1988	185	1990	2
1966	81	1968	1971	116	1973		1980	151	1982	2	1989	186	1991	1
1966	82	1967	1972	117	1974		1980	152	1982	3	1989	187	1990	3
1965	83	1967	1972	118	1976		1981	153	1982	4	1990	188	1991	3
1965	84	1968	1973	119	1976		1983	154	1984	4	1989	189	1990	4
1966	85	1967	1974	120	1976		1981	155	1983	1	1989	190	1992	1
1965	86	1967	1972	121	1974		1980	156	1983	2	1989	191	1992	2
1966	87	1967	1973	122	1975		1981	157	1983	3	1990	192	1992	3
1967	88	1969	1973	123	1975		1981	158	1983	4	1992	193	1993	3
1967	89	1968	1976	124	1978		1983	159	1984	2	1992	194	1993	4
1967	90	1969	1973	125	1975		1982	160	1984	1	1991	196	1992	4
1967	91	1969	1973	126	1975		1983	161	1984	3	1991	197	1993	1
1966	92	1968	1975	127	1977		1983	162	1985	1	1992	198	1994	1
1967	93	1970	1972	128	1974		1984	163	1985	4	1991	200	1993	2
1968	94	1969	1973	129	1976		1983	164	1985	2	1992	201	1994	2
1968	95	1971	1978	130	1979	4	1984	165	1985	3	1993	204	1994	3
1968	96	1970	1974	131	1975		1984	166	1986	1				
1968	97	1969	1973	132	1975		1985	167	1986	2				
1968	98	1971	1973	133	1977		1984	168	1986	3				
1967	99	1968	1975	134	1977		1985	169	1986	4				
1971	100	1974	1975	135	1978		1984	170	1987	1				
1972	101	1975	1975	136	1978		1985	171	1987	2				
1967	102	1970	1976	137	1978		1985	172	1987	3				
1968	103	1970	1975	138	1977		1986	173	1987	4				
1968	104	1971	1980	139	1982	1	1986	174	1988	1				
1969	105	1971	1976	140	1979	1	1986	175	1988	2				
1969	106	1972	1976	141	1979	2	1987	176	1988	3				
1971	107	1976	1976	142	1979	3	1986	177	1988	4				
1968	108	1971	1977	143	1980	1	1986	178	1989	1				

This nonjournal serial does not always appear in strict sequential volume order in the original. Accordingly, the translation will occasionally reflect any gaps resulting thereby, as happens for example with the English imprints of 1984, 1985, and 1990.

CONTENTS

V. V. Arestov, Best approximation of translation invariant unbounded operators by bounded linear operators 1

M.-B. A. Babaev, On the order of approximation of the Sobolev class W_q^r by bilinear forms in L_p for $1 \le q \le 2 \le p \le \infty$ 17

V. M. Badkov, Asymptotic and extremal properties of orthogonal polynomials in the presence of singularities in the weight 37

Yu. S. Vasil′ev, Approximation by splines on an infinite interval 83

S. V. Konyagin, On estimates of Gaussian sums and Waring's problem for a prime modulus 105

O. V. Matveev, Spline interpolation of functions of several variables, and bases in Sobolev spaces 119

A. V. Reztsov, On errors of parallelepiped cubature formulas on classes of differentiable functions 147

V. V. Tarkaev, On divergence of Fourier series in rearranged Price systems ... 155

D. S. Telyakovskiĭ, On asymptotically monogenic functions 177

S. A. Telyakovskiĭ, On approximation of differentiable functions of high smoothness by Fourier sums 183

N. N. Kholshchevnikova, On some thin sets in the theory of functions and topology 203

I. G. Tsar′kov, Widths and an inequality of Jackson type for abstract functions 209

N. I. Chernykh, Jackson's inequality in $L_p(0, 2\pi)$ $(1 \le p < 2)$ with sharp constant 223

V. T. Shevaldin, Lower estimates of the widths of some classes of periodic functions 233

RUSSIAN CONTENTS*

Арестов В. В. Наилучшее приближение неограниченных операторов, инвариантных относительно сдвига, линейными ограниченными операторами 3

Бабаев М.-Б. А. О порядке приближения соболевского класса W_q^r билинейными формами в L_p при $1 \leqslant q \leqslant 2 \leqslant p \leqslant \infty$ 21

Бадков В. М. Асимптотические и экстремальные свойства ортогональных полиномов при наличии особенностей у веса 41

Васильев Ю. С. Приближение сплайнами на бесконечном промежутке 89

Конягин С. В. Об оценках сумм Гаусса и проблеме Варинга по простому модулю 111

Матвеев О. В. Сплайн-интерполяция функций нескольких переменных и базисы в пространствах Соболева 125

Резцов А. В. О погрешностях параллелепипедальных кубатурных формул на классах дифференцируемых функций 153

Таркаев В. В. О расходимости рядов Фурье по переставленным системам Прайса 160

Теляковский Д. С. Об асимптотически моногенных функциях 186

Теляковский С. А. О приближении суммами Фурье дифференцируемых функций высокой гладкости 193

Холщевникова Н. Н. О некоторых тонких множествах в теории функций и топологии 212

Царьков И. Г. Поперечники и неравенства типа Джексона для абстрактных функций 219

Черных Н. И. Неравенство Джексона в $L_p(0,2\pi)$ $(1 \leqslant p < 2)$ с точной константой 232

Шевалдин В. Т. Оценки снизу поперечников некоторых классов периодических функций 242

*The American Mathematical Society scheme for transliteration of Cyrillic may be found at the end of index issues of *Mathematical Reviews*.

Труды Матем. Инст. Стеклова
198 (1992)

Proceedings of the Steklov
Institute of Mathematics
1994, Issue 1

BEST APPROXIMATION OF TRANSLATION INVARIANT UNBOUNDED OPERATORS BY BOUNDED LINEAR OPERATORS

UDC 517.5

V. V. ARESTOV

In this paper we study the problem of best approximation of a translation invariant operator by bounded linear operators on an invariant class of functions in the spaces $L_\gamma(\mathbf{R}^m)$. We shall show that, under certain conditions, this problem reduces to the problem of best approximation of an unbounded functional by bounded functionals on classes of functions in the spaces $L_{r,s} = L_{r,s}(\mathbf{R}^m)$, the duals of which are spaces of multipliers from L_r to L_s. The spaces $L_{r,s}$ are more general than the classical spaces L_γ; for the properties of these spaces see, for example, [1]. Best approximation of functionals is better understood (see [2], and for dual results [3], [4]). On the basis of results from this paper and from [2] we prove that the magnitude of best approximation of the differentiation operator d^k/dt^k in L_γ on the axis $(-\infty, \infty)$ can be expressed in terms of the smallest constant in multiplicative inequalities between the norms of derivatives of functions in $L_{r,s}(-\infty, \infty)$; a solution of both problems is given in a concrete situation.

The problem of best approximation of an unbounded operator (in particular the differentiation operator d^k/dt^k of order k) by bounded operators arose in Stechkin's paper [5]. Subsequently it was studied by L. V. Taĭkov, Yu. N. Subbotin, V. N. Gabushin, V. I. Berdyshev, J. Peetre, B. E. Klots, A. P. Buslaev, the author, and others (see [5]–[8], [2]–[3] and the bibliography therein). A solution of the problem was given in a number of concrete cases, and its connection with other extremal problems was explained. In particular, this problem is connected with the ill-posed problem of best linear recovery of the values of an unbounded operator on the elements of a class given with known error; namely, it offers the possibility of estimating minimal error, above and below, and selecting a good bounded linear method of recovery (for details see [5], [6], [9], [8]).

1

Let $L_\gamma = L_\gamma(\mathbf{R}^m)$, $m \geq 1$, $1 \leq \gamma \leq \infty$, be the space of measurable functions x on $\mathbf{R}^m$ for $1 \leq \gamma < \infty$, with (finite) norm

$$\|x\|_\gamma = \left\{ \int_{\mathbf{R}^m} |x(t)|^\gamma \, dt \right\}^{1/\gamma},$$

and for $\gamma = \infty$ the space $C_0 = C_0(\mathbf{R}^m)$ of continuous functions vanishing at infinity, with the uniform norm $\|x\|_\infty = \|x\|_{C_0} = \max\{|x(t)| : t \in \mathbf{R}^m\}$. In the sequel p, q, r, s will be parameters satisfying the conditions $1 \leq p, q, r, s \leq \infty$; if γ is any one of these parameters, then γ' will be its (Hölder) conjugate, $1/\gamma + 1/\gamma' = 1$. For $1 \leq p, q \leq \infty$ we denote by $\mathscr{L}_p^q$ the set of bounded linear operators from

1991 *Mathematics Subject Classification.* Primary 41A35, 41A50; Secondary 46E30.

L_p to L_q, and by $\mathcal{T}_p^q$ the set of operators $T \in \mathcal{L}_p^q$ invariant with respect to any translation, i.e., operators T such that $\tau_h T = T\tau_h$ for any $h \in \mathbf{R}^m$, where τ_h is a translation operator defined by the relation $(\tau_h x)(t) = x(t-h)$.

Let, further, $\mathcal{S}$ be the space of fast decaying infinitely differentiable functions on $\mathbf{R}^m$ (the space of test functions), and $\mathcal{S}'$ the corresponding space of generalized functions (see, for example, [10], [11], [1]). The value of a functional (generalized function) $\theta \in \mathcal{S}'$ on an element $x \in \mathcal{S}$ will be denoted by $\langle \theta, x\rangle$. The Fourier transform of a function x in L_1 (and, in particular, in $\mathcal{S}$) is defined by the formula

$$\hat{x}(t) = \int x(\eta) e^{-2\pi i \langle \eta, t\rangle}\, d\eta;$$

here and below, in integrals over $\mathbf{R}^m$ the set of integration is not indicated. The inverse Fourier transform is then defined by the formula

$$\tilde{x}(t) = \int x(\eta) e^{2\pi i \langle \eta, t\rangle}\, d\eta.$$

The Fourier transform of a functional $\theta \in \mathcal{S}'$ is a functional $\hat{\theta} \in \mathcal{S}'$ acting by the formula $\langle \hat{\theta}, x\rangle = \langle \theta, \hat{x}\rangle$, $x \in \mathcal{S}$. If $\theta \in L_\gamma$, $1 \le \gamma \le 2$, then $\hat{\theta} \in L_{\gamma'}$, $1/\gamma + 1/\gamma' = 1$. The convolution $\theta * x$ of elements $\theta \in \mathcal{S}'$ and $x \in \mathcal{S}$ is the function $y(\eta) = \langle \theta, \sigma_\eta x\rangle$, where the operation σ_η, $\eta \in \mathbf{R}^m$, is defined by the formula $(\sigma_\eta x)(t) = x(\eta - t)$. For $\theta \in \mathcal{S}'$ we denote by $\check{\theta}$ the functional acting by the formula $\langle \check{\theta}, x\rangle = \langle \theta, \sigma_0 x\rangle = \langle \theta, \check{x}\rangle$, $\check{x}(t) = x(-t)$, $x \in \mathcal{S}$, i.e., this is the operation of change of sign of the argument.

With the aid of an element $a \in \mathcal{S}'$ we define on $\mathcal{S}$ an operator A and its formal adjoint A^* by the formulas

$$Ax = \check{a} * x, \qquad A^* x = a * x, \qquad x \in \mathcal{S}. \tag{1.1}$$

Let us impose on a the condition that $a * x \in \mathcal{S}$ for any $x \in \mathcal{S}$, i.e., that $A^*\mathcal{S} \subset \mathcal{S}$ and $A\mathcal{S} \subset \mathcal{S}$; this is the case, for example, if a has compact support. Let us now extend the operator A to a wider set according to S. L. Sobolev's scheme. Denote by $\Sigma = \Sigma(\mathbf{R}^m)$ the set of measurable, locally integrable functions x on $\mathbf{R}^m$, for each of which there exists an exponent $\mu = \mu(x) \le 0$ with the property

$$\int |x(t)|(1+|t|^2)^\mu\, dt < \infty. \tag{1.2}$$

It is clear that $\Sigma \subset \mathcal{S}'$. For a pair of functions $x, y \in \Sigma$ we say that x belongs to the domain $D(A)$ of the operator A and $y = Ax$, if

$$\int x A^* \varphi\, dt = \int y\varphi\, dt \qquad \forall \varphi \in \mathcal{S}. \tag{1.3}$$

It is not hard to see that the operator A is invariant with respect to translations on $D(A)$. With the aid of yet another element $b \in \mathcal{S}'$ we define a pair (of formally adjoint) operators B and B^* in the same way as for the operators A and A^*; here we also assume that b has the property that $B^*\mathcal{S} \subset \mathcal{S}$ and, hence, $B\mathcal{S} \subset \mathcal{S}$ as well. We introduce the space

$$W_r^p(B) = \{x \in L_r \cap D(B) : Bx \in L_p\} \tag{1.4}$$

of functions x from the domain $D(B)$ of the operator B that lie in L_r and are such that $Bx \in L_p$. In the set $W_r^p(B)$ we select the class

$$Q = Q_r^p(B) = \{x \in W_r^p(B) : \|Bx\|_p \le 1\}. \tag{1.5}$$

In this paper we discuss Stechkin's problem [5] on best approximation of the operator A on the class Q in the metric of the space L_q by the set $\mathscr{L}_r^s(N)$ of bounded linear operators T from L_r to L_s with $\|T\| \le N$, i.e., we investigate the quantity

$$E(N) = E(A, B; r, s; p, q; N) = \inf\{U(T) : T \in \mathscr{L}_r^s(N)\}, \tag{1.6}$$

where

$$U(T) = \sup\{\|Ax - Tx\|_q : x \in Q_r^p(B)\}.$$

In our subsequent work we will have to consider problems of type (1.6) on narrower classes of functions than (1.5), (1.4). Let Φ be a linear subspace in $D(B)$, invariant with respect to translations. We introduce the subspace

$$\Phi_r^p = \Phi_r^p(B) = \{x \in L_r \cap \Phi : Bx \in L_p\} \tag{1.7}$$

of the space $W_r^p(B)$. In $\Phi_r^p(B)$ we distinguish the class of functions

$$Q(\Phi) = Q_r^p(\Phi, B) = \{x \in \Phi_r^p(B) : \|Bx\|_p \le 1\} \tag{1.8}$$

and consider the problem of best approximation of the operator A on this class by bounded operators; more precisely, we study the quantity

$$E(\Phi, N) = E(A, B; \Phi; r, s; p, q; N) = \inf\{U(\Phi, T) : T \in \mathscr{L}_r^s(N)\}, \tag{1.9}$$

where

$$U(\Phi, T) = \sup\{\|Ax - Tx\|_q : x \in Q_r^p(\Phi, B)\}. \tag{1.10}$$

Obviously, (1.6) is a particular case of (1.9).

We assume that Φ, and thus also $Q(\Phi)$, is translation invariant. As was proved in [12] (for generalizations see [8], [13]), under this condition we may restrict our attention to invariant operators T in (1.9), and thus also in (1.6), i.e., the equality

$$E(\Phi, N) = \inf\{U(\Phi, T) : T \in \mathscr{T}_r^s(N)\} \tag{1.11}$$

holds, where

$$\mathscr{T}_r^s(N) = \mathscr{L}_r^s(N) \cap \mathscr{T}_r^s = \{T \in \mathscr{T}_r^s : \|T\| \le N\}.$$

2

It is known (see [10], [11]) that if $s < r$ then the set $\mathscr{T}_r^s$ consists of the single operator $T \equiv 0$, and that if $s \ge r$ then on $\mathscr{S}$ every operator $T \in \mathscr{T}_r^s$ has the form of a convolution:

$$Tx = \theta * x, \qquad x \in \mathscr{S}, \tag{2.1}$$

with a certain element $\theta = \theta(T) \in \mathscr{S}'$ (and can be extended by continuity from $\mathscr{S}$ to L_r). The set $M(r, s)$ of generalized functions $\theta = \theta(T)$ such that the convolution (2.1) defines a bounded operator T from L_r to L_s (i.e., the set of multipliers from L_r to L_s) is a Banach space with the norm $\|\theta\|_{M(r,s)} = \|\theta(T)\|_{M(r,s)} = \|T\|_{L_r \to L_s}$. Below, when speaking of the spaces $M(r, s)$ we assume that $1 \le r \le s \le \infty$. Choose the parameter ρ from the condition $1/r - 1/s = 1 - 1/\rho$; we have $1 \le \rho \le \infty$. For functions $\theta \in L_\rho$, $x \in L_r$ the Young inequality

$$\|\theta * x\|_s \le c(r, s)\|\theta\|_\rho \|x\|_r \tag{2.2}$$

holds with the best constant [14]

$$c(r, s) = \alpha_r \alpha_\rho \alpha_s \le 1, \qquad \alpha_\gamma = (\gamma^{1/\gamma}(\gamma')^{-1/\gamma'})^{m/2}. \tag{2.3}$$

This implies the imbedding

$$L_p \subset M(r, s), \qquad 1/p = 1 - (1/r - 1/s), \tag{2.4}$$

with

$$\|\theta\|_{M(r,s)} \le c(r, s)\|\theta\|_p, \qquad \theta \in L_p. \tag{2.5}$$

Note that, moreover, $M(r, s) = M(s', r')$ (together with equality of the norms of the elements), and if $1/p = (1-\delta)/r + \delta/s'$, $1/q = (1-\delta)/s + \delta/r'$, $0 < \delta < 1$, then there is the imbedding $M(r, s) \subset M(p, q)$. The structure of the spaces $M(2, 2)$ and $M(r, \infty) = M(1, r')$ is known; namely [10], [11],

$$M(2, 2) = \widehat{L}_\infty = \{\hat{\theta} : \theta \in L_\infty\}, \tag{2.6}$$

$$M(r, \infty) = L_{r'} \quad \text{for } 1 \le r < \infty, \qquad M(\infty, \infty) = M(1, 1) = \mathscr{V}, \tag{2.7}$$

where $\mathscr{V}$ is the space of bounded Borel measures on $\mathbf{R}^m$.

The space $M(r, s)$ is the dual of the Banach function space $L(r, s) = L_{r,s}$, constructed as follows (see [1] and the references given there). If $1 \le r \le s < \infty$, then $L_{r,s}$ consists of functions $h \in L_\omega$, $1/\omega = 1/r - 1/s$, of the form

$$h = \sum x_j * y_j, \qquad x_j \in L_r, \quad y_j \in L_{s'}, \quad \sum \|x_j\|_r \|y_j\|_{s'} < \infty, \tag{2.8}$$

and the norm in $L_{r,s}$ is defined by the formula

$$\|h\|_{r,s} = \|h\|_{L_{r,s}} = \inf \sum \|x_j\|_r \|y_j\|_{s'}.$$

Here the infimum is taken over all representations (2.8). If $s = \infty$, then, as is easily realized from (2.7), one should take

$$L_{\infty,\infty} = C_0, \qquad L_{r,\infty} = L_r \quad \text{for } 1 \le r < \infty. \tag{2.9}$$

In all cases, $L_{r,s}$ is a subspace in L_ω, $1/\omega = 1/r - 1/s$, contains the set $L_r * L_{s'} = \{x * y : x \in L_r, y \in L_{s'}\}$, and

$$\|x\|_\omega \le \|x\|_{r,s}, \qquad x \in L_{r,s}, \tag{2.10}$$

$$\|x * y\|_{r,s} \le \|x\|_r \|y\|_{s'}, \qquad x \in L_r, \ y \in L_{s'}. \tag{2.11}$$

Lemma 1. *The set* $\mathscr{S}$ *is dense in* $L_{r,s}$ *for all* r, s $(1 \le r \le s \le \infty)$.

Indeed, if $s = \infty$ this follows from (2.9). For $s < \infty$ the linear hull of $L_r * L_{s'}$ is dense in $L_{r,s}$ by the definition of $L_{r,s}$. In view of (2.11) the set $\mathscr{S} * \mathscr{S} \subset \mathscr{S}$ is dense in $L_r * L_{s'}$ (relative to the norm of $L_{r,s}$). The set $\mathscr{S}$ is imbedded in $L_r * L_{s'}$, which is easily seen by representing the elements $\varphi \in \mathscr{S}$ in the form

$$\varphi = g * G\varphi, \tag{2.12}$$

where $g(t) = e^{-2\pi|t|}$, $t \in \mathbf{R}^m$, and $G\varphi = \chi * \varphi$, $\chi(t) = (\hat{g}(t))^{-1} = c_m(1+|t|^2)^{(m+1)/2}$, where c_m is some constant. Hence we conclude that $\mathscr{S}$ is dense in $L_{r,s}$.

Corollary. *The space* $L_{r,s}$ *is the completion of* $\mathscr{S}$ *with respect to the norm*

$$\|x\|_{r,s} = \sup\{|\langle\theta, x\rangle| : \theta \in M(r, s), \|\theta\| \le 1\}. \tag{2.13}$$

We note several facts that involve the spaces $L_{r,s}$ and are corollaries of the corresponding properties of the spaces $M(r, s)$. The estimate

$$\|x\|_\omega \le c(r, s)\|x\|_{r,s}, \quad x \in L_{r,s}, \tag{2.14}$$

sharper than (2.10), follows from the relations (2.4), (2.5), (2.13). In view of the equality $M(r, s) = M(s', r')$ and Corollary 1 we have also the equality

$$L_{r,s} = L_{s',r'}. \tag{2.15}$$

On the basis of (2.6) we conclude that

$$L_{2,2} = \widetilde{L}_1 = \{x = \tilde{y},\ y \in L_1\}, \tag{2.16}$$

where

$$\|x\|_{2,2} = \|y\|_1, \qquad x = \tilde{y} \in L_{2,2} \quad (y \in L_1). \tag{2.17}$$

If $1 \le p \le r \le s \le \infty$ and the parameter ρ is chosen from the condition $1 - 1/\rho = 1/p - 1/r$, then it follows easily from the definition of the spaces $L_{r,s}$ that the imbedding

$$L_\rho * L_{r,s} \subset L_{p,s} \tag{2.18}$$

and the inequality

$$\|y * x\|_{p,s} \le \|y\|_\rho \|x\|_{r,s}, \qquad y \in L_\rho, \quad x \in L_{r,s}, \tag{2.19}$$

take place.

For convenience of reference, one more property of the spaces $L_{r,s}$ is given in the form of a lemma.

Lemma 2. *Let $x \in L_{r,s}$, $\theta \in L^*_{r,s} = M(r, s)$. Then $\theta * x \in C$, and if $1/r - 1/s < 1$ then $\theta * x \in C_0$.*

Proof. Assume, first, that $x \in \mathscr{S}$. Using the representation (2.12) we may write $\theta * x = g * \theta * Gx$. In this relation $g \in L_1 \cap C_0$, $\theta * Gx \in L_s \cap L_{r'}$, and, therefore, $\theta * x \in C$ if $r = 1$, $s = \infty$, and $\theta * x \in C_0$ in the remaining cases. Let $x \in L_{r,s}$. By Lemma 1 there exists a sequence $\{x_n\}$ of elements $x_n \in \mathscr{S}$ converging in $L_{r,s}$ to x. The assertion of Lemma 2 follows now from the obvious inequality $|\theta * x - \theta * x_n| \le \|\theta\|\, \|x - x_n\|_{r,s}$.

3

The space

$$W_r^p = W_r^p(B) = \{x \in L_r \cap D(B) : Bx \in L_p\} \tag{3.1}$$

is a Banach space with respect to the norm

$$\|x\|_{W_r^p} = \|x\|_r + \|Bx\|_p; \tag{3.2}$$

this follows easily from the relation

$$\int xB^*\varphi\, dt = \int \varphi Bx\, dt, \qquad \varphi \in \mathscr{S}, \tag{3.3}$$

defining the element Bx, and the assumption $B\mathscr{S} \subset \mathscr{S}$. Denote by $\mathring{W}_r^p$ the closure of the set $\mathscr{S}$ in W_r^p.

We introduce, further, the Banach space of functions

$$W_{r,s}^{p,q} = W_{r,s}^{p,q}(B) = \{x \in L_{r,s} \cap D(B) : Bx \in L_{p,q}\} \tag{3.4}$$

with the norm

$$\|x\|_{W_{r,s}^{p,q}} = \|x\|_{r,s} + \|Bx\|_{p,q}, \tag{3.5}$$

and denote by $\mathring{W}_{r,s}^{p,q}$ the closure of $\mathscr{S}$ in $W_{r,s}^{p,q}(B)$.

Let $D_C(A)$ be the set of functions $x \in D(A)$ for which the function Ax is continuous; the assumption $A\mathscr{S} \subset \mathscr{S}$ implies that $\mathscr{S} \subset D_C(A)$. If $x \in \mathscr{S}$, then $\langle a, x\rangle = Ax(0)$. Let us agree to use this relation for the definition of $\langle a, x\rangle$ also in the case when $x \in D_C(A)$. For the linear subspace $\Phi \subset D(B)$ we define the class

$$\Lambda = \Lambda(\Phi) = \{x \in \Phi \cap W_{r,s}^{p;q} : \|Bx\|_{p,q} \leq 1\}$$

and introduce the quantity

$$e(\Phi, N) = \inf\{u(\Phi, \theta) : \theta \in M(r, s),\ \|\theta\| \leq N\}, \tag{3.6}$$

where

$$u(\Phi, \theta) = u(\theta) = \sup\{\langle a, x\rangle - \langle\theta, x\rangle : x \in \Lambda(\Phi)\} \tag{3.7}$$

if $\Lambda(\Phi) \subset D_C(A)$, and $u(\Phi, \theta) = \infty$ otherwise.

In [15] we proved, with the aid of the assertion (1.11), that if $Y = B\mathscr{S}$ is dense in L_p, then

$$E(\mathscr{S}, N) = E(\mathring{W}_r^p(B), N), \tag{3.8}$$

$$e(\mathscr{S}, N) = e(\mathring{W}_{r,s}^{p;q}(B), N), \tag{3.9}$$

$$E(\mathring{W}_r^p(B), N) = e(\mathring{W}_{r,s}^{p;q}(B), N). \tag{3.10}$$

Concerning the density of $B\mathscr{S}$ in L_p, the following rather simple assertion can be made.

Lemma 3. *$B\mathscr{S}$ is dense in L_p if and only if the operator B has the property*

$$h \in L_{p'} \cap C_0,\ Bh = 0 \Rightarrow h = 0 \tag{3.11}$$

if $p > 1$, and the property

$$h \in C,\ Bh = 0 \Rightarrow h = 0 \tag{3.12}$$

if $p = 1$.

Proof. For the linear subspace $B\mathscr{S} \subset L_p$ to be dense in L_p it is necessary and sufficient that for any $f \in L_p^*$ the condition $f * B\eta = 0$, $\eta \in \mathscr{S}$, imply $f = 0$. Taking the convolution of the function $f*B\eta = 0$ with a function $\varphi \in \mathscr{S}$, we obtain the relation $(f * \varphi) * B\eta = 0$, $\eta, \varphi \in \mathscr{S}$. This relation signifies that the function $h = f * \varphi$ satisfies the equality $Bh = 0$. In this circumstance $h = f * \varphi \in L_p^* \cap C^\infty$ and, hence, $h \in L_{p'} \cap C_0$ for $p > 1$, and $h \in C$ for $p = 1$. Under the conditions of the lemma, it now follows that $h = 0$, and, hence, also $f = 0$, i.e., that $B\mathscr{S}$ is dense in L_p.

Let $B\mathscr{S}$ be dense in L_p. Since $L_{p'} \cap C \subset L_p^*$, the properties (3.11), (3.12) are satisfied a fortiori. The lemma is proved.

We give sufficient conditions for $\mathscr{S}$ to be dense in $W_{r,s}^{p;q}(B)$.

Lemma 4. *Let at least one of the following three assumptions be satisfied:*

(α_1) $1 \leq r \leq p \leq q \leq s \leq \infty$,

(α_2) $1/p - 1/q < 1$ *and the operator B has the property*

$$h \in C_0,\ Bh = 0 \Rightarrow h = 0, \tag{3.13}$$

(α_3) *the parameters p, q, r, s are arbitrary, but the operator B has the property*

$$h \in C,\ Bh = 0 \Rightarrow h = 0. \tag{3.14}$$

Then the set $\mathscr{S}$ *is dense in* $W^{p;q}_{r,s}(B)$, *i.e.,*

$$\overset{\circ}{W}{}^{p;q}_{r,s}(B) = W^{p;q}_{r,s}(B). \tag{3.15}$$

Proof. The set $\mathscr{S}$ is a linear subspace in $W = W^{p;q}_{r,s}(B)$. In view of the Hahn-Banach theorem it is dense in W if and only if for a functional $f \in W^*$ the condition $\langle f, \mathscr{S}\rangle = 0$ (i.e., $\langle f, \varphi\rangle = 0$ for all $\varphi \in \mathscr{S}$) implies $f = 0$. A functional $f \in W^*$ is determined by a pair of elements $f_1 \in L^*_{r,s} = M(r, s)$, $f_2 \in L^*_{p,q} = M(p, q)$; here $\langle f, x\rangle = \langle \check{f}_1, x\rangle + \langle \check{f}_2, Bx\rangle$, $x \in W$. The condition $\langle f, \mathscr{S}\rangle = 0$ means that $\langle \check{f}_1, \varphi\rangle + \langle \check{f}_2, B\varphi\rangle = 0$, $\varphi \in \mathscr{S}$, or, which is the same,

$$f_1 * \varphi + f_2 * B\varphi = 0, \qquad \varphi \in \mathscr{S}. \tag{3.16}$$

Let $x \in W$. Put $g = f_1 * x + f_2 * Bx$. By Lemma 2 we have $g \in C$. We need to show that $g = 0$. It suffices to prove that for any function $\varphi \in \mathscr{S}$ the function

$$g * \varphi = (f_1 * x) * \varphi + (f_2 * Bx) * \varphi \tag{3.17}$$

is equal to zero. Let us first verify that

$$(f_1 * x) * \varphi = f_1 * (x * \varphi) = (f_1 * \varphi) * x. \tag{3.18}$$

For $\varepsilon > 0$ we choose a function $\eta \in \mathscr{S}$ such that $\|x - \eta\|_{r,s} < \varepsilon$. Since $f_1 \in L^*_{r,s} \subset \mathscr{S}'$, and $\eta, \varphi \in \mathscr{S}$, it follows that

$$(f_1 * \eta) * \varphi = f_1 * (\eta * \varphi) = (f_1 * \varphi) * \eta. \tag{3.19}$$

In addition, the following relations hold:

$$|f_1 * \eta - f_1 * x| = |f_1 * (\eta - x)| \le \|f_1\| \, \|\eta - x\|_{r,s} \le \varepsilon \|f_1\|,$$
$$\|\eta * \varphi - x * \varphi\|_{r,s} \le \|\varphi\|_1 \|\eta - x\|_{r,s} \le \varepsilon \|\varphi\|_1,$$
$$\|f_1 * \varphi\|_{M(r,s)} \le \|f_1\|_{M(r,s)} \|\varphi\|_1,$$

and therefore the corresponding parts in (3.18) and (3.19) are close, and both equalities in (3.18) are verified. By the same argument, we have the equality

$$(f_2 * Bx) * \varphi = f_2 * (Bx * \varphi).$$

By the definition of the function Bx we have $Bx * \varphi = x * B\varphi$, and thus

$$f_2 * (Bx * \varphi) = f_2 * (B\varphi * x). \tag{3.20}$$

It follows from (3.16) that (for $\varphi \in \mathscr{S}$)

$$f_2 * B\varphi \in L^*_{p,q} \cap L^*_{r,s}, \tag{3.21}$$

and so the expression $(f_2 * B\varphi) * x$ is meaningful. Assume that we have already proved the equality

$$f_2 * (B\varphi * x) = (f_2 * B\varphi) * x. \tag{3.22}$$

Then (3.18), (3.20), (3.22), and (3.16) would imply the equality

$$g * \varphi = (f_1 * \varphi + f_2 * B\varphi) * x = 0,$$

i.e. that $\mathscr{S}$ is dense in W. Thus, to prove the lemma it suffices to establish (3.22).

Put

$$h = f_2 * (B\varphi * x) - (f_2 * B\varphi) * x; \tag{3.23}$$

by Lemma 2 we have $h \in C$. Let us show that $Bh = 0$, i.e., $h * B\eta = 0$ for $\eta \in \mathscr{S}$. Using the relation $x * B\eta = Bx * \eta \in L_{p,q} \cap L_{r,s}$ and (3.21), we obtain

$$(f_2 * (B\varphi * x)) * B\eta = f_2 * ((B\varphi * x) * B\eta) = f_2 * (B\varphi * (x * B\eta))$$
$$= (f_2 * B\varphi) * (x * B\eta) = ((f_2 * B\varphi) * x) * B\eta.$$

The coincidence of the first and the last function in this relation signifies that $Bh = 0$. From this we can make the deduction that if the third assumption in the lemma is satisfied, then $\mathscr{S}$ is dense in W.

Let $1/p - 1/q < 1$. Then, by Lemma 2, the first function on the right in (3.23) belongs to C_0. The second function there also has that same property. Indeed, if $\psi \in \mathscr{S}$, then $f_2 * B\varphi * \psi \in C_0$, since $f_2 * B\varphi \in L_{p,q}$. Approximating the function x by the functions $\psi \in \mathscr{S}$ in the norm of $L_{r,s}$ and using the fact that $f_2 * B\varphi \in L^*_{r,s}$, we see that $(f_2 * B\varphi) * x \in C_0$. Thus, if $1/p - 1/q < 1$, then $h \in C_0$. Consequently, if the second assumption in the lemma is satisfied, again $\mathscr{S}$ is dense in W.

Now let $\psi = \eta * \zeta$, $\eta, \zeta \in \mathscr{S}$. Then $(f_2 * (B\varphi * x)) * \psi = (f_2 * \psi) * (B\varphi * x)$. If $r \le p \le q \le s$, then $f_2 * \psi = \eta * f_2 * \zeta \in L^*_{r,s} = M(r, s)$, and therefore

$$(f_2 * \psi) * (B\varphi * x) = (f_2 * \psi * B\varphi) * x = \psi * ((f_2 * B\varphi) * x).$$

This relation means that $h * \psi = 0$, and so $h = 0$. Thus the assertion of the lemma holds under the first assumption as well. Lemma 4 is completely proved.

Since $W^{1,p'}_{1,r'} = W^{p,\infty}_{r,\infty} = W^p_r$, Lemma 4 provides sufficient conditions for $\mathscr{S}$ to be dense in $W^p_r(B)$. At present, many results of this kind are known (see the relevant material in the books [16] and [17]). We make two more comments.

Remark 1. If the condition

(α_4) $1/p - 1/r \le 1/m$ and B is a differentiation operator with constant coefficients

holds, then, as is easily verified [15], the set $\mathscr{S}$ is dense in $W^p_r(B)$.

Remark 2. In the general case $\mathscr{S}$ is not dense in $W^p_r(B)$, and so the conclusion of Lemma 4 (the equality (3.15)) does not follow without additional assumptions.

Assume, for example, that the operator B has the following properties: the characteristic function κ of the operator B belongs to L_γ, $1/\gamma = 1-(1/p-1/r)$, $r \ge p$, and the homogeneous equation $Bz = 0$ has a nonzero solution $z \in L_r$. Then for $x \in \mathscr{S}$ we have $x = \kappa * Bx$, and therefore the inequality

$$\|x\|_r \le \|\kappa\|_\gamma \|Bx\|_p \tag{3.24}$$

holds on the set $\mathscr{S}$, and so also on the set $\overset{\circ}{W}{}^p_r(B)$.

The function z belongs to $W^p_r(B)$, but (3.24) is violated on the function z; hence $z \notin \overset{\circ}{W}{}^p_r(B)$ and (3.15) does not hold. Let us give a specific example of such a situation. Let Δ be the Laplace operator and I the identity operator. Let us show that the operator

$$B = (\Delta - I)^l (\Delta + I) \tag{3.25}$$

possesses the indicated properties for

$$m > 1, \quad l > m/2, \quad r > 2m/(m-1), \quad 1/p - 1/r > (m+1)/2m. \tag{3.26}$$

Let J_ν, Y_ν be the Bessel functions of the first and second kind, $\nu = m/2 - 1$. On the radial functions $x(t) = f(|t|)$ the equation $(\Delta + I)x = 0$ has the form

$$f''(\rho) + \frac{m-1}{\rho} f'(\rho) + f(\rho) = 0, \qquad \rho = |t|.$$

Therefore, the function $z(t) = |t|^{-\nu} J_\nu(|t|)$ is a solution of the equation $(\Delta + I)z = 0$, and thus of the equation $Bz = 0$ as well. Since $J_\nu(\rho) \sim \rho^{-1/2}$ as $\rho \to \infty$, it follows that $z \in L_r$ for $r > 2m/(m-1)$. The characteristic function κ_0 of the operator $\Delta + I$ coincides, up to a constant, with the function $|t|^{-\nu} J_{-\nu}(|t|)$ or $|t|^{-\nu} Y_\nu(|t|)$ for m even or m odd, respectively (see [18], Chapter III, §11.8). The characteristic function g_l of the operator $(\Delta - I)^l$ for $l > m/2$ is continuous and behaves at infinity like $e^{-c|t|}$, $c > 0$. The function $\kappa = g_l * \kappa_0$ will be a fundamental function for the operator B. This function is bounded, its behavior as $t \to \infty$ is determined by that of κ_0, and so $\kappa(t) \sim |t|^{-(m-1)/2}$, $t \to \infty$, and $\kappa \in L_\gamma$, $\gamma > 2m/(m-1)$. It follows that if condition (3.26) is fulfilled, then the set $\mathscr{S}$ is not dense in $W_r^p(B)$ for B from (3.25).

Theorem 1. *If the condition* (α_2) *or* (α_3) *in Lemma* 4 *is satisfied, then*

$$E(\mathscr{S}, N) = E(W_r^p(B), N), \tag{3.27}$$

$$e(\mathscr{S}, N) = e(W_{r,s}^{p;q}(B), N), \tag{3.28}$$

$$e(W_{r,s}^{p;q}(B), N) = E(W_r^p(B), N). \tag{3.29}$$

Indeed, if (α_2) or (α_3) is satisfied, then (3.15) holds. Under these conditions the assumptions (3.11), (3.12) in Lemma 3 are a fortiori satisfied, and hence $B\mathscr{S}$ is dense in L_p. Now Theorem 1 follows from (3.8)–(3.10).

4

We assume that $A\mathscr{S} \subset \mathscr{S}$ and therefore we may define the convolution $\zeta * a$ of elements $\zeta \in \mathscr{S}'$ with a by $\langle \zeta * a, \varphi \rangle = \langle \zeta, \check{a} * \varphi \rangle$, $\varphi \in \mathscr{S}$. Suppose that there is an element $h \in \mathscr{S}'$ such that

$$a = h * b. \tag{4.1}$$

Under this supposition we have

$$A = H \circ B, \tag{4.2}$$

where the linear operator H is the convolution with the element h. The decomposition (4.1) holds, for example, in the case when there is a fundamental function κ for the operator B^*, i.e., there exists an element $\kappa \in \mathscr{S}'$ such that $\kappa * b$ is a δ-function; in this case one can take

$$h = a * \kappa = A^* \kappa \tag{4.3}$$

in (4.1).

Denote by $M(N) = M_{r,s}(B^*, N)$ the set of functionals $\zeta \in \mathscr{S}'$ such that $B^*\zeta = \zeta * b \in M(r, s)$ and $\|B^*\zeta\| \le N$. Finally, we define the quantity

$$\varepsilon(N) = \varepsilon(h; r, s; p, q; N) = \inf\{\|h - \zeta\|_{M(p,q)} : \zeta \in M_{r,s}(B^*, N)\}; \tag{4.4}$$

this is the best approximation of the element h by the set $M(N)$.

Lemma 5.

$$\varepsilon(N) = e(\mathscr{S}, N). \tag{4.5}$$

Proof. We first prove the inequality $e(\mathscr{S}, N) \le \varepsilon(N)$. We may consider $\varepsilon(N)$ finite. Let a functional $\zeta \in \mathscr{S}'$ have the property $B^*\zeta = \zeta * b \in M(r, s)$, $h - \zeta \in M(p, q)$. Then for $x \in \mathscr{S}$ we have

$$\langle h - \zeta, Bx \rangle = \langle h - \zeta, \check{b} * x \rangle = \langle h * b, x \rangle - \langle \zeta * b, x \rangle = \langle a, x \rangle - \langle \theta, x \rangle,$$

where $\theta = \zeta * b = B^*\zeta$. It follows that

$$u(\mathscr{S}, \theta) \le \|h - \zeta\|_{M(p,q)},$$

and this implies the inequality $e(\mathscr{S}, N) \le \varepsilon(N)$.

Let us prove the reverse inequality. We consider $e(\mathcal{S}, N)$ finite. Let a functional $\theta \in M(r, s)$ be such that $u(\mathcal{S}, \theta) < \infty$. Then the following representation ([15], Lemma 2.1) holds:

$$(4.6) \qquad \langle a, x\rangle - \langle \theta, x\rangle = \langle f, Bx\rangle, \qquad x \in \mathcal{S},$$

where f is a bounded linear functional on $B\mathcal{S}$ with the norm of $L_{p,q}$ and $\|f\| = u(\mathcal{S}, \theta)$. Denote by f also the extension of this functional to the whole of $L_{p,q}$ with preservation of the norm. It follows from (4.1) and (4.6) that

$$\langle \theta, x\rangle = \langle h - f, \check{b} * x\rangle = \langle (h - f) * b, x\rangle,$$

i.e., $\theta = B^*\zeta$, $\zeta = h - f$. We have $\|h - \zeta\| = \|f\| = u(\mathcal{S}, \theta)$ and, consequently, $\varepsilon(N) \le e(\mathcal{S}, N)$. The lemma is proved.

Lemma 5 and Theorem 1 permit us to express the following claim.

Theorem 2. *If condition* (4.1) *and one of the conditions* (α_2) *or* (α_3) *are satisfied, then*

$$(4.7) \qquad E(W_r^p(B), N) = e(W_{r,s}^{p;q}(B), N) = \varepsilon(N).$$

Denote by $\mathcal{E}(N)$ the common value of the quantities in (4.7). In view of the definition (4.4) the following assertion is valid.

Corollary 2. *Under the conditions of Theorem 2, for any functional* $\zeta \in \mathcal{S}'$ *with the property* $B^*\zeta \in M(r, s)$ *the following (sharp) inequality holds:*

$$(4.8) \qquad \|h - \zeta\|_{M(p,q)} \ge \mathcal{E}(\|B^*\zeta\|_{M(r,s)}).$$

Remark 3. If $B\mathcal{S}$ is dense in L_p (an assumption weaker than that of Theorem 2), then Lemma 5 and the equalities (3.27), (3.28) imply the inequality

$$(4.9) \qquad \varepsilon(N) \le E(N) = E(W_r^p(B), N).$$

We give an example in which the inequality (4.9) is strict; this, in particular, means that in the general case Theorem 2 does not hold. We make use of the example of Remark 2. Suppose that the characteristic function κ for the operator B belongs to L_γ, $1/\gamma = 1 - (1/p - 1/q)$, and the homogeneous equation $Bz = 0$ has a nonzero solution $z \in L_r \cap L_s$. Consider the problem of best approximation of the identity operator $A = I$ on the class $Q_r^p(B)$. Thus, let

$$(4.10) \qquad E(N) = \inf_{\|T\|_{L_r \to L_s} \le N} \ \sup_{x \in Q_r^p(B)} \|x - Tx\|_q.$$

In this case we have the representation (4.1) with $h = \kappa$, so that

$$(4.11) \qquad \varepsilon(N) = \inf\{\|\kappa - \zeta\|_{M(p,q)} : \zeta \in M_{r,s}(B^*, N)\}.$$

For any N we have $\varepsilon(N) \le \|\kappa\|_{M(p,q)} \le \|\kappa\|_\gamma < \infty$. On the other hand, for sufficiently small N, $E(N) = \infty$. Indeed, the function z belongs to L_r and $Bz = 0$; therefore, $cz \in Q_r^p(B)$ for any c. It follows from this that if the deviation $U(T) = \sup\{\|x - Tx\|_q : x \in Q_r^p(B)\}$ is finite, then $z - Tz = 0$ and, therefore,

$$\|T\|_{L_r \to L_s} \ge N(z) = \|z\|_s \|z\|_r^{-1};$$

consequently, $E(N) = \infty$ for $N < N(z)$. Thus, in this case $N < N(z)$ and the inequality (4.9) is strict. More concretely, it would be possible to take the operator (3.25), and assume that $q = s = r$ and that conditions (3.26) hold. If we assume, in addition, that $p \ge 2$, then $B\mathcal{S}$ will be dense in L_p (see [15], the remark to Theorem 2.1).

5

We consider one more problem connected with the (functional) problem (3.6). We distinguish the class

$$\Lambda = \Lambda_{r,s}^{p;q}(B) = \{x \in W_{r,s}^{p;q}(B) : \|Bx\|_{p,q} \leq 1\} \tag{5.1}$$

in the space $W_{r,s}^{p;q}(B) = \{x \in L_{r,s} \colon Bx \in L_{p,q}\}$ and define a function ω on the semiaxis $[0, \infty)$ by setting

$$\omega(\delta) = \sup\{|\langle a, x\rangle| : x \in \Lambda_{r,s}^{p;q}\}(B), \|x\|_{r,s} \leq \delta\} \tag{5.2}$$

if $\Lambda \subset D_C(A)$, and $\omega(\delta) = \infty$ if $\Lambda \not\subset D_C(A)$. This function is a modulus of continuity of the functional a on the class Λ. Finally, let

$$\Omega(N) = \sup\{\omega(\delta) - N\delta : \delta \geq 0\}. \tag{5.3}$$

The equality

$$e(W_{r,s}^{p;q}(B), N) = \Omega(N)$$

is a consequence of the results of [2]. This fact and Theorem 2 imply the next assertion.

Corollary 3. *Under the conditions of Theorem* 2,

$$E(W_r^p(B), N) = \Omega(N). \tag{5.4}$$

6

We consider in more detail the problem of best approximation of the operator of differentiation of order k on the class of n-times $(0 \leq k < n)$ differentiable functions by bounded linear operators in the spaces $L_\gamma = L_\gamma(-\infty, \infty)$ on the real line, i.e., the problem of studying the quantities

$$E(N) = E_{k,n}(r, s; p, q; N) = \inf\{U(T) \colon T \in \mathscr{L}_r^s(N)\}, \tag{6.1}$$

where

$$U(T) = \sup\{\|x^{(k)} - Tx\|_q : x \in Q_r^p(n)\},$$
$$Q_r^p(n) = \{x \in L_r : x^{(n)} \in L_p, \|x^{(n)}\|_p \leq 1\}.$$

This problem, as mentioned above, has been studied by S. B. Stechkin, L. V. Taĭkov, Yu. N. Subbotin, V. N. Gabushin, V. I. Berdyshev, A. P. Buslaev, the author, and others (see the references in [5], [12], [15], [7], [8]); we note that in the majority of the papers (with the exception of [12], [15]) the case $s = q$ was considered.

If $s < r$, $n - k + 1/q - 1/p > 0$ or $q < p$, $k - 1/s + 1/r > 0$, then $E(N) = \infty$ for any $N \geq 0$. But if $s \geq r$, $q \geq p$, then (excluding the degenerate case $k = 0$, $s = r$) the quantity $E(N)$ is finite and

$$E(N) = E(1)N^{-\gamma}, \qquad \gamma = (n - k + 1/q - 1/p)/(k + 1/r - 1/s). \tag{6.2}$$

The finiteness of $E(N)$ and the formula (6.2) for $s = q \geq \max(p, q)$ were given in [5], [19]. The fact that $E(N) = \infty$ for $s = q < \max(p, r)$ was proved in [20], and for $s \neq q$, $s < r$ or $q < p$, in [12].

Problem (6.1) is a particular case of (1.6) for the operators

$$A = D^k = d^k/dt^k, \qquad B = D^n = d^n/dt^n, \tag{6.3}$$

which are convolution operators and have the property $A\mathscr{S} \subset \mathscr{S}$, $B\mathscr{S} \subset \mathscr{S}$. For the operator $B = D^n$ the set $W_r^p = W_r^p(D^n) = W_r^p(n)$ consists of the functions

$x \in L_r$ for which (the classical derivative $x^{(n-1)}$ of order $n-1$ exists and is locally absolutely continuous, and) $x^{(n)} \in L_p$.

Let $s \geq r$, $q \geq p$. The set $W_{r,s}^{p,q} = W_{r,s}^{p,q}(n)$ consists of the functions $x \in L_{r,s} \subset L_{\rho_1}$, $1/\rho_1 = 1/r - 1/s$, for which $x^{(n)} \in L_{p,q} \subset L_{\rho_r}$, $1/\rho_2 = 1/p - 1/q$; we have $W_{r,s}^{p,q} \subset W_{\rho_1}^{\rho_2}$. In the present case $\omega(\delta)$ and $\Omega(N)$ take the form

$$\omega(\delta) = \omega_{k,n}(\delta) = \sup\{|x^{(k)}(0)| : \|x\|_{r,s} \leq \delta,\ \|x^{(n)}\|_{p,q} \leq 1\}, \tag{6.4}$$

$$\Omega(N) = \Omega_{k,n}(N) = \sup\{\omega_{k,n}(\delta) - N\delta : \delta \geq 0\}. \tag{6.5}$$

In the same way as in [15] in the proof of Theorem 3.1, it is not difficult to see that the following assertion holds.

Lemma 6. *If* $s \geq r$, $q \geq p$, *and*

$$n + 1/q - 1/p + 1/r - 1/s > 0, \tag{6.6}$$

then

$$K = \omega(1) < \infty, \tag{6.7}$$

$$\omega(\delta) = K\delta^{\alpha}, \tag{6.8}$$

$$\begin{gathered} \Omega(N) = \beta\alpha^{\alpha/\beta} K^{1/\beta} N^{-\gamma}, \\ \alpha = (n - k + 1/q - 1/p)/(n + 1/q - 1/p + 1/r - 1/s), \\ \beta = 1 - \alpha, \qquad \gamma = \alpha/\beta = (n - k + 1/q - 1/p)/(k + 1/r - 1/s). \end{gathered} \tag{6.9}$$

Assertions (6.7) and (6.8) make it possible to derive the following.

Corollary 4. *If* $0 \leq k < n$, $s \geq r$, $q \geq p$, $n + 1/q - 1/p + 1/r - 1/s > 0$, *then the functions in* $W_{r,s}^{p,n}(n)$ *satisfy the inequality*

$$\|x^{(k)}\|_C \leq K \|x\|_{r,s}^{\alpha} \|x^{(n)}\|_{p,q}^{\beta} \tag{6.10}$$

with the finite (and smallest possible) constant $K = \omega(1)$.

The operator $B = D^n$ has a classical characteristic function

$$\kappa_n(t) = t_+^{n-1}/(n-1)!, \qquad t_+ = \max(0, t), \tag{6.11}$$

and, therefore, the representation (4.1) takes place, in which the function h coincides up to sign with the function

$$\kappa_n^{(k)}(t) = \kappa_{n-k}(t) = t_+^{n-k-1}/(n-k-1)!.$$

Consequently, in this case the quantity (4.4) has the form

$$\begin{aligned} \varepsilon(N) &= \varepsilon_{k,n}(N) \\ &= \inf\{\|\kappa_{n-k} - \zeta\|_{M(p,q)} : \zeta \in \mathscr{S}',\ \zeta^{(n)} \in M(r,s),\ \|\zeta^{(n)}\|_{M(r,s)} \leq N\}. \end{aligned} \tag{6.12}$$

Finally, we put

$$\begin{gathered} e_{k,n}(N) = \inf\{u_{k,n}(\theta) : \theta \in M(r,s),\ \|\theta\| \leq N\}, \\ u_{k,n}(\theta) = \sup\{x^{(k)}(0) - \langle\theta, x\rangle : x \in L_{r,s},\ \|x^{(s)}\|_{p,q} \leq 1\}. \end{gathered} \tag{6.13}$$

Theorem 3. *If* $s \geq r$, $q \geq p$, $p > 1$, *with* $s > r$ *for* $k = 0$, *then for any* $N > 0$

$$E_{k,n}(N) = \varepsilon_{k,n}(N) = e_{k,n}(N) = \beta\alpha^{\alpha/\beta} K^{1/\beta} N^{-\alpha/\beta}, \tag{6.14}$$

where K *is the smallest constant in inequality* (6.10).

Proof. If $\varphi \in C_0$ and $\varphi^{(n)} = 0$, then $\varphi = 0$; moreover, we require that $p > 1$. This means that in this case condition (α_2) of Lemma 4 is satisfied. Applying (4.7), (5.4), and (6.9), we obtain the assertion of Theorem 3.

The first equality in (6.14) was proved (directly, without using Theorem 2) in our earlier paper [12]; the analog of this equality for the periodic case was obtained in [21].

For functions in the class $W_r^p(n) = \{x \in L_r, x^{(n)} \in L_p\}$ the following inequality holds:

$$\|x^{(k)}\|_p \le G\|x\|_r^\alpha \|x^{(n)}\|_p^\beta,$$
$$\alpha = (n-k+1/q-1/p)/(n-1/p+1/r), \qquad \beta = 1-\alpha. \tag{6.15}$$

Such inequalities have been studied by G. H. Hardy, J. E. Littlewood, E. Landau, J. Hadamard, B. Sz.-Nagy, A. N. Kolmogorov, S. B. Stechkin, L. V. Taĭkov, V. N. Gabushin, N. P. Kuptsov, A. P. Buslaev, G. G. Magaril-Il'yaev, V. M. Tikhomirov, and others (see the bibliography in [22]). Stechkin [5] observed that $E_{k,n}(N)$ (with $s = q$) gives an upper bound for the best constant G in (6.15); more precisely,

$$G \le \left(\frac{N}{\alpha}\right)^\alpha \left(\frac{E_{k,n}(N)}{\beta}\right)^\beta. \tag{6.16}$$

It was proved in [23] that

$$G < \infty \Leftrightarrow \frac{n-k}{r} + \frac{k}{p} \ge \frac{n}{q}.$$

On the other hand, as noted above,

$$E_{k,n}(N) < \infty \Leftrightarrow s \ge r, \ q \ge p.$$

Thus, the finiteness conditions for G and $E_{k,n}(N)$ are different. It is known that in the following three cases: $q = \infty$ [2], $p = q = r = 2$ [24], $p = q = r = 1$ [25], the inequality (6.16) becomes an equality. However, this is not always so, even if $s = q \ge \max(p, r)$, when both sides in (6.16) are finite. The inequality (6.16) is probably always strict, except for the above-listed cases. We note that [26] is devoted to the question of compatibility of the problems (6.1) and (6.15) in the periodic case.

In view of (6.14), the right-hand side of (6.16) is equal to K, and, therefore, (6.16) is equivalent to the inequality $G \le K$ between the best constants in (6.10) and (6.15) for $s = q$. The inequality (6.15) with constant K, i.e., the inequality

$$\|y^{(k)}\|_q \le K\|y\|_r^\alpha \|y^{(n)}\|_p^\beta, \qquad y \in W_r^p(n), \tag{6.17}$$

can be obtained directly from (6.10). In fact, if $q = \infty$, then (6.10) and (6.15) coincide. We suppose that $q < \infty$. Let $y \in W_r^p(n)$, $y \not\equiv 0$. Set $z = |y^{(k)}|^{q-1} \operatorname{sgn} y^{(k)}$, $x = y * \check{z}$. Then

$$z \in L_{q'}, \qquad \|z\|_{q'} = \|y^{(k)}\|_q^{q-1}, \qquad x \in W_{r,q}^{p,q}(n),$$
$$x^{(n)} = y^{(n)} * \check{z}, \qquad x^{(k)}(0) = \|x^{(k)}\|_C = \|y^{(k)}\|_q^q,$$

and

$$\|x\|_{r,q} \le \|z\|_{q'}\|y\|_r, \tag{6.18}$$
$$\|x^{(n)}\|_{p,q} \le \|z\|_{q'}\|y^{(n)}\|_p. \tag{6.19}$$

Substituting the function x in (6.10) and applying the last three relations, we obtain the inequality (6.17) for y.

It was proved in [27] that if $(n-k)/r+k/p > n/q$ and $p > 1$, then there exists a (nonzero) extremal function. It follows from the proof of (6.17) that if $G = K$ (i.e., the constants in (6.10), (6.15) coincide) or, which is the same, the inequality (6.16) becomes an equality, then both (6.18) and (6.19) become equalities on the extremal function y for (6.15).

Theorem 4. *The inequality* (6.16) *is strict in the following three cases*:

1) $q = r = 2,\ 1 \le p < 2,\ 0 < k < n,$
2) $q = p = 2,\ 1 \le r < 2,\ 0 \le k < n,$
3) $r = 1,\ 1 \le q < \infty,\ 1 < p \le \infty,\ 0 \le k < n.$

To prove the theorem it suffices to show that in each of the cases listed at least one of the inequalities (6.18), (6.19) is strict on any function $y \in W_r^p(n)$. By (2.16) we have $\|x\|_{2,2} = \|\hat{x}\|_1 = \|\hat{y}\hat{z}\|_1 \le \|\hat{y}\|_2\|\hat{z}\|_2$. The last inequality becomes an equality only when $\hat{y}$ and $\hat{z}$ are proportional. Since for $q = 2$ we have $z = y^{(k)}$, it follows that $\hat{z}(t) = (2\pi i t)^k \hat{y}(t)$, and this implies the nonexistence of a function $y \in L_2$, $y \not\equiv 0$, such that $\hat{y}$ and $\hat{z}$ are proportional. Thus, for $q = r = 2$ the inequality (6.18) is strict on any function $y \in W_2^2(n)$, and the assertion of the theorem is true in the first case. It can be verified similarly that if $q = p = 2$, then (6.19) is strict on any function $y \in W_r^2(n)$. It remains to consider the third case. By (2.9) we have $\|x\|_{1,q} = \|x\|_{q',\infty} = \|x\|_{q'}$, and therefore (6.18) takes the form $\|y * \check{z}\|_{q'} \le \|y\|_1\|\check{z}\|_{q'}$. It is known that the last inequality is strict on functions $y \in L_1$, $z \in L_{q'}$. Theorem 4 is proved.

To conclude of this section we present the analog of Corollary 3. Namely, (6.12), (6.14) imply

Corollary 5. *If* $s \ge r$, $q \ge p$, $p > 1$, *and* $s > r$ *for* $k = 0$, *then for any* $\zeta \in \mathcal{S}'$ *the following* (*sharp*) *inequality holds*:

$$\|\zeta^{(n)}\|^{\alpha}_{M(r,s)}\|\kappa_{n-k} - \zeta\|^{\beta}_{M(p,q)} \ge \alpha^{\alpha}\beta^{\beta}K = (\varepsilon_{k,n}(1))^{\beta}, \tag{6.20}$$

where K *is the smallest constant in* (6.10).

An inequality of this type (for $q = s = \infty$, $p = r = 2$) was first given by Taĭkov [28], and then (for $q = s = \infty$ and arbitrary p, r) by the author [7]; we note that for $s = q = \infty$ K is the smallest constant in (6.15).

7

Let us calculate all the quantities in (6.14) for

$$k = 0, \quad n = 2, \quad s = 4/3, \quad r = 1, \quad p = q = 2. \tag{7.1}$$

It is easy to see with the aid of (2.7), (2.8) that if $p = q = 2$, $r = 1$, $1 \le s \le 2$, then the following formula is valid for (6.12) (for a detailed proof, see [15]):

$$\varepsilon(N) = (2\pi)^{k-n} \inf\left\{\left\|\frac{t^k - \hat{\eta}(t)}{t^n}\right\|_{\infty} : \|\eta\|_s \le (2\pi)^{-k}N\right\}. \tag{7.2}$$

From this moment on we assume that the parameters satisfy the conditions (7.1). We first find an upper bound for $\varepsilon(N)$. For the function

$$\eta(t) = \frac{1}{3h^2\pi^3}\left(\frac{\sin 2\pi h t}{t}\right)^3, \qquad h > 0,$$

we have (see [29], p. 19, formula (10))

$$\hat{\eta}(t) = \frac{8}{2h^2\pi}\int_0^{\infty}\left(\frac{\sin hu}{u}\right)^3 \cos ut\, du = \begin{cases} 1 - t^2/3h^2, & 0 \le t \le h, \\ (3h-t)^2/6h^2, & h \le t \le 3h, \\ 0, & t \ge 3h; \end{cases}$$

in addition, the function η is even. It is not difficult to obtain from here that the function $\chi(t) = (1-\hat{\eta}(t))t^{-2}$ has the value $1/3h^2$ on $[-h, h]$ and that $|\chi(t)| \le 1/3h^2$ for all values of t, so that $\|(1-\hat{\eta}(t))t^{-2}\|_\infty \le 1/3h^2$. We calculate $\|\eta\| = \|\eta\|_{4/3}$. We have

$$\|\eta\| = \frac{1}{3h^2\pi}\left(2\int_0^\infty \left(\frac{\sin 2\pi t h}{t}\right)^4 dt\right)^{3/4},$$

and since (see [29], p. 20)

$$\int_0^\infty \left(\frac{\sin t}{t}\right)^4 dt = \frac{\pi}{3}, \tag{7.3}$$

we have $\|\eta\| = \frac{8}{9}(3h)^{1/4}$. Using (7.2), we obtain the estimate

$$\varepsilon(N_h) \le \frac{1}{12h^2\pi^2}, \qquad N_h = \tfrac{8}{9}(3h)^{1/4}, \qquad h > 0.$$

Let us estimate $\varepsilon(N_h)$ from below. Define the function $y = y_h$ by setting $y(t) = 3/(2h^3)$ for $|t| \le h$ and $y(t) = 0$ for $|t| > h$. We have

$$\|t^2 y\|_1 = 1, \qquad \|y\|_1 = 3h^{-2}, \qquad \hat{y}(t) = \frac{3}{2h^3\pi}\frac{\sin 2\pi h t}{t}$$

and, by (7.3), $\|\hat{y}\|_4 = (3h^{-3})^{3/4}$. For any function η with $\|\eta\|_{4/3} \le N_h$

$$\begin{aligned}\left\|\frac{1-\hat{\eta}(t)}{t^2}\right\|_\infty &\ge \left|\int_{-\infty}^\infty \frac{1-\hat{\eta}(t)}{t^2} t^2 y(t)\,dt\right| \ge \int_{-\infty}^\infty y(t)\,dt - \left|\int_{-\infty}^\infty \hat{\eta}(t) y(t)\,dt\right| \\ &= 3h^{-2} - \left|\int_{-\infty}^\infty \eta(t)\hat{y}(t)\,dt\right| \\ &\ge 3h^{-2} - \|\eta\|_s \|\hat{y}\|_{s'} \ge 3h^{-2} - N_h\|\hat{y}\|_4 = 1/3h^2.\end{aligned}$$

Consequently, we can make the following deduction:

If $k = 0$, $n = 2$, $s = 4/3$, $r = 1$, $p = q = 2$, *then*

$$E(N_h) = \varepsilon(N_h) = \frac{1}{12h^2\pi^2}, \qquad N_h = \frac{8}{9}(3h)^{1/4}, \qquad h > 0. \tag{7.4}$$

By (2.10), (2.16), (2.17) the set $W^{2,2}_{1,4/3} = W^{2,2}_{4,\infty}(2)$ consists of the functions $x \in L_4$ such that $x'' = \tilde{y}$, $y \in L_1$. The inequality (6.10) in this case takes the form

$$\|x\|_C \le K(\|x\|_4^8 \|\widehat{x''}\|_1)^{1/9}. \tag{7.5}$$

It can be established on the basis of Theorem 3 and (7.4) that the smallest constant in (7.5) is $K = 3^{1/3}(2\pi)^{-2/9}$. It is easy to verify directly that the inequality (7.5) turns into an equality on the functions $x(t) = c(\sin ht)/t$, where $h > 0$ and c are arbitrary constants.

BIBLIOGRAPHY

1. Ronald Larsen, *An introduction to the theory of multipliers*, Springer-Verlag, Berlin, 1971.
2. V. N. Gabushin, *Best approximation of functionals on certain sets*, Mat. Zametki **8** (1970), 551–562; English transl. in Math. Notes **8** (1970).
3. Jaak Peetre, *Approximation of linear operators*, Constructive Theory of Functions (Proc. Internat. Conf., Varna, 1970; B. Penkov and D. Vachov, editors), Izdat. Bolgar. Akad. Nauk, Sofia, 1972, pp. 245–263.
4. C. A. Micchelli and T. J. Rivlin, *A survey of optimal recovery*, Optimal Estimation in Approximation Theory (Proc. Internat. Sympos., Freudenstadt, 1976; C. A. Micchelli and T. J. Rivlin, editors), Plenum Press, New York, 1977, pp. 1–54.

5. S. B. Stechkin, *Best approximation of linear operators*, Mat. Zametki **1** (1967), 137–148; English transl. in Math. Notes **1** (1967).

6. V. V Arestov, *The uniform regularization of the problem of computing the values of an operator*, Mat. Zametki **22** (1977), 231–244; English transl. in Math. Notes **22** (1977).

7. ———, *On some extremal problems for differentiable functions of one variable*, Trudy Mat. Inst. Steklov. **138** (1975), 3–28; English transl. in Proc. Steklov Inst. Math. **138** (1975).

8. V. N. Gabushin, *Optimal methods of computing the values of the operator Ux if x is given with an error. Differentiation of functions defined with an error*, Trudy Mat. Inst. Steklov. **145** (1980), 63–78; English transl. in Proc. Steklov Inst. Math. **1981**, no. 1 (145).

9. V. K. Ivanov, V. V. Vasin, and V. P. Tanana, *Theory of linear ill-posed problems and its applications*, "Nauka", Moscow, 1978. (Russian)

10. Lars Hörmander, *Estimates for translation invariant operators in L^p-spaces*, Acta Math. **104** (1960), 93–140.

11. Elias M. Stein and Guido Weiss, *Introduction to Fourier analysis on Euclidean spaces*, Princeton Univ. Press, Princeton, NJ, 1971.

12. V. V. Arestov, *Approximation of translation-invariant operators*, Trudy Mat. Inst. Steklov. **138** (1975), 43–70; English transl. in Proc. Steklov Inst. Math. **138** (1975).

13. ———, *Approximation of invariant operators*, Mat. Zametki **34** (1983), 9–29; English transl. in Math. Notes **34** (1983).

14. William Beckner, *Inequalities in Fourier analysis of R^n*, Proc. Nat. Acad. Sci. U.S.A. **72** (1975), 638–641.

15. V. V. Arestov, *Approximation of operators of convolution type by bounded linear operators*, Trudy Mat. Inst. Steklov. **145** (1980), 3–19; English transl. in Proc. Steklov Inst. Math. **1981**, no. 1 (145).

16. Elias M. Stein, *Singular integrals and differentiability properties of functions*, Princeton Univ. Press, Princeton, NJ, 1970.

17. O. V. Besov, V. P. Il′in, and S. M. Nikol′skiĭ, *Integral representations of functions and imbedding theorems*, "Nauka", Moscow, 1975; English transl., Vols. 1, 2, Wiley, New York, 1979.

18. V. S. Vladimirov, *Equations of mathematical physics*, 4th ed., "Nauka", Moscow, 1981; English transl. of 1st ed., Marcel Dekker, New York, 1971.

19. V. V. Arestov, *On the best uniform approximation of differentiation operators*, Mat. Zametki **5** (1969), 273–284; English transl. in Math. Notes **5** (1969).

20. V. N. Gabushin, *On the best approximation of the differentiation operator in the L_p-metric*, Mat. Zametki **12** (1972), 531–538; English transl. in Math. Notes **12** (1972).

21. B. E. Klots, *Linear deviations of the classes $\widetilde{W}_p^\alpha$ and approximations in spaces of multipliers*, Mat. Zametki **18** (1975), 97–108; English transl. in Math. Notes **18** (1975).

22. V. M. Tikhomirov and G. G. Magaril-Il′yaev, *Inequalities for derivatives*, in A. N. Kolmogorov, *Selected works*. Vol. I: *Mathematics and mechanics*, "Nauka", Moscow, 1985, pp. 387–390; English transl., Kluwer, Dordrecht, 1991.

23. V. N. Gabushin, *Inequalities for norms of a function and its derivatives in L_p-metrics*, Mat. Zametki **1** (1967), 291–298; English transl. in Math. Notes **1** (1967).

24. Yu. N. Subbotin and L. V. Taĭkov, *Best approximation of a differentiation operator in the space L_2*, Mat. Zametki **3** (1968), 157–164; English transl. in Math. Notes **3** (1968).

25. V. V. Arestov, *On the best approximation of differentiation operators in the uniform metric*, Candidate's dissertation, Moscow, 1969. (Russian)

26. B. E. Klots, *Best linear and nonlinear approximations of smooth functions*, Funktsional. Anal. i Prilozhen. **12** (1978), no. 1, 15–24; English transl. in Functional Anal. Appl. **12** (1978).

27. A. P. Buslaev, G. G. Magaril-Il′yaev, and V. M. Tikhomirov, *The existence of extremal functions in inequalities for derivatives*, Mat. Zametki **32** (1982), 823–834; English transl. in Math. Notes **32** (1982).

28. L. V. Taĭkov, *Inequalities of Kolmogorov type and best formulas for numerical differentiation*, Mat. Zametki **4** (1968), 233–238; English transl. in Math. Notes **4** (1968).

29. A. Erdélyi et al., *Tables of integral transforms*, Vol. 1, McGraw-Hill, New York, 1954.

Translated by S. IVANOV

Труды Матем. Инст. Стеклова
198 (1992)

Proceedings of the Steklov
Institute of Mathematics
1994, Issue 1

ON THE ORDER OF APPROXIMATION OF THE SOBOLEV CLASS W_q^r BY BILINEAR FORMS IN L_p FOR $1 \le q \le 2 \le p \le \infty$

UDC 517.51

M.-B. A. BABAEV

In this paper we study approximation of functions of several variables by sums of pair products of functions of fewer variables. First we introduce the necessary notation and definitions. Consider the space $L_p = L_p(K)$ of functions $f = f(x)$, $x = (x_1, \dots, x_l) \in \mathbf{R}^l$, defined on the cube $K = I^l$, $I = [0, 1]$, such that

$$\|f\|_p = \left(\int_K |f(x)|^p\,dx\right)^{1/p} < \infty, \qquad 1 \le p < \infty,$$
$$\|f\|_\infty = \operatorname*{ess\,sup}_{x\in K} |f(x)| < \infty, \qquad p = \infty.$$

Let $\Gamma = \{u, v\}$, $u = (x_1, \dots, x_d)$, $v = (x_{d+1}, \dots, x_l)$, $1 \le d < l$. Denote

$$G^M(\Gamma) = G^M(\Gamma, K) = \left\{ g \,\middle|\, g = \sum_{k=1}^M \varphi_k(u)\psi_k(v),\ \varphi_k \in L_p(I^d),\ \psi_k \in L_p(I^{l-d}) \right\}.$$

Let Q be a compact set in L_p. We call the quantity

$$\tau_M = \tau_{M\Gamma}(Q)_p = \tau(Q, G^M(\Gamma), L_p) = \sup_{f\in Q} \inf_{g\in G^M(\Gamma)} \|f - g\|_p$$

the deviation of the compact set Q from the manifold $G^M(\Gamma)$ in L_p. Let r be a positive integer, and $1 \le q \le \infty$. Denote by $W_q^r = W_q^r(K)$ the Sobolev class of functions f on K satisfying the inequalities $\|f\|_q \le 1$ and

$$\|f\|_{L_q^r}^q \overset{\text{def}}{=} \sum_{|\tilde r| = r} \|f^{(\tilde r)}(x)\|_q^q \le 1,$$

where

$$f^{(\tilde r)}(x) = \frac{\partial^{|\tilde r|}}{\partial x_1^{r_1} \cdots \partial x_l^{r_l}} f, \qquad \tilde r = (r_1, \dots, r_l), \qquad |\tilde r| = r_1 + \dots + r_l,$$

and the summation is over all nonnegative integer vectors $\tilde r$ for which $|\tilde r| = r$. Without loss of generality we shall assume that $d \ge l - d \ge 1$. Further, the order equality $\tau_M \asymp \theta_M$ and the inequality $\tau_M \ll \theta_M$ are equivalent, respectively, to the estimates $C_1\theta_M \le \tau_M \le C_2\theta_M$ and $\tau_M \le C_2\theta_M$ ($M \in \mathbf{N}$) with constants $0 < C_1 \le C_2 < \infty$ independent of M.

The main result of this paper is the following theorem.

Theorem. *Let* $1 \le q \le 2 \le p \le \infty$, $d \ge l - d \ge 1$, $r > l/q - d/p$. *Then*

$$\tau(W_q^r, G^M(\Gamma), L_p) \asymp M^{-\frac{1}{l-d}(r - \frac{d}{q} + \frac{d}{p}) + \frac{1}{p} - \frac{1}{2}}.$$

1991 *Mathematics Subject Classification.* Primary 41A25, 41A63; Secondary 46E35.

In a number of articles V. N. Temlyakov (see, for example, [1], [2], and the monograph [3]) studied the approximation $\tau_{M\Gamma}(F)_p$ of classes of periodic functions of several variables by bilinear forms when $d = l - d$ in $\Gamma = \{u, v\}$. These are the classes $W^{\bar{r}}_{q,\alpha}$, $SW^r_{q,\alpha}$, $H^{\bar{r}}_q$, and NH^r_q of periodic functions, defined by restrictions on the corresponding partial derivatives or by restrictions on the corresponding differences. In particular, for $1 \leq q \leq 2 \leq p \leq \infty$ and $d = l - d$, he found the orders of $\tau_M(F)_p$. Denote by $|\mathfrak{S}|$ the number of variables in the group $\mathfrak{S}$. We give one of those results (in our notation) for isotropic classes.

Theorem T (Temlyakov [1]). *Let F be one of the classes $SW^r_{q,\alpha}$ or NH^r_q of functions of $l = 2d$ variables and let $|u| = |v| = d$. Then for $1 \leq q < 2 < p \leq \infty$, $r > l/q$ the following relation holds*:

$$\tau_{M\Gamma}(F) \asymp M^{-r/d+1/q-1/2}.$$

In [4] Temlyakov obtained such estimates in various mixed norms for anisotropic Sobolev and Nikol′skiĭ classes, and applied them to estimating Kolmogorov widths of the images of the unit balls of L_q spaces in the L_p metric for integral operators with kernels in these classes.

In the theorem concerning the class $W^r_q(I^l)$ the groups of variables u and v are arbitrary. As seen from the formulation of the theorem, the specific character of the groups of variables of different length is manifested also in that the order of $\tau_M(W^r_q)_p$ depends on p.

Proof of the theorem. In the case $p = 2$ the assertion of the theorem follows from our earlier result [5]: for $1 \leq q \leq p \leq 2$ and $r > l/q - l/p$

$$\tau(W^r_q, G^M(\Gamma), L_p) \asymp M^{-\frac{1}{l-d}(r-\frac{d}{q}+\frac{d}{p})}.$$

For that reason we assume further that $1 \leq q \leq 2 \leq p \leq \infty$ and $p \neq q$.

Let us agree that throughout the proof $(\int_S |f(x)|^p\,dx)^{1/p}$ and $(\sum_{i\in\sigma} |a_i|^p)^{1/p}$ for $p = \infty$ mean $\operatorname{ess\,sup}_{x\in S} |f(x)|$ and $\sup_{i\in\sigma} |a_i|$, respectively. We shall denote by $\rho(a, \mathscr{L}, X)$ the distance from an element $a \in X$ to a set $\mathscr{L} \subset X$ in the Banach space X:

$$\rho(a, \mathscr{L}, X) = \inf_{y\in\mathscr{L}} \|a - y\|_X.$$

Recall that the Kolmogorov widths of a subset W of a Banach space X are the quantities

$$d_n(W, X) = \inf_{\mathscr{L}_n} \sup_{a\in W} \rho(a, \mathscr{L}_n, X),$$

where the infimum is taken over all subspaces of X of dimension n.

To be able to estimate τ_M from below, we shall need two lemmas. Let B^m_q be the unit ball in the space l^m_q.

Lemma 1. *If $n \leq m$ and $2 \leq p \leq \infty$, then*

$$d_n(B^m_1, l^m_p) \geq m^{1/p-1/2}(1 - n/m)^{1/2}. \tag{1}$$

This estimate follows directly from Stechkin's equality [6]

$$d_n(B^m_1, l^m_2) = (1 - n/m)^{1/2} \qquad (n \leq m)$$

and the inequality

$$\|\overline{x}\|_{l^m_2} \leq m^{1/2-1/p} \|\overline{x}\|_{l^m_p} \qquad (2 \leq p \leq \infty).$$

In the case $n/m \leq \theta < 1$, (1) implies the estimate

$$d_n(B^m_1, l^m_p) \geq C(\theta) m^{1/p-1/2} \qquad (2 \leq p \leq \infty) \tag{2}$$

with constant $C(\theta) > 0$ independent of m and n. The last inequality may also be found in Kashin's paper [7].

Let $\mu = m^{l-d}$. Choose a positive integer $m \geq 2$ from the condition $\alpha\mu < m < \beta\mu$, where $\alpha = 1/2^{l-d+1}$, $\beta = 1/2$ (this is always possible, since for such α, β the set of intervals $(\alpha\mu, \beta\mu)$ covers the interval $(1/2, \infty)$. Denote by J the set m^l of multi-indices

$$J = J_l = \{\bar{i} = (i_1, \dots, i_l) \mid 1 \leq i_j \leq m, \ j = 1, \dots, l\}.$$

Partition the cube $K = I^l$ $(I = [0, 1])$ into m^l cubes

$$K_{\bar{i}} = \left\{ x \,\middle|\, \frac{i_j - 1}{m} \leq x_j \leq \frac{i_j}{m}, \ j = 1, \dots, l \right\}, \qquad \bar{i} = (i_1, \dots, i_l) \in J,$$

and consider the m^l functions

$$s_{\bar{i}}(x) = \begin{cases} \prod_{j=l}^m \sin^\rho(\pi m x_j), & x \in K_{\bar{i}}, \\ 0, & x \in \mathbf{R}^l \setminus K_{\bar{i}}, \end{cases}$$

in $\mathbf{R}^l$.

Lemma 2. *For arbitrary* $1 \leq p \leq \infty$, $\bar{i} \in J$ *and integer* $r_j \geq 0$ $(j = 1, \dots, l)$ *with* $\sum_{j=1}^l r_j = r < \rho$ *the equality*

$$\left(\int_{K_{\bar{i}}} |s_{\bar{i}}^{(\tilde{r})}(x)|^p \, dx \right)^{1/p} = C_{\tilde{r},p} m^{r-l/p} \tag{3}$$

holds, where $C_{\tilde{r},p}$ *is a constant dependent only on* l, ρ, $\tilde{r}$, *and* p.

The proof of Lemma 2 is carried out by the change of variables $mx_j = t_j$ $(j = 1, \dots, l)$ in the integral in (3), after which we obtain the equality (3) with

$$C_{\tilde{r},p} = \left(\int_0^1 \cdots \int_0^1 \left| \left(\prod_{j=1}^l \sin^\rho \pi t_j \right)^{(\tilde{r})} \right|^p dt_1 \cdots dt_l \right)^{1/p}.$$

A lower estimate for $\tau_{M\Gamma}(W_q^r)_p$

Consider $\mu = m^{l-d}$ cubes $K_{\bar{i}}$ with indices of the form

$$\bar{i} = (i_1, i_2, \dots, i_{l-d}; \ 1, 1, \dots, 1; \ i_1, i_2, \dots, i_{l-d}). \tag{4}$$

These are cubes containing the $(l-d)$-dimensional diagonal plane in the $2(l-d)$-dimensional base $x_{l-d+1} = \cdots = x_d = 1$ of the cube I^l defined by the system $x_1 = x_{d+1}, \dots, x_{l-d} = x_l$. Define on K a function $s(x)$, setting it equal to $b_m s_{\bar{i}}(x)$ on every such $K_{\bar{i}}$ and zero at the remaining points. Here $b_m = C^* m^{-r+d/q}$ (C^* will be determined later). Analytically this can be written as follows: denote by $J^* \subset J$ the set of indices of the form (4), where $i_j = 1, \dots, m$ $(j = 1, \dots, l-d)$. Then

$$s(x) = b_m \sum_{\bar{i} \in J^*} s_{\bar{i}}(x) = b_m \sum_{i_{\overline{1,l-d}}=1} s_{\bar{i}}(x), \qquad \bar{i} \in J^*,$$

where

$$i_{\overline{1,l-d}} = (i_1, \dots, i_{l-d}), \qquad \sum_{i_{\overline{1,l-d}}=1}^m = \sum_{i_1=1}^m \cdots \sum_{i_{l-d}=1}^m.$$

Let us show that $s(x) \in W_q^r(K)$. Since for $|\tilde{r}| < \rho$ the derivatives $s_{\bar{i}}^{(\tilde{r})}(x)$ are continuous everywhere in K, for this it suffices to verify the norm conditions of the class. Using Lemma 2 and the fact that the supports of the functions $s_{\bar{i}}^{(\tilde{r})}(x)$ do not intersect, for $C^* \leq (\sum_{|\tilde{r}|=r} C_{\tilde{r},q}^q)^{-1/q}$ in b_m we obtain

$$\begin{aligned}\|s(x)\|_{L_q^r(K)}^q &= \sum_{|\tilde{r}|=r} b_m^q \sum_{\bar{i}\in J^*} \|s_{\bar{i}}^{(\tilde{r})}(x)\|_{L_q(K)}^q = b_m^q \sum_{|\tilde{r}|=r}\sum_{\bar{i}\in J^*} \int_{K_{\bar{i}}} |s_{\bar{i}}^{(\tilde{r})}(x)|^q\,dx \\ &= b_m^q m^{rq-l} \sum_{|\tilde{r}|=r} C_{\tilde{r},q}^q \sum_{i_{\overline{1,l-d}}=1} 1 \leq m^{rq-l} m^{-rq+d} m^{l-d} = 1.\end{aligned}$$

Analogously, if $C^{*q} \leq C_{0,q}^{-q}$ then

$$\|s(x)\|_{L_q(K)}^q = b_m m^{-l} m^{l-d} C_{0,q}^q = m^{-rq+d} m^{-d} C_{0,q}^q C^{*q} \leq m^{-rd} \leq 1.$$

Consequently, for

$$C^* \leq \min\left\{ C_{0,q}^{-1}, \left(\sum_{|\tilde{r}|=r} C_{\tilde{r},q}^q\right)^{-1/q}\right\}$$

in the definition of b_m and $s(x)$, we have $s(x) \in W_q^r(K)$.

Let us now estimate the distance from $s(x)$ to the set $G^M(\Gamma)$ in $L_p(K)$. Denote by $\xi_{\bar{i}} = (\xi_{i_1}, \ldots, \xi_{i_l})$ the corner of the cube $K_{\bar{i}}$ closest to the point $(0, \ldots, 0)$ (i.e., $\xi_{i_s} = (i_s - 1)/m$, $s = 1, \ldots, l$). For each $g \in G^M(\Gamma)$, denoting $K_{\bar{1}} = K_{(1,1,\ldots,1)}$, we have

$$\tag{5} \begin{aligned}\|s-g\|_p &= \left(\sum_{\bar{i}\in J} \int_{K_{\bar{i}}} |s(x) - g(x)|^p\,dx\right)^{1/p} \\ &= \left(\sum_{\bar{i}\in J} \int_{K_{\bar{1}}} |s(x+\xi_{\bar{i}}) - g(x+\xi_{\bar{i}})|^p\,dx\right)^{1/p} = \left(\int_{K_{\bar{1}}} I^p(x)\,dx\right)^{1/p},\end{aligned}$$

where $I(x) = (\sum_{\bar{i}\in J} |s(x+\xi_{\bar{i}}) - g(x+\xi_{\bar{i}})|^p)^{1/p}$ $(1 \leq p \leq \infty)$. Using the discretization applied in [8], we estimate $I(x)$ for each fixed $x \in K_{\bar{1}}$ (although $g(x), I(x) \in L_p(K)$, they may be considered defined on the whole of K, extending, if necessary, their definitions arbitrarily on a set of measure zero). Clearly,

$$\tag{6} I(x) \geq \left(\sum_{i_{\overline{1,l-d}}=1}^{m} \left[\sum_{i_{\overline{d+1,l}}=1}^{m} |s(x+\xi_{\bar{i}'}) - g'(x+\xi_{\bar{i}'})|^p\right]\right)^{1/p},$$

where

$$\bar{i}' = (i_1, \ldots, i_{l-d},\ 1, \ldots, 1,\ i_{d+1}, \ldots, i_l).$$

Since

$$|\sin^\rho \pi m(x_j + \xi_{i_j})| = |\sin^\rho \pi m x_j| \qquad \text{for all } j = 1, \ldots, l,$$

$s(x+\xi_{\bar{i}'}) = b_m s_{\bar{1}}(x) = s(x)$ for $x \in K_{\bar{1}}$ and $\bar{i}' \in J^*$. Therefore, in accordance with the definition of the function $s(x)$ for $x \in K_{\bar{1}}$ we have

$$s(x+\xi_{\bar{i}'}) = \begin{cases} s(x) & \text{if } \bar{i} \in J^*, \text{ i.e., if } i_k = i_{d+k} \text{ in } \bar{i}' \ (k = 1, \ldots, l-d); \\ 0 & \text{in the remaining cases, i.e., when} \\ & \exists k_0 \in \{1, \ldots, l-d\} \text{ such that } i_{k_0} \neq i_{d+k_0} \text{ in } \bar{i}'. \end{cases}$$

Thus, we have $s(x+\xi_{\bar{i}'}) = s(x)\delta_{\bar{i}'}$, where $\delta_{\bar{i}'} = \delta_{i_1 i_{d+1}} \cdots \delta_{i_{l-d} i_l}$ and δ_{ij} is the Kronecker symbol. Taking this into account, from (6) we obtain

(7)
$$\begin{aligned}I(x) &\ge \left(\sum_{i_{\overline{1,l-d}}=1}^{m}\left[\sum_{i_{\overline{d+1,l}}=1}^{m} |s(x)\delta_{\bar{i}'} - g(x+\xi_{\bar{i}'})|^p\right]\right)^{1/p} \\ &= \left(\sum_{i_{\overline{1,l-d}}=1}^{m}\left[\sum_{i_{\overline{d+1,l}}=1}^{m} \left|s(x)\delta_{\bar{i}'} \right.\right.\right. \\ &\qquad \left.\left.\left. - \sum_{k=1}^{M}\varphi_k(x_{\overline{1,d}} + \xi_{i_{\overline{1,l-d}}1_{\overline{1,2d-l}}})\psi_k(x_{\overline{d+1,l}} + \xi_{i_{\overline{d+1,l}}})\right|^p\right]\right)^{1/p},\end{aligned}$$

where

$$x_{\overline{1,d}} = u = (x_1, \dots, x_d), \qquad 1_{\overline{1,2d-l}} = (\underbrace{1, \dots, 1}_{2d-l});$$
$$\xi_{i_{\overline{1,l-d}}1_{\overline{1,2d-l}}} = (\xi_{i_1}, \dots, \xi_{i_{l-d}}, 1, \dots, 1),$$
$$x_{\overline{d+1,l}} = v = (x_{d+1}, \dots, x_l), \qquad \xi_{i_{\overline{d+1,l}}} = (\xi_{i_{d+1}}, \dots, \xi_l), \qquad \xi_i = \frac{i-1}{m}.$$

In the space l_p^μ, $\mu = m^{l-d}$, the vectors $\{e_j\}_{j=1}^\mu$ ($e_j = (0, \dots, 1, \dots, 0)$, with 1 at the jth position) of the natural basis and the coordinates of any vector $y = \{y_k\}_{k=1}^\mu$ in this basis are enumerated by the multi-indices

$$(i_1, \dots, i_{l-d}) = i_{\overline{1,l-d}} \quad \text{and} \quad (i_{d+1}, \dots, i_l) = i_{\overline{d+1,l}}$$
$$(i_j = 1, \dots, m; \; j = 1, \dots, l-d \quad \text{and} \quad j = d+1, \dots, l)$$

respectively, considering $e_{i_{\overline{1,l-d}}}$ to be the vector in l_p^μ whose coordinate with multi-index $i_{\overline{d+1,l}} = i_{\overline{1,l-d}}$ is equal to 1 and the rest are zeros. Then for each fixed multi-index $i_{\overline{1,l-d}}$ the collection of numbers $\delta_{\bar{i}'}$ (where $\bar{i}' = (i_{\overline{1,l-d}}, 1_{\overline{l-d+1,d}}, i_{\overline{d+1,l}})$ and only the multi-index $i_{\overline{d+1,l}}$ changes) coincides with the collection of the coordinates of the vector $e_{i_{\overline{1,l-d}}}$. Therefore, denoting by $\overline{\psi}_k$ ($k = 1, \dots, M$) the vector in l_p^μ whose coordinates with multi-index $i_{\overline{d+1,l}}$ ($i_j = 1, \dots, m$, $j = d+1, \dots, l$) coincide with $\psi_k(v + \xi_{i_{\overline{d+1,l}}})$, the estimate (7) can be rewritten in the form

$$I(x) \ge \left(\sum_{i_{\overline{1,l-d}}=1} \left\| s(x)e_{i_{\overline{1,l-d}}} - \sum_{k=1}^{M}\varphi_k(u + \xi_{i_{\overline{1,l-d}}1_{\overline{1,2d-l}}})\overline{\psi}_k\right\|_{l_p^\mu}^p\right)^{1/p} \qquad (1 \le p \le \infty).$$

Denote by $\Phi_M = \Phi_M(x)$ the linear hull of the vectors $\overline{\psi}_1, \dots, \overline{\psi}_M$. For each $x \in K_{\bar{1}}$, Φ_M is a subspace of dimension $\nu \le M$ in l_p^μ. Since for any $\overline{a} \in \Phi_M$

$$\|s(x)e_{i_{\overline{1,l-d}}} - \overline{a}\|_{l_p^\mu} \ge \rho(e_{i_{\overline{1,l-d}}}, \Phi_M, l_p^\mu)s(x) \qquad (s(x) \ge 0),$$

from the previous estimate we get

$$I(x) \ge s(x)\left(\sum_{i_{\overline{1,l-d}}=1}^{m} \rho^p(e_{i_{\overline{1,l-d}}}, \Phi_M, l_p^\mu)\right)^{1/p}$$

or, returning to the usual enumeration of the natural basis in l_p^μ,

$$(8)\qquad I(x) \geq s(x)\left(\sum_{j=1}^{\mu}\rho^p(e_j, \Phi_M, l_p^\mu)\right)^{1/p} \qquad (x\in K_{\bar{1}},\ 1\leq p\leq\infty),$$

where $\rho(e, \Phi_M, l_p^\mu)$ is the distance from e to Φ_M in l_p^μ. From (5) and (8) we have

$$\|s-g\|_{L_p(K)} \geq \left(\int_{K_{\bar{1}}}\left(\left(\sum_{j=1}^{\mu}\rho^p(e_j, \Phi_M, l_p^\mu)\right)^{1/p}|s(x)|\right)^p dx\right)^{1/p},$$

whence

$$(9)\qquad \tau(W_q^r, G^M(\Gamma), L_p) \geq \left(\int_{K_{\bar{1}}}\left(\sum_{j=1}^{\mu}\rho^p(e_j, \Phi_M, l_p^\mu)\right)^{1/p\cdot p}|s(x)|^p\,dx\right)^{1/p}.$$

Fix an arbitrary $x\in K_{\bar{1}}$. Then, without loss of generality, we may assume that the sequence $\{\rho(e_j, \Phi_M(x), l_p^\mu)\}_{j=1}^{\mu}$ ($\mu = m^{l-d}$) does not increase. Then, since $2M<\mu$ by the definition of the number m,

$$R_{mM} \overset{\text{def}}{=} \left(\sum_{j=1}^{M}\rho^p(e_j, \Phi_M, l_p^\mu)\right)^{1/p} \geq \left(\sum_{j=1}^{[M/2]+1}\rho^p(e_j, \Phi_M, l_p^\mu)\right)^{1/p}$$

$$\geq \rho(e_{[M/2]+1}, \Phi_M, l_p^\mu)\left(\frac{M}{2}\right)^{1/p} \qquad (1\leq p\leq\infty).$$

Denote by $l_{p,M}^\mu$ the subspace in l_p^μ spanned by the vectors $\bar{a} = (a_1, a_2, \dots, a_\mu)$ with $a_1 = a_2 = a_{[M/2]} = 0$, and by $\bar{a}'$ and Φ_M' the projections of the vectors $\bar{a}\in l_p^\mu$ and the subspaces $\Phi_M\subset l_p^\mu$ onto $l_{p,M}^\mu$. The subspace $l_{p,M}^\mu$ with the metric of the space l_p^μ, as usual, may be identified with the space l_p^N ($N=\mu-[M/2]$). Then we have

$$\|e_j-\bar{a}\|_{l_p^\mu} \geq \|e_j'-\bar{a}'\|_{l_p^\mu} = \|e_j'-\bar{a}'\|_{l_p^N},$$

where in the last expression, for $\bar{a} = (a_1, \dots, a_\mu)$, by $\bar{a}'$ we mean $(a_{[M/2]+1}, \dots, a_\mu)$. Therefore,

$$\rho(e_{[M/2]+1}, \Phi_M, l_p^\mu) = \max_{[M/2]+1\leq j\leq\mu}\rho(e_j, \Phi_M, l_p^\mu) \geq \max_{[M/2]+1\leq j\leq\mu}\rho(e_j', \Phi_M', l_p^N).$$

Thus,

$$R_{mM} \geq \left(\frac{M}{2}\right)^{1/p}\max_{[M/2]+1\leq j\leq\mu}\rho(e_j', \Phi_{M/2}', l_p^N),$$

where $\Phi_M' = \Phi_M'(x)$ is a subspace of dimension ν' ($\nu'\leq\nu\leq M$) of the space l_p^N of dimension $N = \mu-[M/2] > 2M-M/2 = \frac{3}{2}M$, and so $\nu'/N < \theta = 2/3$.

For further estimation we use Lemma 1. The ball B_1^N ($\subset l_p^N$) is a polyhedron with vertices at the points $\pm e_j'\in l_p^N$ ($j=[M/2]+1, \dots, \mu$). Therefore, $\{\pm e_j'\}_{j=[M/2]+1}^{\mu}$ is the total collection of extreme points of the set B_1^N. From this by the Krein-Mil′man theorem (see [9], p. 21), for any subspace $L_{\nu'}\subset l_p^N$,

$$\sup_{x\in B_1^N}\rho(x, L_{\nu'}, l_p^N) = \max_{i=[M/2]+1,\dots,\mu}\rho(e_i', L_{\nu'}, l_p^N).$$

Therefore

$$\inf_{L_{\nu'}} \max_{i=[M/2]+1,\dots,\mu} \rho(e'_i, L_{\nu'}, l_p^N) = d_{\nu'}(B_1^N, l_p^N).$$

From here, taking into account that $\nu' < \theta N$ $(\theta = 2/3)$ $(N < \mu < 2^{l-d+1}M)$, for $2 \le p \le \infty$ by (2) we obtain the following: for each $x \in K_{\bar{1}}$

$$R_{mM} \ge \left(\frac{M}{2}\right)^{1/p} d_{\nu'}(B_1^N, l_p^N) \ge \tfrac{1}{2}C(\theta)(N)^{1/p-1/2}M^{1/p} \ge C_{l,p}(\theta)M^{2/p-1/2}. \tag{10}$$

From (9) and (10) we have

$$\tau(W_q^r, G^M(\Gamma), L_p) \gg M^{1/p-1/2}M^{1/p}\|s(x)\|_{L_p(K_{\bar{1}})}.$$

Using the definition of the function $s(x)$ and Lemma 2, we have

$$\begin{aligned} \|s(x)\|_{L_p(K_{\bar{1}})} &= \left(\int_{K_{\bar{1}}} \left| b_m \sum_{i_{\overline{1,l-d}}=1}^{m} s_{\bar{i}'}(x)\right|^p dx\right)^{1/p} \\ &= b_m \|s_{\bar{1}}(x)\|_{L_p(K_{\bar{1}})} = b_m m^{-l/p} C_{0,p} = Cm^{-r+d/q-l/p}. \end{aligned} \tag{11}$$

Using (11) and $\alpha\mu \le M \le \beta\mu$ we extend the estimate:

$$\begin{aligned} \tau(W_q^r, G^M(\Gamma), L_p) &\gg M^{2/p-1/2}m^{-r+d/q-l/p} \\ &\gg M^{2/p-1/2}M^{(-r+d/q-l/p)/(l-d)} = M^{-(r-d/q+d/p)/(l-d)+1/p-1/2}, \end{aligned}$$

where $2 \le p \le \infty$.

The lower estimate for $\tau_M(W_q^r)_p$ is proved.

We now give some auxiliary assertions needed in the estimation of τ_M from above.

Lemma 3 (B.S. Kashin and E. D. Gluskin; see [10]). *For $1 \le n < m$ the two-sided estimate*

$$d_n(B_q^m, l_p^m) \asymp \begin{cases} (\min\{1, m^{1/p}n^{-1/2}\})^{(1/q-1/p)(1/2-1/p)}, & 2 \le q < p < \infty, \\ \max\{m^{1/p-1/q}, \min\{1, m^{1/p}n^{-1/2}\}\sqrt{1-\tfrac{n}{m}}\}, & \\ & 1 \le q < 2 \le p < \infty, \end{cases}$$

holds with constants independent of m and n.

We shall need the following obvious corollary to this lemma: since $\sqrt{1-n/m} < 1$,

$$d_n(B_q^m, l_p^m) \le C\min\{1, m^{1/p}n^{-1/2}\} \qquad (1 \le q \le 2 \le p < \infty,\ p \ne q), \tag{12}$$

where C does not depend on m or n.

Lemma 3′ (A. Yu. Garnaev and E. D. Gluskin [11]). *Let $2 \le t < \infty$. Then for $1 \le n < m$ with some absolute constant C the following inequality is valid:*

$$d_n(B_t^m, l_\infty^m) \le \min\left\{1, C\left(\frac{\log(1+m/n)}{n}\right)^{1/t}\right\}.$$

In [16] an order equality was established for $d_n(B_t^m, l_\infty^m)$. We use the upper estimate for $t = 2$: since $B_q^m \subset B_2^m$ $(q \le 2)$, we have

$$d_n(B_q^m, l_\infty^m) \le C\left(\frac{\log(1+m/n)}{n}\right)^{1/2}. \tag{12′}$$

Lemma 4. *Let $1 \le p, q \le \infty$ and let $\Delta \subset I^l$ be a cube with sides parallel to the coordinate axes. If the condition $r > l/q - l/p$ is satisfied, then there exists a linear* *projection operator* P_Δ, *mapping the space* $W_q^2(\Delta)$ *onto the finite-dimensional subspace of polynomials of degree* $\le r-1$ *in* l *variables, such that for each* $f \in W_q^r(\Delta)$

$$\|f - P_\Delta f\|_{L_p(\Delta)} \le C|\Delta|^{r/l-1/q+1/p}\|f\|_{L_q^r(\Delta)}, \tag{13}$$

where $|\Delta|$ *is the Lebesgue volume of the cube* Δ, *and the constant* C *does not depend on the cube* Δ *or on* f.

The assertion of this lemma for $p=\infty$, $r > l/q$ and for $l/q - l/p < r \le l/q$ follows directly from Lemmas 3.1 and 3.2 in [12], respectively, which lemmas are based on Sobolev's imbedding theorem [13]. The remaining case $r > l/q$, $p = \infty$ is obtained simply by integrating the pth power of the inequality

$$|f(x) - (P_\Delta f)(x)| \le C|\Delta|^{r/l-1/q}\|f\|_{L_q^r(\Delta)} \qquad (x\in\Delta,\ r > l/q)$$

(equivalent to the inequality (13) for $p=\infty$).

In particular, under the conditions of Lemma 4

$$\|f - P_\Delta f\|_{L_p(\Delta)} \le C|\Delta|^{r/l-1/q+1/p}\|f\|_{W_q^r(\Delta)}, \tag{14}$$

where

$$\|f\|_{W_q^r(\Delta)} = (\|f\|^q_{L_q(\Delta)} + \|f\|^q_{L_q^r(\Delta)})^{1/q}.$$

We partition the cube I^l into 2^{kl}, $k = 0, 1, \ldots$, equal cubes. Denote by R_k the set thus obtained of cubes Δ. We construct the operator $P_{R_k} = P_k$ by setting $(P_k f)(x) = (P_\Delta f)(x)$ inside each cube $\Delta \in R_k$. To ensure single-valuedness of the operator $P_k f$ we agree to choose the values of either one of the polynomials $P_\Delta f$ or $P_{\Delta'} f$ as $P_k f$ on the planes belonging to common faces of different Δ and Δ'; $(P_k f)(x)$ is a piecewise polynomial function almost everywhere on I^l, where each polynomial is of total degree $\le r-1$ with respect to the collection of its variables.

Lemma 5. *Let* p, q, r *satisfy the hypothesis of Lemma* 4. *Then*

$$f \in W_q^r(K) \Rightarrow \|f - P_k f\|_{L_p(K)} \le C2^{(-r+l/q-l/p)k}\left(\sum_{\Delta\in R_k}\|f\|^p_{W_q^r(\Delta)}\right)^{1/p}, \tag{15}$$

where C *is the constant from Lemma* 4.

Proof. According to Lemma 4, for each $\Delta \in R_k$ we have

$$\begin{aligned}\|f(x) - (P_k f)(x)\|_{L_p(\Delta)} &= \|f(x) - (P_\Delta f)(x)\|_{L_p(\Delta)}\\ &\le C(2^{-kl})^{r/l-1/q+1/p}\|f\|_{W_q^r(\Delta)} = C2^{(-r+l/q-l/p)k}\|f\|_{W_q^r(\Delta)}.\end{aligned}$$

This implies (15) for $p=\infty$. For $p<\infty$ we raise this inequality to power p and sum over all $\Delta\in R_k$:

$$\sum_{\Delta\in R_k}\int_\Delta |f(x) - (P_k f)(x)|^p\,dx \le C^p 2^{(-r+l/q-l/p)kp}\sum_{\Delta\in R_k}\|f\|^p_{W_q^r(\Delta)},$$

whence (15) follows.

Lemma 6. *Let* $1 \le q \le p \le \infty$, $r > l/q - l/p$. *For* $f \in W_q^r(I^l)$ *put* $T_0 f = P_0 f$, $T_k f = (P_k - P_{k-1})f = P_k f - P_{k-1} f$ $(k = 1, 2, \ldots)$, *where* $P_k f$ *is the operator from Lemma* 5. *Then*

$$\|T_k f\|_{L_p(I^l)} \le 4C2^{(-r+l/q-l/p)(k-1)}$$

for any $f \in W_q^r(I^l)$, *where* C *is the constant from Lemma* 4, *and the expansion*

$$f = \sum_{k=0}^{\infty} T_k f$$

holds, where the convergence is in the sense of $L_p(I^l)$.

Proof. We have

$$f - \sum_{k=0}^{n} T_k f = f - P_n f.$$

Denote the right-hand side of (15) by α_k. Then from (15) we have

$$\|f - P_n f\|_{L_p(I^l)} \le \alpha_n.$$

Taking into account that

$$q \le p \Rightarrow \left(\sum_{i=1}^{m} |a_i|^p \right)^{1/p} \le \left(\sum_{i=1}^{m} |a_i|^q \right)^{1/q},$$

we obtain

$$\begin{aligned}
\left(\sum_{\Delta \in R_n} \|f\|^p_{W_q^r(\Delta)} \right)^{1/p} &\le \left(\sum_{\Delta \in R_n} \|f\|^q_{W_q^r(\Delta)} \right)^{1/q} \\
&= \left(\sum_{\Delta \in R_n} \left[\int_\Delta |f|^q\, dx + \sum_{|\bar{r}|=r} \int_\Delta |f^{(\bar{r})}(x)|^q\, dx \right] \right)^{1/q} \\
&= \left(\int_{I^l} |f|^q\, dx + \sum_{|\bar{r}|=r} \int_{I^l} |f^{(\bar{r})}(x)|^q\, dx \right)^{1/q} = \|f\|_{W_q^r(I^l)} \le 2^{1/q}.
\end{aligned}$$

Then, setting $\omega = r - l/q + l/p$, we have

$$\alpha_n = C 2^{-\omega n} \left(\sum_{\Delta \in R_n} \|f\|^p_{W_q^r(\Delta)} \right)^{1/p} \le C 2^{1/q} 2^{-\omega n} \le 2C 2^{-\omega n}.$$

Since $\omega > 0$, it follows that $\alpha_n \to 0$ $(n \to \infty)$, which proves the expansion in the lemma.

We use the estimate for α_n $(n = 0, 1, \dots)$ just obtained to estimate $\|T_k f\|_{L_p(I^l)}$. For $k \ge 1$ we have

$$\begin{aligned}
\|T_k f\|_{L_p(I^l)} &\le \|P_k f - f\|_{L_p(I^l)} + \|f - P_{k-1} f\|_{L_p(I^l)} \le \alpha_k + \alpha_{k-1} \\
&< 2C(2^{-\omega k} + 2^{-\omega(k-1)}) = 2C(1 + 2^{-\omega}) 2^{-\omega(k-1)} \le 4C 2^{-\omega(k-1)}.
\end{aligned}$$

Lemma 6 is proved.

Lemma 7. *Let* $M_k \in \{0, 1, \dots\}$ *and* $\sum_{k=0}^{\infty} M_k \le N$. *Then, setting* $G^0 = \{\varphi \equiv 0\}$, *for* $1 \le p \le \infty$ *we have*

$$\rho\left(\sum_{k=0}^{\infty} T_k f, G^N, L_p \right) \le \sum_{k=0}^{\infty} \rho(T_k f, G^{M_k}, L_p). \tag{16}$$

Proof. Let us agree to assume that $\sum_{j=1}^{0} a_j = 0$. Then

$$\begin{aligned}\rho\left(\sum_{k=0}^{\infty} T_k f, G^N, L_p\right) &= \inf_{\varphi_j, \psi_j} \left\| \sum_{k=0}^{\infty} (T_k f)(x) - \sum_{j=1}^{N} \varphi_j(u)\psi_j(v) \right\|_{L_p} \\ &\le \inf_{\varphi_{k_j}, \psi_{k_j}} \left\| \sum_{k=0}^{\infty} T_k f - \sum_{k=0}^{\infty} \sum_{j=1}^{M_k} \varphi_{k_j}(u)\psi_{k_j}(v) \right\|_{L_p} \\ &\le \inf_{\varphi_{k_j}, \psi_{k_j}} \sum_{k=0}^{\infty} \left\| T_k f - \sum_{j=1}^{M_k} \varphi_{k_j}\psi_{k_j} \right\|_{L_p} \\ &= \sum_{k=0}^{\infty} \inf_{\varphi_{k_j}, \psi_{k_j}} \left\| T_k f - \sum_{j=1}^{M_k} \varphi_{k_j}\psi_{k_j} \right\|_{L_p} = \sum_{k=0}^{\infty} \rho(T_k f, G^{M_k}, L_p).\end{aligned}$$

The lemma is proved.

Note that since $M_k = 0$ for all k from some place on, the series in Lemma 7 converges for $\omega = r - l/q + l/p > 0$ by Lemma 6. Since for $m_k = r \cdot 2^k$ the piecewise polynomial function $(T_k f)(x) \in G^{m_k^{l-d}}$, here we have $\rho(T_k f, G^{M_k}, L_p) = 0$ for $M_k \ge m_k^{l-d}$. For $M_k = 0$, $G^0 = \{0\}$ we obtain the quantities

$$\rho(T_k f, G^{M_k}, L_p) = \|T_k f\|_p,$$

the estimates of which were given in Lemma 6. The estimates of the remaining terms in (16) (for $0 < M_k < m_k^{l-d}$) with be based on the following lemma and its corollary.

Consider the net Q on the cube I^l:

$$Q = \{\xi = (\xi_1, \dots, \xi_l) \mid \xi_s = (j-1)/m, \ j = 1, \dots, m, \ s = 1, \dots, l\},$$

and to each net function $V(\xi)$ put in correspondence the vector $(V(\xi))_{\xi \in Q} = V \in \mathbf{R}^{m^l}$, the coordinates of which, enumerated by the multi-index ξ running over the set of its values in Q, coincide with the corresponding values of $V(\xi)$. Let

$$V \in B_q^{m^l}, \quad \text{i.e.,} \ \|V(\xi)\|_{l_q^{m^l}} = \left(\sum_{\xi \in Q} |V(\xi)|^q\right)^{1/q} \le 1,$$

$$\xi_u = \xi_{\overline{1,d}} = (\xi_1, \dots, \xi_d), \quad \xi_v = \xi_{\overline{d+1,l}}; \qquad \xi = (\xi_1, \dots, \xi_l) = (\xi_u, \xi_v).$$

Lemma 8. *Let $1 \le q < p \le \infty$. For any positive integers m, $\mu = m^{l-d}$ there exists an operator H_{mM} on the class of net functions $V(\xi) \in B_q^{m^l}$ $(\xi \in Q)$ of the form*

$$(H_{mM}V)(\xi) = g_{2M}(\xi; V) = \sum_{j=1}^{2M} \varphi_j(\xi_u; V)\psi_j(\xi_v; V)$$

such that

$$\left\{\sum_{\xi \in Q} |V(\xi) - (H_{mM}V)(\xi)|^p\right\}^{1/p} \le CM^{1/p-1/q} \min\{1, m^{(l-d)/p} M^{-1/2}\}$$

with a constant $0 < C < \infty$ independent of m, M, and V.

Proof. Denote by $\widetilde{Q}_u$ and $\widetilde{Q}_v$ the projections of the net Q onto the spaces of the variables $u = (x_1, \dots, x_d)$ and $v = (x_{d+1}, \dots, x_l)$, respectively. We divide the net Q into m^d columns

$$Q_u = Q_{\xi_u} = \{\xi = (\xi_u, \xi_v) \mid \xi_v \in \widetilde{Q}_v\} \quad (\xi_u \in \widetilde{Q}_u).$$

We partition the coordinates of each vector $V = (V(\xi))_{\xi \in Q}$ into groups (vector components) $V_u = (V(\xi))_{\xi \in Q_u}$ $(u \in \widetilde{Q}_u)$. By the definition of width

$$d_M(B_q^\mu, l_p^\mu) \stackrel{\text{def}}{=} d^*_{mM} \quad (\mu = m^{l-d}),$$

for any positive integer $M < \mu$ there exists a subspace $L^*_M \subset l_p^\mu$ $(\dim L^*_M = M)$ such that

$$\rho(a, L^*_M, l_p^\mu) \le d^*_{mM} \qquad \text{for all } a \in B_q^\mu. \tag{17}$$

Denote by A the operator of metric projection in l_p^μ onto L^*_M. Then, taking the homogeneity of A into account, the inequality (17) may be rewritten in the form

$$\|a - Aa\|_{l_p^\mu} \le d^*_{mM} \|a\|_{l_q^\mu} \qquad \forall a \in \mathbf{R}^\mu.$$

In particular, for any $V \in B_q^{m^l}$ and any vector component V_u

$$\|V_u - AV_u\|_{l_p^\mu} \le d^*_{mM} \|V_u\|_{l_q^\mu} = d^*_{mM} \left(\sum_{\xi \in Q_u} |V(\xi)|^q \right)^{1/q} \stackrel{\text{def}}{=} d^*_{mM} \beta_u^{1/q}. \tag{18}$$

We renumber the elements of the set $\{\beta_u\}$ in decreasing order:

$$\beta_1 \ge \beta_2 \ge \dots \ge \beta_i \ge \dots \ge \beta_{m^d}. \tag{19}$$

Then, for any $j = 1, \dots, m^d$,

$$1 \ge \sum_{\xi \in Q} |V(\xi)|^q = \sum_{i=1}^{m^d} \beta_i \ge \sum_{i=1}^{j} \beta_i \ge \beta_j \sum_{1}^{j} 1 = j\beta_j \Rightarrow \beta_j \le 1/j.$$

From here (since $q < p \le \infty$)

$$\begin{aligned} \left(\sum_{j=M+1}^{m^d} \beta_j^{p/q} \right)^{1/p} &\le \left(\sum_{j=M+1}^{\infty} \frac{1}{j^{p/q}} \right)^{1/p} \le \left(\int_M^\infty \frac{dx}{x^{p/q}} \right)^{1/p} \\ &= \left(\frac{1}{p/q - 1} M^{1-p/q} \right)^{1/p} = M^{1/p-1/q} \left(\frac{q}{p-q} \right)^{1/p} \le C_1 M^{1/p-1/q}, \end{aligned} \tag{20}$$

where

$$C_1 = C_{p,q} = \begin{cases} 1, & p = \infty, \\ (q/(p-q))^{1/p}, & p < \infty. \end{cases}$$

In particular, we have also proved that if $\beta_j > 1/M$, then $j \le j_M < M$. Denote $\bigcup_{\beta_u > 1/M} Q_u$ by Q^M, and the projection of Q^M onto the space of the variables $u = (x_1, \dots, x_d)$ by Q_u^M. Using (20) we obtain

$$\begin{aligned} \left(\sum_{\xi_u \in \widetilde{Q}_u \setminus Q_u^M} \beta_u^{p/q} \right)^{1/p} &= \left(\sum_{i=j_M+1}^{m^d} \beta_j^{p/q} \right)^{1/p} = \left(\sum_{i=j_M+1}^{M} + \sum_{i=M+1}^{m^d} \right)^{1/p} \\ &\le \left(\left(\frac{1}{M} \right)^{p/q} M + (C_1 M^{1/p-1/q})^p \right)^{1/p} < C_2 M^{1/p-1/q}. \end{aligned}$$

Therefore, for any $M < \mu$ and $1 \le q < p \le \infty$, in accordance with (18)

$$\left(\sum_{\xi_u \in \widetilde{Q}_\mu \setminus Q_u^M} \|V_u - AV_u\|_{l_p^\mu}^p \right)^{1/p} \le C_2 M^{1/p-1/q} \, d_{mM}^*. \tag{21}$$

We now return to the initial vector $V = (V(\xi))_{\xi \in Q} \in B_q^{m^l}$ with vector components $V_u = (V(\xi))_{\xi \in Q_u}$. To this end, we put in correspondence to each vector $V = (V_u)_{u \in \widetilde{Q}_u} \in B_q^{m^l}$ the vector

$$H_{mM} V = ([HV]_u)_{\xi_u \in \widetilde{Q}_u} \in B_q^{m^l}$$

with vector components $[HV]_u \in l_q^\mu$, setting

$$[HV]_u = \begin{cases} V_u & \text{if } \beta_u = \|V_u\|_{l_q^\mu}^q > 1/M, \\ AV_u & \text{otherwise.} \end{cases}$$

Then the left-hand side of (21) coincides with $\|V - H_{mM}V\|_{l_p^{m^l}}$, and we obtain

$$\rho(V, H_{mM}V)_{l_p^{m^l}} \le C_2 M^{1/p-1/q} d_{mM}^*, \tag{22}$$

i.e.,

$$\left(\sum_{\xi \in Q} |V(\xi) - (H_{mM}V)(\xi)|^p \right)^{1/p} < C_2 M^{1/p-1/q} \, d_{mM}^*. \tag{22'}$$

Let $\{\overline{\psi}_1, \overline{\psi}_2, \dots, \overline{\psi}_M\}$ be a basis in $L_M^* \subset l_p^\mu$. Then

$$[H_{mM}V]_u = \sum_{j=1}^M C_j(\xi_u) \overline{\psi}_j \qquad \forall u \in \widetilde{Q}_u \setminus Q_u^M. \tag{23}$$

The coordinates of any vector $\overline{a} \in l_p^\mu$ in the natural basis will be enumerated by multi-indices $\xi_v \in \widetilde{Q}_v$, i.e., $\overline{a} \in l_p^\mu$ will be represented in the form $\overline{a} = (a(\xi_v))_{\xi_v \in \widetilde{Q}_v}$.

Suppose that each vector $\overline{\psi}_j$ has the coordinates $\psi_j(\xi_v)$ in the natural basis, i.e., $\overline{\psi}_j = (\psi_j(\xi_v))_{\xi_v \in \widetilde{Q}_v}$. Then, going over in (23) to coordinatewise notation in the natural basis, we get

$$[H_{mM}V]_u(\xi_v) = \sum_{j=1}^M C_j(\xi_u) \psi_j(\xi_v), \qquad \xi_u \in Q_u \setminus Q_u^M,$$

i.e., considering $(H_{mM}V)(\xi)$ as a net function on Q, we conclude that for all $\xi = (\xi_u, \xi_v) \in Q \setminus Q^M$

$$(H_{mM}V)(\xi) = \sum_{j=1}^M C_j(\xi_u) \psi_j(\xi_v).$$

Setting

$$\varphi_j(\xi_n) = \begin{cases} C_j(\xi_u) & \text{if } \xi_u \in \widetilde{Q}_u \setminus Q_u^M, \\ 0 & \text{if } \xi_u \in Q_u^M, \end{cases}$$

we obtain a bilinear form

$$g_M(\xi, V) = \sum_{j=1}^M \varphi_j(\xi_u; V) \psi_j(\xi_v) \qquad (\xi \in Q)$$

such that

$$g_M(\xi, V) = \begin{cases} (H_{mM}V)(\xi), & \xi \in Q \backslash Q^M, \\ 0, & \xi \in Q^M. \end{cases} \tag{24}$$

We show that the function

$$U(\xi) = \begin{cases} 0, & \xi \in Q \backslash Q^M, \\ (H_{mM}V)(\xi), & \xi \in Q^M, \end{cases} \tag{25}$$

also belongs to $G^M(\Gamma)$. Since Q^M consists of $j_M < M$ columns Q_u, it is seen from the structure of the function $U(\xi)$ that in each determinant

$$\left(\overset{M+1}{\nabla} U\right)(u_1, \dots, u_{M+1}, v_1, \dots, v_{M+1}) \overset{\text{def}}{=} \begin{vmatrix} U(u_1, v_1) \cdots U(u_{M+1}, v_1) \\ \dots\dots\dots\dots\dots\dots \\ U(u_1, v_{M+1}) \cdots U(u_{M+1}, v_{M+1}) \end{vmatrix},$$

where $u_i = \xi_u^{(i)}$, $v_j = \xi_v^{(j)}$, $(u_i, v_j) \in Q$, at least one column consists of zeros. Consequently, this determinant is zero.

In [14] a criterion was proved for the representability of a function $f(u, v)$ as a bilinear form in $G^M(\Gamma)$; the net analog of this criterion consists in the following: a function $V(\xi)$ on the net Q is representable in the form $\sum_{k=1}^M \varphi_k(\xi_u)\psi_k(\xi_v)$ if and only if for any $(u_i, v_j) \in Q$ $(i, j = 1, \dots, M+1)$

$$\left(\overset{M+1}{\nabla}(V)\right)(u_i, v_j)_{i,j=1,\dots,M+1} = 0.$$

By the criterion $U(\xi) = \tilde{g}_M(\xi; V) \in G^M(\Gamma)$.

Taking (24) and (25) into account, we obtain

$$(H_{mM}V)(\xi) = g_M(\xi, V) + \tilde{g}_M(\xi, V) \in G^{2M}(\Gamma).$$

This and (22′) imply the assertion of the lemma. Lemma 8 is proved.

Let Q be a net on I^l and let ξ, ξ_u, ξ_v be the discrete variables defined before Lemma 8.

Lemma 9 (Corollary). *For $1 \le q \le 2 \le p \le \infty$ and any positive integers $m, M < \mu = m^{l-d}$, there exists an operator $g_M(\xi, V)$ of the form*

$$g_M(\xi; V) = \sum_{j=1}^M \varphi_j(\xi_u; V)\psi_j(\xi_v; V)$$

$$(\xi_u = (\xi_1, \dots, \xi_d), \quad \xi_v = (\xi_{d+1}, \dots, \xi_l), \quad \xi = (\xi_u, \xi_v)),$$

on the class of net functions $V(\xi)$ $(\xi = (\xi_1, \dots, \xi_l) \in Q)$, such that

$$\left(\sum_{\xi \in Q} |V(\xi) - g_M(\xi; V)|^p\right)^{1/p} \le CM^{1/p-1/q} d_{mM} \|V(\xi)\|_{l_q^{m^l}},$$

where

$$d_{mM} = \begin{cases} \min(1, m^{(l-d)/p} M^{-1/2}), & \text{for } p < \infty, \\ \min(1, [\frac{1}{M}\log(1 + \frac{m^{l-d}}{M})]^{1/2}), & \text{for } p = \infty, \end{cases}$$

and the constant $C < \infty$ does not depend on m and M.

Indeed, $M < \mu \Rightarrow [M/2] < \mu$; in addition, $[M/2] \ge M/3$ and, for $M \ge 2$, $[M/2] \ge 1$. Therefore, by Lemma 8 for $M \ge 2$ and $g_M(\xi; V) = H_{m,[M/2]}(\xi; V)$, using the estimates (12), (12′), we get

$$\|V(\xi) - g_M(\xi; V)\|_{l_p^M} \le C[M/2]^{1/p-1/q} d^*_{m,[M/2]}$$
$$\le \widetilde{C}[M/2]^{1/p-1/q} d_{m,[M/2]} \le 3^2\widetilde{C}M^{1/p-1/q} d_{mM}.$$

Since $2[M/2] \le M$, $g_M(\xi, V)$ has the desired form. For $M = 1$ the lemma is trivial $(g_M(\xi; V) \equiv 0)$.

An upper estimate for $\tau_M(W_q^r)_p$

We first estimate the terms $\rho(T_k f, G^{M_k}, L_p)$ on the right-hand side of the inequality (16) in Lemma 7, for $0 < M_k < m_k^{l-d}$ $(m_k = r2^k)$. To this end, we partition every cube $\Delta \in R_k$ into r^l equal cubes. As a result of this we get a partition of I^l into $(r2^k)^l$ equal cubes and the corresponding net $Q = Q(r, k)$, coinciding with that considered in Lemmas 8 and 9 for $m = m_k = r2^k$. Let

$$g_{M_k}(\xi) = \sum_{j=1}^{M_k} \varphi_j(\xi_u)\psi_j(\xi_v)$$

be any bilinear form on the net $Q(r, k)$. Let Δ_u (Δ_v) and $Q_u(r, k)$ $(Q_v(r, k))$ denote the projections of Δ and $Q(r, k)$ onto the space of the variables $u = (x_{\overline{1,d}})$ $(v = (x_{\overline{d+1,l}}))$. Let $P_{j,\Delta_u}(U)$ $(P_{i,\Delta_v}(V))$ be the Lagrange interpolation polynomial in each of the variables $x_1, \dots, x_d$ $(x_{d+1}, \dots, x_l)$ interpolating the function $\varphi_j(\xi_u)$ $(\psi_j(\xi_v))$ on the net

$$Q_u(r, k) \cap \overline{\Delta}_u (Q_v(r, k) \cap \overline{\Delta}_v)$$

(see, for example, [15]). Setting

$$\varphi_j(u) = \sum_{\Delta_u \subset I^d} \chi_{\Delta_u}(u) P_{j,\Delta_u}(u),$$
$$\psi_j(v) = \sum_{\Delta_v \subset I^{l-d}} \chi_{\Delta_v}(v) P_{j,\Delta_v}(v),$$

where $\chi_E(t)$ is the characteristic function of the set E, we extend $g_{M_k}(\xi)$ to I^l:

$$g_{M_k}(x) = \sum_{j=1}^{M_k} \varphi_j(u)\psi_j(v).$$

The function $g_{M_k}(x)$ inside each $\Delta \in R_k$ is a polynomial of degree r in each variable $x_1, \dots, x_l$. Extending the definition of the functions $\varphi_j(u)$, $\psi_j(v)$ to the faces of Δ_u, Δ_v, for example by continuity from the left in each variable, we obtain a bilinear form $g_{M_k}(x)$ on I^l interpolating the initial function $g_{M_k}(\xi)$ on the net $Q(r, k)$.

Put

$$\|(T_k f)(\xi)\|_{l_q} \stackrel{\text{def}}{=} \left(\sum_{\xi \in Q(r,k)} |(T_k f)(\xi)|^q \right)^{1/q}.$$

We apply Lemma 9 to the net function $V(\xi, f) = (T_k f)(\xi)/\|(T_k f)(\xi)\|_{l_q}$ for $m = m_k$ and $M = M_k \le \mu_k = m_k^{l-d}$. As a result, for $M_k \le \mu_k = m_k^{l-d}$ we put in correspondence to the function $(T_k f)(\xi)$ the function $g_{M_k}(\xi; f) \in G^{M_k}(\Gamma, Q(r, k))$, which, for $1 \le q \le 2 \le p \le \infty$ $(p \ne q)$, satisfies the inequality

$$\left(\sum_{\xi \in Q(r,k)} |(T_k f)(\xi) - g_{M_k}(\xi; f)|^p \right)^{1/p} \le C M_k^{1/p-1/q} d_{m_k M_k} \|(T_k f)(\xi)\|_{l_q}, \tag{26}$$

where

$$(26') \qquad d_{m_k M_k} = \begin{cases} m_k^{(l-d)/p} M^{-1/2}, & \text{for } p < \infty, \\ [M_k^{-1} \log(1 + m_k^{l-d}/M_k)]^{1/2} & \text{for } p = \infty. \end{cases}$$

We extend $g_{M_k}(\xi; f)$ to I^l as indicated above. Since on each $\Delta \in R_k$, $(T_k f)(x)$ is a polynomial of total degree $\le r$ in the collection of its variables and $g_{M_k}(x; f)$ is, by construction, a polynomial of degree $\le r$ in each variable, it follows that

$$(T_k f)(x) - g_{M_k}(x; f) \stackrel{\text{def}}{=} \widetilde{P}_\Delta(x) \qquad (x \in \Delta)$$

is a polynomial of degree r in each variable. Then, in view of Lemma 1.2 in [16],

$$\left\| \sum_{\Delta \in R_k} \chi_\Delta(x) \widetilde{P}_\Delta(x) \right\|_p \le C_{r,l} |\Delta|^{1/p} \left(\sum_{\Delta \in R_k} \sum_{\xi \in \bar{\Delta} \cap Q(r,k)} |\widetilde{P}_\Delta(\xi)|^p \right)^{1/p}.$$

Taking into account the equivalence of the norms in l_p on $\mathbf{R}^{(r+1)^l}$ $(1 \le p \le \infty$, $(r+1)^l$ the cardinality of $\overline{\Delta} \cap Q(r, k))$, we see that the proof of this inequality reduces to that of the inequality

$$\|\widetilde{P}_\Delta(x)\|_{L_p(\Delta)} \le C \|\widetilde{P}_\Delta(\xi)\|_{l_1(\xi \in \bar{\Delta} \cap Q(r,k))} |\Delta|^{1/p}, \qquad C = C(r, l),$$

which follows directly from the representation of $\widetilde{P}_\Delta(x)$ as the Lagrange interpolation polynomial of degree r on Δ:

$$\widetilde{P}_\Delta(x) = \sum_{\xi_{\overline{1,l}} \in \bar{\Delta} \cap Q(r,k)} \widetilde{P}_\Delta(\xi_{\overline{1,l}}) l_{\xi_1}(x_1) l_{\xi_2}(x_2) \cdots l_{\xi_l}(x_l),$$

where the $l_{\xi_i}(x_i)$—polynomials of degree r in x_i—are the Lagrange multipliers. Then

$$C|\Delta|^{1/p} = C(r, l)|\Delta|^{1/p} = \max_{\xi_{\overline{1,l}} \in \bar{\Delta} \cap Q(r,k)} \|l_{\xi_1}(x_1) l_{\xi_2}(x_2) \cdots l_{\xi_l}(x_l)\|_{L_p(\Delta)}$$

$$= |\Delta|^{1/p} \left\{ \max_{0 \le j \le r} \left(\int_0^1 \left| \prod_{\substack{s=0 \\ s \ne j}}^{r} \frac{(t - s/r)}{(j/r - s/r)} \right|^p dt \right)^{1/p} \right\}^l.$$

Therefore,

$$\|(T_k f)(x) - g_{M_k}(x)\|_{L_p(I^l)} \le C_{r,k} |\Delta|^{1/p} \left(\sum_{\xi \in Q(r,k)} |(T_k f)(\xi) - g_{M_k}(\xi)|^p \right)^{1/p}.$$

In view of (26), for $1 \le q \le 2 \le p \le \infty$ $(p \ne q)$ we obtain

$$(27) \qquad \|(T_k f)(x) - g_{M_k}(x)\|_{L_p(I^l)} \le \widetilde{C} M_k^{1/p - 1/q} 2^{-kl/p} d_{r \cdot 2^k, M_k} \|(T_k f)(\xi)\|_{l_q}.$$

Let us estimate $\|(T_k f)(\xi)\|_{l_q}$. We have

(28)

$$\|(T_k f)(\xi)\|_{l_q} = \left(\sum_{\Delta \in R_k} \sum_{\xi \in \Delta} |(T_k f)(\xi)|^q \right)^{1/q} \le \left(\sum_{\Delta \in R_k} \sum_{\xi \in \Delta} (\|(T_k f)(x)\|_{L_\infty(\Delta)})^q \right)^{1/q}.$$

Since $T_kf = P_kf - P_{k-1}f$ is a polynomial of degree $\leq k$, according to the well-known inequality of S. M. Nikol'skiĭ (for the homogeneous case see, for example, [17])

$$\|(T_kf)(x)\|_{L_\infty(\Delta)} \leq C(k, p)|\Delta|^{-1/p}\|(T_kf)(x)\|_{L_p(\Delta)}, \qquad p \geq 1. \tag{29}$$

Further, using Lemma 4, we obtain

$$\|(T_kf)(x)\|_{L_p(\Delta)} \leq \|f - P_kf\|_{L_p(\Delta)} + \|f - P_{k-1}f\|_{L_p(\Delta)} \ll |\Delta|^{r/l-1/q+1/p}\|f\|_{W_q^r(\Delta)}.$$

By this and (29), from (28) we have

$$\begin{aligned}\|(T_kf)(\xi)\|_{l_q} &\ll \left[\sum_{\Delta\in R_k}\sum_{\xi\in\Delta}(|\Delta|^{-1/p}|\Delta|^{r/l-1/q+1/p}\|f\|_{W_q^r(\Delta)})^q\right]^{1/q} \\ &= |\Delta|^{r/l-1/q}r^l\left(\sum_{\Delta\in R_k}\|f\|^q_{W_q^r(\Delta)}\right)^{1/q} = |\Delta|^{r/l-1/q}r^l(\|f\|^q_{W_q^r(I^l)})^{1/q} \\ &\ll |\Delta|^{r/l-1/q} = 2^{-(r/l-1/q)kl}.\end{aligned} \tag{30}$$

Applying the estimate (30) in (27), we obtain

$$\begin{aligned}\rho(T_kf, G^{M_k}, L_p) &\leq \|(T_kf)(x) - g_{M_k}(x; f)\|_{L_p(I^l)} \\ &\leq \widetilde{C}M_k^{1/p-1/q}\, d_{r2^k, M_k}2^{(-r/l+1/q-1/p)kl},\end{aligned} \tag{31}$$

where $1 \leq q \leq 2 \leq p \leq \infty$.

To complete the upper estimate for $\tau_M(W_q^r)_p$ we specify the numbers M_k in Lemma 7 as follows. For $r > l/q - d/p$ set

$$M_k = \begin{cases} (r2^k)^{l-d}, & k \leq k' = \frac{1}{l-d}\log_2 M, \\ [M2^{\alpha(k'-k)}], & k' < k \leq k^* = \frac{\alpha+1}{a}k', \\ 0, & k > k^*, \end{cases}$$

where the number α is determined from the relation

$$0 < \alpha < \min\left\{\frac{r - l/q + d/p}{1/2 + 1/q - 1/p}, \frac{r - l/q + l/p}{(l-d)(1/2 + 1/q - 2/p)}\right\}.$$

Then

$$\begin{aligned}\sum_{k=0}^\infty M_k &\leq \sum_{k=0}^{k'}(r2^k)^{l-d} + \sum_{k=k'}^{k^*}[M2^{\alpha(k'-k)}] \leq r^{l-d}\frac{2^{l-d}(2^{k'})^{l-d} - 1}{2^{l-d} - 1} + M\sum_{k=k'}^{k^*}2^{\alpha(k'-k)} \\ &< r^{l-d}\frac{2^{l-d}}{2^{l-d}-1}M + M\frac{1 - 2^{\alpha(k'-k^*)-\alpha}}{1 - 2^{-\alpha}}.\end{aligned}$$

But since

$$l - d \geq 1 \Rightarrow \frac{2^{l-d}}{2^{l-d}-1} \leq 2 \quad\text{and}\quad \alpha(k' - k^*) = \alpha k'\left(1 - \frac{\alpha+1}{\alpha}\right) = -k',$$

it follows that

$$\sum_{k=0}^\infty M_k \leq M\left[2r^{l-d} + \frac{1 - 2^{-\alpha-k'}}{1 - 2^{-\alpha}}\right] \leq M\theta \stackrel{\text{def}}{=} N \quad \left(\theta = 2r^{l-d} + \left[\frac{1}{1 - 2^{-\alpha}}\right] + 1\right).$$

Thus, by Lemma 7 and the remark following its proof, we have, for any M and for $N = \theta M$ (θ does not depend on M):

$$\forall f \in W_q^r$$

$$\rho(f, G^N, L_p) \le \sum_{k>k'} \rho(T_k f, G^{M_k}, L_p) \tag{32}$$

$$\le \sum_{k'<k\le k^*} \|T_k f - g_{M_k}(x; f)\|_{L_p(I^l)} + \sum_{k>k^*} \|T_k f\|_{L_p(I^l)}.$$

Let us start by estimating the first sum on the right in (32) for $p < \infty$.

Set $\beta = 1/p - 1/q - 1/2$ for $1 \le q \le 2 \le p < \infty$ $(p \ne q)$. By (31) and (26), for $m_k = r2^k$ we have

$$\sum_{k'<k\le k^*} \|T_k f - g_{M_k}(x; f)\|_{L_p(I^l)}$$

$$\le \widetilde{C} \sum_{k'\le k\le k^*} M_k^{1/p-1/q} 2^{-\omega k} \min\{1, m_k^{(l-d)/p}, M_k^{-1/2}\} \tag{33}$$

$$\le \widetilde{C} \sum_{k'\le k\le k^*} M_k^{1/p-1/q} 2^{-\omega k} m_k^{(l-d)/p} M_k^{-1/2} \overset{\text{def}}{=} J_1.$$

Since $[\lambda] \ge \frac{1}{2}\lambda$ for $\lambda \ge 1$,

$$J_1 \le C \sum_{k\ge k'} (M 2^{\alpha(k'-k)})^\beta 2^{-\omega k} 2^{k(l-d)/p} = C_1 M^\beta 2^{\alpha k'\beta} \sum_{k\ge k'} 2^{[-\alpha\beta+(-r+l/q-d/p)]k}.$$

By the definition of α the last sum is the sum of a decreasing geometric progression. Therefore,

$$J_1 \ll M^\beta 2^{\alpha k'\beta} \frac{2^{[-\alpha\beta+(-r+l/q-d/p)]k'}}{1 - 2^{-\alpha\beta+(-r+l/q-d/p)}} \ll M^\beta 2^{\alpha k'\beta} 2^{-\alpha k'\beta} 2^{k'(-r+l/q-d/p)} \tag{34}$$

$$= M^{1/p-1/q-1/2} M^{(-r+l/q-d/p)/(l-d)} = M^{-(r-d/q+d/p)/(l-d)+1/p-1/2}.$$

With this in mind, from (33) we obtain

$$\sum_{k'\le k\le k^*} \|T_k f - g_{M_k}(x; f)\|_{L_p(I^l)} \le M^{-(r-d/q+d/p)/(l-d)+1/p-1/2} \tag{35}$$

$$(1 \le q \le 2 \le p < \infty, \ p \ne q).$$

For $p = \infty$ by (31) and (26′) we have

$$J^{\mathrm{I}} \overset{\text{def}}{=} \sum_{k'<k\le k^*} \|(T_k f)(x) - g_{M_k}(x; f)\|_{L_\infty(I^l)}$$

$$\le C M^{-1/q} \sum_{k'<k\le k^*} 2^{(-r+l/q)k} M_k^{-1/2} \left[\log\left(1 + \frac{m_k^{l-d}}{M_k}\right)\right]^{1/2}$$

$$= C_1 M^{-1/q} M^{-1/2} \sum_{k'<k\le k^*} 2^{(-r+l/q)^k} 2^{-(1/2)\alpha(k'-k)} \left[\log\left(1 + \frac{m_k^{l-d}}{M_k}\right)\right]^{1/2} \tag{36}$$

$$= C M^{-1/q-1/2} 2^{-(1/2)\alpha k'} \sum_{k'<k\le k^*} 2^{(-r+l/q+\alpha/2)k} \left[\log\left(1 + \frac{m_k^{l-d}}{M_k}\right)\right]^{1/2}.$$

Taking into account that $k > k'$, we obtain

$$\log\left(1+\frac{m_k^{l-d}}{M_k}\right) \le \log\left(1+\frac{2r^{l-d}2^{k(l-d)}}{M2^{\alpha(k'-k)}}\right) = \log(1+2r^{l-d}2^{k(l-d)+\alpha(k-k')}/M),$$

where, by the definition of the number $k' = k'(M)$, $1/M = 2^{-k'(l+d)}$. Therefore,

$$\log_2\left(1+\frac{m_k^{l-d}}{M_k}\right) \le \log_2(1+2r^{l-d}2^{k(l-d)+\alpha(k-k')-k'(l-d)})$$
$$\le C_{r,l}\log_2 2^{(\alpha+l-d)(k-k')} \le C_{r,l}(\alpha+l)(k-k').$$

Keeping this in mind, from (36) we obtain

$$\begin{aligned} J^{\mathrm{I}} &\ll M^{-1/q-1/2}2^{-(1/2)\alpha k'} \sum_{k'<k\le k^*} 2^{(-r+l/q+\alpha/2)k}(\alpha+l)(k-k') \\ &\le M^{-1/q-1/2}2^{-(1/2)\alpha k'} \sum_{k=[k']+1} 2^{(-r+l/q+\alpha/2)k}(\alpha+l)(k-k') \\ &\le M^{-1/q-1/2}2^{-(1/2)\alpha k'}2^{(-r+l/q+\alpha/2)k'}(\alpha+l)\sum_{j=1}^{\infty}2^{(-r+l/q+\alpha/2)j}j. \end{aligned} \tag{37}$$

Since $-r+l/q+\alpha/2<0$ by the condition of the theorem and the definition of α, the series $\sum_{j=1}^{\infty}2^{(-r+l/q+\alpha/2)j}j$ converges. Therefore, from (37) we have

$$\begin{aligned} J^{\mathrm{I}} \ll M^{-1/q-1/2}2^{(-r+l/q)k'} &= M^{-1/q-1/2}M^{(1/(l-d))(-r+l/q)} \\ &= M^{-(1/(l-d))(r-d/q)-1/2}. \end{aligned} \tag{38}$$

In order to estimate the remaining terms in (32) we use Lemma 6. We have

$$J^{\mathrm{II}} \overset{\mathrm{def}}{=} \sum_{k\ge k^*}\|T_kf\|_{L_p(I^l)} \le 4C\sum_{k\ge k^*}2^{-\omega(k-1)} \ll \sum_{k\ge k^*}2^{-\omega k}.$$

Taking into account that $-\omega<0$ by the condition of the theorem, we extend the estimate:

$$J^{\mathrm{II}} \ll \frac{2^{-\omega k^*}}{1-2^{-\omega}} \ll 2^{-\omega k^*}.$$

Considering the definition of k^*, we obtain

$$J^{\mathrm{II}} \ll 2^{-\frac{\alpha+1}{\alpha}\frac{\omega}{l-d}\log_2 M} = M^{-\frac{\alpha+1}{\alpha}\frac{\omega}{l-d}}.$$

By the definition of α

$$0<\alpha<\frac{\omega}{(l-d)(1/q+1/2-2/p)}.$$

Then

$$\frac{\alpha+1}{\alpha} = 1+\frac{1}{\alpha} > 1+\frac{(l-d)(1/q+1/2-2/p)}{\omega}$$
$$\Rightarrow -\frac{\alpha+1}{\alpha}\frac{\omega}{l-d} \le -\frac{1}{l-d}\left(r-\frac{d}{q}+\frac{d}{p}\right)+\frac{1}{p}-\frac{1}{2}.$$

Therefore,

$$J^{\mathrm{II}} \ll M^{-(r-d/q+d/p)/(l-d)+1/p-1/2}. \tag{39}$$

Thus, from (32), (35), (38), and (39) we have

$$\rho(f, G^N, L_p) \ll M^{-(r-d/q+d/p)/(l-d)+1/p-1/2},$$

where $N = \theta M$, whence the needed upper estimate of $\tau_M(W_q^r)_p$ for $1 \leq q \leq 2 \leq p \leq \infty$ follows. The theorem is completely proved.

Remark. The assertion of the theorem remains valid for the space $W_q^r(I^l)$ also in the case of fractional r (Sobolev-Slobodetskiĭ spaces). For a noninteger r the norm $\|f\|_{L_p^r(\Delta)}$ is given by

$$\|f\|^p_{L_p^r(\Delta)} = \sum_{|\bar{r}|=[r]} \int_\Delta \int_\Delta \frac{|f^{(\bar{r})}(x) - f^{(\bar{r})}(y)|^p}{|x-y|^{p\theta+l}}\, dx\, dy,$$

where $[r]$ is the integer part of r, $\theta = r - [r]$ (see, for example, [12] and [18]).

A comparison of our theorem with Theorem T also shows that the condition on r is somewhat weakened: instead of $r > l/q$ we now need $r > l/q - d/p$.

The author expresses his gratitude to S.V.Konyagin, V. E. Maĭorov, and N. I. Chernykh for useful discussions.

BIBLIOGRAPHY

1. V. N. Temlyakov, *On best bilinear aproximations of periodic functions of several variables*, Dokl. Akad. Nauk SSSR **286** (1986), 301–304; English transl. in Soviet Math. Dokl. **33** (1986).
2. ———, *Approximation of periodic functions of several variables by combinations of functions depending on fewer variables*, Trudy Mat. Inst. Steklov **173** (1986), 243–252; English transl. in Proc. Steklov Inst. Math. **1987**, no. 4 (173).
3. ———, *Approximation of functions with bounded mixed derivative*, Trudy Mat. Inst. Steklov. **178** (1986); English transl., Proc. Steklov Inst. Math. **1989**, no. 1 (178).
4. ———, *Bilinear approximation and applications*, Trudy Mat. Inst. Steklov. **187** (1989), 191–215; English transl. in Proc. Steklov Inst. Math. **1990**, no. 3 (187).
5. M.-B. A. Babaev, *On the degree of approximation of the Sobolev class W_q^r by bilinear forms in L_p for $1 \leq q \leq p \leq 2$*, Mat. Sb. **182** (1991), 122–129; Englisn transl. in Math. USSR Sb. **72** (1992).
6. S. B. Stechkin, *On best approximations of given classes of functions by arbitrary polynomials*, Uspekhi Mat. Nauk **9** (1954), no. 1 (59), 133–134. (Russian)
7. B. S. Kashin, *Widths of some finite-dimensional sets and classes of smooth functions*, Izv. Akad. Nauk USSR Ser. Mat. **41** (1977), 334–351; English transl. in, Math. USSR Izv. **11** (1977).
8. V. E. Maĭorov, *Discretization of a problem on widths*, Uspekhi Mat. Nauk **30** (1975), no. 6 (186), 179–180. (Russian)
9. V. M. Tikhomirov, *Some questions of approximation theory*, Izdat. Moskov. Gos. Univ., Moscow, 1976. (Russian)
10. E. D. Gluskin, *Norms of random matrices and widths of finite-dimensional sets*, Mat. Sb. **120 (162)** (1983), 180–189; English transl. in Math. USSR Sb. **48** (1984).
11. A. Yu. Garnaev and E. D. Gluskin, *On widths of the Euclidean ball*, Dokl. Akad. Nauk SSSR **277** (1984), 1048–1052; English transl. in Soviet Math. Dokl. **30** (1984).
12. M. Sh. Birman and M. Z. Solomyak, *Piecewise-polynomial approximations of functions of the classes W_p^α*, Mat. Sb. **73 (115)** (1967), 331–355; English transl. in Math. USSR Sb. **2** (1967).
13. S. L. Sobolev, *Applications of functional analysis in mathematical physics*, Izdat. Leningrad. Univ., Leningrad, 1950; English transl., Amer. Math. Soc., Providence, RI, 1963.
14. M.-B. A. Babaev, *Best approximation by bilinear forms*, preprint, Inst. Mat. Mekh. Akad. Nauk Azerbaĭdzhan. SSR, Baku, 1987 = Manuscript No. 8624-B87, deposited at VINITI, 1987. (Russian) R. Zh. Mat. 1988, 3Б70.
15. V. E. Maĭorov, *Best approximation of the classes $W_1^r(I^s)$ in the space $L_\infty(I^s)$*, Mat. Zametki **19** (1976), 699–706; English transl. in Math. Notes **19** (1976).
16. ———, *Representations and best approximations on classes of functions of several variables*, Candidate's dissertation, Moscow, 1976. (Russian)

17. A. F. Timan, *Theory of approximation of functions of a real variable*, Fizmatgiz, Moscow, 1960; English transl., Pergamon Press, Oxford, and Macmillan, New York, 1963.

18. S. M. Nikol'skiĭ, *Imbedding, continuation and approximation theorems for differentiable functions of several variables*, Uspekhi Mat. Nauk **16** (1961), no. 5 (101), 63–114; English transl. in Russian Math. Surveys **16** (1961).

Translated by S. IVANOV

Труды Матем. Инст. Стеклова
198 (1992)

Proceedings of the Steklov
Institute of Mathematics
1994, Issue 1

ASYMPTOTIC AND EXTREMAL PROPERTIES OF ORTHOGONAL POLYNOMIALS IN THE PRESENCE OF SINGULARITIES IN THE WEIGHT

UDC 517.5

V. M. BADKOV

1. Introduction

In this paper we establish pointwise two-sided estimates, and sometimes also asymptotic formulas, for polynomials orthogonal on the circle and their derivatives of arbitrary order, on the entire contour of orthogonality, including the singular points of the weight. The weight is constructed from moduli of continuity. It turns out that in many cases best possible order is achieved on such orthogonal polynomials for the analogs obtained here of known inequalities for the norms of the polynomials and their derivatives. As consequences we obtain certain results for trigonometric orthogonal polynomials and for polynomials orthogonal on a segment.

Everywhere below, $\mathbf{C}$ is the complex plane, $\mathbf{R} = (-\infty, \infty)$, $\mathbf{Z}_+ = \{0, 1, \dots\}$, $\mathbf{N} = \{1, 2, \dots\}$. For a complex-valued function F Lebesgue integrable on $[a, b]$, for $0 < r < \infty$ we set

$$\|F\|_{L^r[a,b]} = \left\{\int_a^b |F(\tau)|^r \, d\tau\right\}^{1/r}, \qquad \|F\|_\infty = \operatorname*{ess\,sup}_{a \le \tau \le b} |F(\tau)|.$$

For $0 < r \le \infty$, by $L^r[a, b]$ we mean the space of functions F such that $\|F\|_{L^r[a,b]} < \infty$. For $r \ge 1$ this quantity is a norm, and for $r < 1$ its rth power defines a metric in $L^r[a, b]$. If F is 2π-periodic and $0 < r \le \infty$, we set

$$L^r = L^r[0, 2\pi], \qquad \|F\|_r = (2\pi)^{-1/r} \|F\|_{L^r[0, 2\pi]}.$$

The quantity $\|F\|_r$ is nondecreasing in r; if $F \in L^\infty$, then $\|F\|_r \to \|F\|_\infty$ $(r \to \infty)$, and if $F \in L^r$ for some $r > 0$, then

$$\lim_{r \to +0} \|F\|_r = \|F\|_0 \stackrel{\text{def}}{=} \exp\left\{\frac{1}{2\pi} \int_0^{2\pi} \ln|F(\tau)| \, d\tau\right\}. \tag{1.1}$$

We shall also consider Hardy spaces H^r of functions analytic in the disc $D = \{z : |z| < 1\}$ for which the supremum with respect to $\rho \in (0, 1)$ of

$$M_r(f; \rho) = \|f(\rho e^{i\tau})\|_r \tag{1.2}$$

is finite. By a modulus of continuity we mean a nondecreasing, continuous, semi-additive function $\omega(\delta)$ on $[0, \infty)$ with $\omega(0) = 0$. If, in addition, for any $a, b \ge 0$

$$2\omega(a/2 + b/2) \ge \omega(a) + \omega(b),$$

then we say that $\omega(\delta)$ is a concave modulus of continuity. We define $\omega(F; \delta)_r = \sup_{|\lambda| \le \delta} \|F(\lambda + \cdot) - F(\cdot)\|_r$—the modulus of continuity in L^r of the function F.

1991 *Mathematics Subject Classification*. Primary 41A10, 41A17, 41A60, 42A10.

Let $\sigma(\tau)\colon [0, 2\pi] \to \mathbf{R}$ be a nondecreasing bounded function with an infinite set of values, $\{\varphi_{\sigma,n}(z)\}_{n=0}^{\infty}$ a system of polynomials orthonormal on the unit circle $\Gamma_1 = \partial D$ relative to $d\sigma(\tau)$ [1], [2], i.e.,

$$\varphi_{\sigma,n}(z) = \kappa_{\alpha,n} z^n + \cdots, \qquad \deg \varphi_{\sigma,n} = n, \quad x_{\sigma,n} > 0 \quad (n \in \mathbf{Z}_+),$$
$$\frac{1}{2\pi}\int_0^{2\pi} \varphi_{\sigma,n}(e^{i\tau})\overline{\varphi_{\sigma,m}(e^{i\tau})}\,d\sigma(\tau) = \delta_{m,n} \qquad (m, n \in \mathbf{Z}_+).$$

If $\sigma \in AC[0, 2\pi]$ (i.e., σ is absolutely continuous on $[0, 2\pi]$), then $\{\varphi_{\sigma,n}\}_{n=0}^{\infty}$ is orthonormal with a 2π-periodic weight coinciding with $\sigma'(\tau)$ on $[0, 2\pi]$. Generally, we call a function φ a weight if it is 2π-periodic, measurable, nonnegative, and not equivalent to zero. When we speak of orthogonal polynomials with weight φ, we additionally require that $\varphi \in L^1$. We shall say that a point θ_0 is regular for the weight φ if for some $\varepsilon > 0$ the essential suprema of φ and $1/\varphi$ on $[\theta_0 - \varepsilon, \theta_0 + \varepsilon]$ are finite. Otherwise, φ has a singularity at θ_0.

If $Q_n(z)$ is a polynomial and $l_n = \deg Q_n \le n$, we set

$$Q_n^*(z) = z^{l_n}\overline{Q_n(\bar{z}^{-1})}. \tag{1.3}$$

Consider the system $\{T_{\sigma,n}(\tau)\}_{n=0}^{\infty}$ of trigonometric polynomials obtained by the Sonin-Schmidt orthogonalization process with respect to the measure $d\sigma$ from the trigonometric system 1, $\cos\tau$, $\sin\tau$, $\cos 2\tau$, $\sin 2\tau$, ... It is known [3] that

$$\begin{aligned} T_{\sigma,2n-1}(\tau) &= c_n \operatorname{Re}\{e^{-i(n-1)\tau}\varphi_{\sigma,2n-1}(e^{i\tau})\}, \\ T_{\sigma,2n}(\tau) &= c_n \operatorname{Im}\{e^{-in\tau}\varphi_{\sigma,2n}(e^{i\tau})\}, \end{aligned} \tag{1.4}$$

where

$$c_n = (2\kappa_{\sigma,2n})^{1/2}[\kappa_{\sigma,2n} - \operatorname{Re}\varphi_{\sigma,2n}(0)]^{-1/2} \qquad (n \in \mathbf{N}). \tag{1.5}$$

The two types of systems of orthogonal polynomials considered, and also polynomials $\{p_{\alpha,n}(t)\}_{n=0}^{\infty}$ orthogonal on $[-1, 1]$ with respect to a measure $d\alpha(t)$: $p_{\alpha,n}(t) = \kappa_{\alpha,n}t^n + \cdots$, $\deg p_{\alpha,n} = n$, $k_{\alpha,n} > 0$ $(n \in \mathbf{Z}_+)$,

$$\int_{-1}^{1} p_{\alpha,n}(t)p_{\alpha,m}(t)\,d\alpha(t) = \delta_{m,n} \qquad (m, n \in \mathbf{Z}_+),$$

are conveniently studied in the framework of a unified theory in which the system $\{\varphi_{\sigma,n}(z)\}_{n=0}^{\infty}$ plays the main role. The founder of this theory is Gábor Szegő [1], [4], who was the first to consider the polynomials $\varphi_{\sigma,n}$ and to establish the formula

$$P_{\alpha,n}\left(\frac{1}{2}\left(z + \frac{1}{z}\right)\right) = \frac{[\varphi_{\sigma,2n}(z) + \varphi_{\sigma,2n}^*(z)]z^{-n}}{\sqrt{2\pi[1 + \varphi_{\sigma,2n}(0)/\kappa_{\sigma,2n}]}} \qquad (n \in \mathbf{Z}_+,\ z \in \mathbf{C}) \tag{1.6}$$

under the assumption that $\alpha(-1) = 0$, $\sigma(\tau) = \alpha(\cos\tau)\operatorname{sgn}(\tau - \pi)$, as well as formulas analogous to (1.4), in which in the role of $\{T_{\sigma,n}\}_{n=0}^{\infty}$ there appears some other system of trigonometric polynomials orthonormal with respect to $d\sigma$, in general not coinciding with $\{T_{\sigma,n}\}_{n=0}^{\infty}$. Szegő established that if $\sigma \in AC[0, 2\pi]$ (subsequently, Ya. L. Geronimus showed that this restriction can be dropped) and

$$\ln\sigma'(\tau) \in L^1[0, 2\pi], \tag{1.7}$$

then

$$\varphi_{\sigma,n}^*(z) = \pi(\sigma'; z) + o(1) \qquad (n \to \infty) \tag{1.8}$$

uniformly inside D, where the function

$$\pi(\sigma'; z) = \exp\left\{-\frac{1}{4\pi}\int_0^{2\pi}\frac{e^{i\tau}+z}{e^{i\tau}-z}\ln\sigma'(\tau)\,d\tau\right\} \qquad (|z|<1) \tag{1.9}$$

is nowadays often called the Szegő function.

Many authors have searched for conditions guaranteeing the validity of (1.8) for $z \in \Gamma_1$ (for more details see [1], [2], [5]–[8]). By $\pi(\sigma'; z)$ in that case the radial limit value is meant. If $\sigma'(\tau)$ has a zero, the relation (1.8) cannot hold uniformly in $z \in \Gamma_1$, since then, on one hand, $\pi(\sigma'; z)$ would have to be continuous in $\overline{D}$, and, on the other hand, because

$$|\pi(\sigma'; e^{i\tau})| = \{\sigma'(\tau)\}^{-1/2} \quad \text{a.e.} \tag{1.10}$$

(see [2], formula (2.4)), $\pi(\sigma'; z)$ is not bounded. In [9], a uniform asymptotic representation (u.a.r.) on Γ_1 for $\varphi^*_{\sigma,n}(z)$ was sought in a new form:

$$\varphi^*_{\sigma,n}(z) = \pi(\sigma'; (1-cn^{-1})z)[1+o(1)], \tag{1.11}$$

where $c \in (0, \infty)$ is fixed. Formula (1.11) is equivalent to the formula

$$\varphi_{\sigma,n}(z) = z^n\overline{\pi(\sigma'; (1-cn^{-1})z)}[1+o(1)] \qquad (n\to\infty). \tag{1.12}$$

The result obtained in [9] is generalized by the following theorem, announced in [10].

Theorem 1.1. *Suppose* $j \in \mathbf{Z}_+$, *condition* (1.7) *is fulfilled, and*

$$\delta_{\sigma,n} \overset{\text{def}}{=} [\pi(\sigma'; 0)]^{-1}\{|\varphi_{\sigma,n+1}(0)|^2 + |\varphi_{\sigma,n+2}(0)|^2 + \cdots\}^{1/2} = o(n^{-1/2}), \tag{1.13}$$

$$\max_{|z|=\rho}|\pi'(\sigma'; z)/\pi(\sigma'; z)| = o((1-\rho)^{-1}) \qquad (\rho\to 1-0). \tag{1.14}$$

Then, for a fixed $c \in (0, \infty)$ *the u.a.r.*

$$\varphi^{(j)}_{\sigma,n}(z) = n^j z^{n-j}\overline{\pi(\sigma'; (1-cn^{-1})z)}[1+o(1)] \qquad (n\to\infty) \tag{1.15}$$

holds on Γ_1.

If in (1.14) o is replaced by O, Theorem 1.1 is no longer valid. The case $j=0$ was proved in [9]. In the present paper we prove Theorem 1.1 in full.

We note that the asymptotic behavior of $\varphi^{(j)}_{\sigma,n}(e^{i\theta})$ as $n\to\infty$ and for fixed $j \in \mathbf{N}$ was investigated in [11]–[14], but exclusively for regular points θ of the function $\sigma'(\tau)$. The quantity (1.13) has an approximative and probabilistic meaning, and plays an important role in the investigation of the behavior of $\varphi_{\sigma,n}(z)$. About this and investigations of the behavior of $\delta_{\sigma,n}$ as $n\to\infty$, consult, for example, [15]–[17]. Below, in §§3 and 4, we shall obtain certain new sufficient conditions guaranteeing the estimates $\delta_{\sigma,n} = o(n^{-1/2})$ and $\delta_{\sigma,n} = O(n^{-1/2})$, which play an important supporting role in the subsequent constructions. In [9] and [16] we found a wide class of functions $\sigma \in AC[0, 2\pi]$ satisfying the conditions (1.7), (1.13), and (1.14) and such that $\sigma'(\tau)$ has singularities at separate points of the segment $[0, 2\pi]$. In this paper we shall further widen this class. Theorem 1.1 fails for weights with power singularity. In the case of a generalized Jacobi weight, a u.a.r. on the entire set of orthogonality for each of the three above-mentioned systems was obtained in [18] in terms of Jacobi polynomials. In particular, the corresponding results of Bernstein [19] were strengthened and generalized. Concerning explicit expressions of orthogonal polynomials via Jacobi polynomials, see [20].

One of the main goals of this paper is the proof of the following three theorems.

Theorem 1.2. *Let* $j \in \mathbf{Z}_+$, $d\sigma(\tau) = \varphi(\tau)\,d\tau$, *and let the weight* φ *have the form*

$$\varphi(\tau) = h(\tau)\prod_{\nu=1}^{m} w_\nu\left(\left|\sin\frac{\tau-\theta_\nu}{2}\right|\right) \qquad (\tau \in \mathbf{R},\ -\pi < \theta_1 < \cdots < \theta_m \le \pi), \tag{1.16}$$

where

$$w_\nu(u) = \prod_{\mu=1}^{l_\nu}[g_{\mu,\nu}(u)]^{\alpha(\mu,\nu)} \in L^1[0,1], \tag{1.17}$$

$m, l_\nu \in \mathbf{N}$, $\alpha(\mu,\nu) \in \mathbf{R}$, $g_{\mu,\nu}(u)$ *are concave moduli of continuity* ($\mu = 1, \dots, l_\nu$; $\nu = 1, \dots, m$), *and the function* h *satisfies the conditions*

$$0 \le h(\tau), \quad h \text{ and } 1/h \in L^\infty, \tag{1.18}$$

$$\omega(h;\delta)_2 = O(\sqrt{\delta}) \qquad (\delta \to +0). \tag{1.19}$$

Then there is a constant $C_1(\varphi; j)$ *such that*

$$|\varphi_{\sigma,n}^{(j)}(e^{i\theta})| \le C_1(\varphi;j)n^j\left\{\prod_{\nu=1}^{m} w_\nu\left(\left|\sin\frac{\theta-\theta_\nu}{2}\right| + \frac{1}{n}\right)\right\}^{-1/2} \tag{1.20}$$

for all $\theta \in \mathbf{R}$ *and* $n \in \mathbf{N}$.

Theorem 1.3. *The conclusion of Theorem* 1.2 *remains in force if in its formulation condition* (1.19) *is replaced by the conditions*

$$\omega(h;\tau)_\infty \tau^{-1} \in L^1[0,1], \tag{1.21}$$

$$\int_0^\theta w_\nu(\tau)\,d\tau = O(\theta w_\nu(\theta)) \qquad (\theta \to +0;\ \nu = 1, \dots, m). \tag{1.22}$$

Theorem 1.4. *Suppose* $j \in \mathbf{Z}_+$ *and* $d\sigma(\tau) = \varphi(\tau)d\tau$, *where* φ *is defined by the formulas* (1.15)–(1.17), *in which* m, l_ν, $\alpha(\mu,\nu)$, $g_{\mu,\nu}(u)$ *have the same meaning as in Theorem* 1.2, *for* w_ν *the relations* (1.22) *are fulfilled, and* h *satisfies the condition* (1.18) *and, for* $j = 0$, *also one of the conditions* (1.19) *or* (1.21). *Then there is a constant* $C_2(\varphi; j) > 0$ *such that*

$$|\varphi_{\sigma,n}^{(j)}(e^{i\theta})| \ge C_2(\varphi;j)n^j\left\{\prod_{\nu=1}^{m} w_\nu\left(\left|\sin\frac{\theta-\theta_\nu}{2}\right| + \frac{1}{n}\right)\right\}^{-1/2} \tag{1.23}$$

for all $\theta \in \mathbf{R}$ *and* $n \ge j$.

We note that Golinskiĭ [7] proved the uniform boundedness on Γ_1 of a sequence of polynomials $\{h_n(z)\}_{n=0}^\infty$ orthonormal on Γ_1 with a continuous positive weight h satisfying the Lebesgue condition

$$E_n(h)\ln n \to 0 \qquad (n \to \infty), \tag{1.24}$$

where $E_n(h)$ is the nth best uniform trigonometric approximation for h. This fact raises the hope that condition (1.21) in Theorem 1.3 could be replaced by (1.24). To completely drop (1.21) from the formulation of Theorem 1.3, without replacing it by some other condition, is not possible. This follows from the results of Rakhmanov [21], [22] and Ambroladze [23], who established that the condition (1.18) does not guarantee the boundedness of the sequence $\|h_n(e^{i\theta})\|_\infty$ even for h continuous.

In §7 we establish certain weight analogs of the A. A. Markov, S. N. Bernstein, and S. M. Nikol′skiĭ inequalities. One particular case of these results is used in the proof of Theorems 1.1–1.4 (in later sections), which then, in turn, are used to prove that some of the inequalities obtained in §7 are best possible in order. The method used in that section is new and allows us, in particular, to prove, with the aid of the refined Hardy-Littlewood inequality (§5), that the smallest constant c, independent of q, r $(0 < q \leq r \leq \infty)$ and $n \in \mathbf{N}$, in the Nikol'skiĭ inequality [24]

$$\|t_n\|_r \leq cn^{1/q-1/r}\|t_n\|_q \tag{1.25}$$

for a trigonometric polynomial t_n of order $\leq n$ is $c = 4$. We note that in [24], $\|F\|_r$ is $(2\pi)^{1/r}$ times larger than here. Therefore, to the constant $c = 2$ in the inequality (1.25) proved in [24], there corresponds $c = 2(2\pi)^{1/q-1/r}$ in our notation. We also note that (1.25) was established in [24] in the m-dimensional case.

2. Two-sided estimates of the modulus of the Szegő function

In this section we establish two-sided estimates, dependent on ρ, θ, of the quantity $|\pi(\varphi; \rho e^{i\theta})|$ for certain weights φ defined by means of functions of the type of moduli of continuity of various orders.

Lemma 2.1. *If the weight φ coincides with $\omega(|\tau|)$ on $[-\pi, \pi]$, where $\omega(\delta)$ is a modulus of continuity, then there exist positive constants $C_1(\varphi)$ and $C_2(\varphi)$ such that*

$$C_1(\varphi) \leq |\pi(\varphi; \rho e^{i\theta})|\,[\omega(1-\rho+|\sin(\theta/2)|)]^{1/2} \leq C_2(\varphi) \tag{2.1}$$

for all $\rho \in [\frac{1}{2}, 1)$ and $\theta \in \mathbf{R}$.

Proof. The inequality ([25], §3.2.4, formula (6))

$$\omega(\tau_2)\tau_2^{-1} \leq 2\omega(\tau_1)\tau_1^{-1} \qquad (0 < \tau_1 < \tau_2) \tag{2.2}$$

implies that $1 \leq \omega(\pi)/\omega(|\tau|) \leq 2\pi/|\tau|$ on $[-\tau, \tau]$. Therefore, $\ln\varphi \in L^1$, so that $\pi(\varphi; z)$ (see (1.9)) is meaningful. Setting $A(\tau, \theta, \delta) = \delta^2 + 4(1-\delta)\sin^2[(\tau-\theta)/2]$, $\delta = 1-\rho$, we have

$$\ln|\pi(\varphi; \rho e^{i\theta})| = -\frac{(2-\delta)\delta}{4\pi}\int_{\theta-\pi}^{\theta+\pi} \frac{\ln\varphi(\tau)}{A(\tau, \theta, \delta)}\,d\tau. \tag{2.3}$$

Using the 2π-periodicity and the evenness with respect to θ of the right-hand side of (2.3), we assume that $0 \leq \theta \leq \pi$. If $0 < \varepsilon < \pi$, $\varepsilon \leq \theta \leq \pi$, $|\tau| \leq \varepsilon/2$, then $\varepsilon/2 \leq \tau + \theta \leq \pi + \varepsilon/2$ and, consequently, $|\ln\varphi(\tau+\theta)| \leq C_3(\varphi; \varepsilon)$. Therefore the modulus of the right-hand side of (2.3) does not exceed

$$\begin{aligned}\frac{(2-\delta)\delta}{4\pi}\int_{-\pi}^{\pi}\frac{|\ln\varphi(\tau)|\,d\tau}{\delta^2+4(1-\delta)\sin^2(\tau/2)} &\leq C_4(\varphi;\varepsilon)\delta\int_0^{\varepsilon/2}\frac{d\tau}{(\tau+\delta)^2} + C_5(\varepsilon)\delta\|\ln\varphi(\cdot)\|_1 \\ &\leq C_6(\varphi; \varepsilon),\end{aligned}$$

whence (2.1) follows.

Now let $0 \leq \theta \leq \varepsilon$, ε sufficiently small. Let us rewrite the right-hand side of (2.3) in the form

$$I_0 = -\frac{(2-\delta)\delta}{4\pi}\int_{-\pi}^{\pi}\frac{\ln\varphi(\tau)\,d\tau}{\delta^2+4(1-\delta)\sin^2[(\tau-\theta)/2]}. \tag{2.4}$$

Here $4\sin^2[(\tau-\theta)/2] = (\tau-\theta)^2 + O((\tau-\theta)^4)$, since $-\pi-\varepsilon \leq \tau-\theta \leq \pi$. Therefore

$$\begin{aligned}&|\delta(2-\delta)/A(\tau,\theta,\delta) - 2\delta/[\delta^2+(\tau-\theta)^2]| \\ &\quad\leq \delta\left|\frac{2-\delta}{\delta^2+4(1-\delta)\sin^2[(\tau-\theta)/2]} - \frac{2}{\delta^2+4\sin^2[(\tau-\theta)/2]}\right| \\ &\quad\quad+\delta\left|\frac{2}{\delta^2+4\sin^2[(\tau-\theta)/2]} - \frac{2}{\delta^2+(\tau-\theta)^2}\right| \leq C_7(\varepsilon),\end{aligned}$$

in view of which

$$(2.5)\quad -2\pi I_0 = \delta\int_0^\pi \frac{\ln\omega(\tau)\,d\tau}{\delta^2+(\tau-\theta)^2} + \delta\int_0^\pi \frac{\ln\omega(\tau)\,d\tau}{\delta^2+(\tau+\theta)^2} + O(1) = I_1+\cdots+I_4+O(1)$$

uniformly with respect to $\theta\in[0,\varepsilon]$ and $\delta\in(0,\frac{1}{2}]$, where

$$I_1 = \int_{\theta/\delta}^{(\pi-\theta)/\delta} \frac{\ln\omega(\delta u-\theta)}{1+u^2}\,du, \qquad I_2 = \int_{(\pi-\theta)/\delta}^{(\pi+\theta)/\delta} \frac{\ln\omega(\delta u-\theta)}{1+u^2}\,du,$$

$$I_3 = \int_0^{\theta/\delta} \frac{\ln\omega(\theta-\delta u)}{1+u^2}\,du, \qquad I_4 = \int_0^{(\pi-\theta)/\delta} \frac{\ln\omega(\delta u+\theta)}{1+u^2}\,du.$$

The numerator in the integral I_2 is $\le C_8(\varepsilon)$ in modulus, since $\pi-2\varepsilon\le\delta u-\theta\le\pi$. Consequently,

$$(2.6)\qquad |I_2|\le C_8(\varepsilon)\pi/2.$$

Consider the integral I_4. By (2.2) and the fact that $\omega(\delta)$ is nondecreasing,

$$1\le\omega(\delta u+\theta)/\omega(\delta+\theta)\le 2u \qquad (u\ge 1),$$

$$1\le\frac{\omega(\delta+\theta)}{\omega(\delta u+\theta)}\le\frac{2(\delta+\theta)}{\delta u+\theta}\le\frac{2(\delta+\theta)}{\delta u+\theta u}=\frac{2}{u} \qquad (0<u<1).$$

Therefore,

$$(2.7)\qquad 0\le\ln[\omega(\delta u+\theta)/\omega(\delta+\theta)]\le\ln(2u) \qquad (u\ge 1),$$

$$(2.8)\qquad 0\le\ln[\omega(\delta+\theta)/\omega(\delta u+\theta)]\le\ln(2/u) \qquad (0<u<1).$$

It follows from (2.7) and (2.8) that

$$\left|I_4-\int_0^{(\pi-\theta)/\delta}\frac{du}{1+u^2}\ln\omega(\delta+\theta)\right|\le\int_0^\infty(1+u^2)^{-1}(\ln 2+|\ln u|)\,du=C_9<\infty.$$

But then

$$\begin{aligned} I_4 &= \ln\omega(\delta+\theta)\int_0^{(\pi-\theta)/\delta}(1+u^2)^{-1}\,du+O(1)\\ &= \int_0^\infty(1+u^2)^{-1}\,du\ln\omega(\delta+\theta)+O(1)+O(\delta)\ln\omega(\delta+\theta) \qquad (2.9)\\ &= (\pi/2)\ln\omega(\delta+\theta)+O(1)+O(\delta+\theta)\ln\omega(\delta+\theta)\end{aligned}$$

uniformly in $\theta\in[0,\varepsilon]$ and $\delta\in(0,1/2]$. Using (2.2), we conclude that

$$0\le(\theta+\delta)\ln[\omega(\pi)/\omega(\theta+\delta)]\le(\theta+\delta)\ln[2\pi/(\theta+\delta)].$$

Therefore, it follows from (2.9) that

$$(2.10)\qquad I_4=(\pi/2)\ln\omega(\delta+\theta)+O(1) \qquad (\delta\to+0)$$

uniformly in θ.

To estimate I_3 we apply the inequalities

$$0\le\ln[\omega(\theta+\delta u)/\omega(\theta-\delta u)]\le\ln[2(\theta+\delta u)/(\theta-\delta u)]$$

(see (2.2)). Then we obtain

$$(2.11)\qquad \begin{aligned} &\left|I_3-\int_0^{\theta/\delta}\frac{\ln\omega(\theta+\delta u)\,du}{1+u^2}\right| = \int_0^{\theta/\delta}\ln\frac{\omega(\theta+\delta u)}{\omega(\theta-\delta u)}\frac{du}{1+u^2}\\ &\quad\le\int_0^{\theta/\delta}\ln\frac{2(\theta+\delta u)}{\theta-\delta u}\frac{du}{1+u^2}=\frac{\theta}{\delta}\int_0^1\ln\frac{2(1+v)}{1-v}\frac{dv}{1+(\theta v/\delta)^2}=I_5.\end{aligned}$$

If $\theta < \delta$, then

$$I_5 \leq \int_0^1 \ln[2(1+v)/(1-v)]\,dv = C_{10} < \infty. \tag{2.12}$$

And if $\theta > \delta$, then

$$\begin{aligned} I_5 &\leq \frac{\theta}{\delta}\int_0^{1/2} \frac{\ln 6}{1+(\theta v/\delta)^2}dv + \frac{\delta}{\theta}\int_{1/2}^1 \left[\left(\frac{\delta}{\theta}\right)^2 + v^2\right]^{-1} \ln\frac{2(1+v)}{1-v}\,dv \\ &\leq \ln 6 \int_0^\infty \frac{du}{1+u^2} + \int_{1/2}^1 \ln\frac{2(1+v)}{1-v}\frac{dv}{v^2} = C_{11} < \infty. \end{aligned} \tag{2.13}$$

From (2.11)–(2.13) we find that

$$I_3 = \int_0^{\theta/\delta} (1+u^2)^{-1} \ln\omega(\theta+\delta u)\,du + O(1) \qquad (\delta \to +0) \tag{2.14}$$

uniformly in θ. Using the inequalities (see (2.2))

$$0 \leq \ln[\omega(\delta u+\theta)/\omega(\delta u-\theta)] \leq \ln[2(\delta u+\theta)/(\delta u-\theta)],$$

we obtain

$$\begin{aligned} \left| I_1 - \int_{\theta/\delta}^{(\pi-\theta)/\delta} \frac{\ln\omega(\delta u+\theta)}{1+u^2}du \right| &\leq \int_{\theta/\delta}^{(\pi-\theta)/\delta} \ln\frac{\omega(\delta u+\theta)}{\omega(\delta u-\theta)}\frac{du}{1+u^2} \\ &\leq \int_{\theta/\delta}^{(\pi-\theta)/\delta} \ln\frac{2(\delta u+\theta)}{\delta u-\theta}\frac{du}{1+u^2} = \frac{\theta}{\delta}\int_1^{(\pi-\theta)/\theta} \ln\frac{2(v+1)}{v-1}\frac{dv}{1+(\theta v/\delta)^2} \\ &\leq \frac{\theta}{\delta}\int_1^2 \ln\frac{2(v+1)}{v-1}\frac{dv}{1+(\theta v/\delta)2} + \frac{\theta}{\delta}\int_2^\infty \frac{\ln 6\,dv}{1+(\theta v/\delta)^2} = I_6 + I_7. \end{aligned} \tag{2.15}$$

It is obvious that

$$I_7 \leq \ln 6 \int_0^\infty (1+u^2)^{-1}\,du = (\pi/2)\ln 6. \tag{2.16}$$

If $\theta \leq \delta$, then

$$I_6 \leq \int_1^2 \ln[2(v+1)/(v-1)]\,dv, \tag{2.17}$$

and for $\theta > \delta$

$$I_6 \leq \frac{\delta}{\theta}\int_1^2 \ln\frac{2(v+1)}{v-1}\frac{dv}{v^2} < \int_1^2 \ln\frac{2(v+1)}{v-1}\frac{dv}{v^2}. \tag{2.18}$$

It follows from (2.15)–(2.18) that

$$I_1 = \int_{\theta/\delta}^{(\pi-\theta)/\delta} (1+u^2)^{-1}\ln\omega(\delta u+\theta)\,du + O(1) \qquad (\delta \to +0) \tag{2.19}$$

uniformly in θ.

By (2.5), (2.6), (2.10), (2.14), and (2.19)

$$I_0 = -(2\pi)^{-1}[2I_4 + O(1)] = -2^{-1}\ln\omega(\delta+\theta) + O(1) \tag{2.20}$$

uniformly in $\theta \in [0, \varepsilon]$ and $\delta \in (0, 1/2]$. The validity of Lemma 2.1 for $\theta \in [0, \varepsilon]$ follows from (2.3), (2.4), and (2.20).

Corollary 2.1. *Lemma* 2.1 *remains in force if in its formulation* $\omega(\delta)$ *is replaced by* $\omega_k(\delta)$, *where* $\omega_k(\delta)$ *is a function of* kth $(k \in \mathbf{N})$ *modulus of continuity type, that is,* $\omega_k(\delta)$ *is continuous, does not decrease,* $\omega_k(0) = 0$, *and for all* $n \in \mathbf{N}$

$$\omega_k(n\delta) \leq n^k \omega(\delta) \qquad (0 \leq \delta \leq \pi/n).$$

Indeed, in this case $g(\delta) = \{\omega_k(\delta)\}^{1/k}$ is a function of the first modulus of continuity type, and for it by the N. K. Bari and S. B. Stechkin theorem [26] (see also [27], p. 170) there is a modulus of continuity $\omega(\delta) \asymp g(\delta)$ $(0 \leq \delta \leq \pi)$. Let us define for $-\pi \leq \tau \leq \pi$ weights ψ, ξ, h by the equalities

$$\psi(\tau) = \omega(|\tau|), \qquad \xi(\tau) = g(|\tau|), \qquad h(\tau) = \xi(\tau)/\psi(\tau). \tag{2.21}$$

We note that if u, v are weights with integrable logarithms, then (see (1.9))

$$\pi(uv\,;\,z) = \pi(u\,;\,z)\pi(v\,;\,z) \qquad (z \in D). \tag{2.22}$$

By (2.21), (2.22), $\pi(\xi\,;\,z) = \pi(\psi\,;\,z)\pi(h\,;\,z)$, and, as is known (see [2], §1.8),

$$\|h\|_\infty^{-1/2} \leq |\pi(h\,;\,z)| \leq \|1/h\|_\infty^{1/2} \qquad (z \in D), \tag{2.23}$$

so that, applying Lemma 2.1 to the weight ψ, we have

$$\begin{aligned} |\pi(\xi\,;\,\rho e^{i\theta})| &\asymp |\pi(\psi\,;\,\rho e^{i\theta})| \asymp [\omega(1-\rho+|\sin(\theta/2)|)]^{-1/2} \\ &\asymp [g(1-\rho+|\sin(\theta/2)|)]^{-1/2} = [\omega_k(1-\rho+|\sin(\theta/2)|)]^{-1/2k}. \end{aligned}$$

It remains to put $\varphi(\tau) = \omega_k(|\tau|)$ on $[-\pi, \pi]$ and to take into account that then $\pi(\varphi\,;\,z) = [\pi(\xi\,;\,z)]^k$.

Corollary 2.2. *Let the weight* φ *be given by the formula*

$$\varphi(\tau) = \omega_k(|\sin(\tau/2)|) \qquad (-\pi \leq \tau \leq \pi), \tag{2.24}$$

where $\omega_k(u)$ *is a function of the* kth $(k \in \mathbf{N})$ *modulus of continuity type. Then constants* $C_{12}(\varphi)$ *and* $C_{13}(\varphi) > 0$ *can be found such that, for all* $\theta \in \mathbf{R}$ *and* $\rho \in [\frac{1}{2}, 1]$,

$$C_{12}(\varphi) \leq |\pi(\varphi\,;\,\rho e^{i\theta})|\,[\omega_k(|\sin(\theta/2)| + 1 - \rho)]^{1/2} \leq C_{13}(\varphi). \tag{2.25}$$

Proof. Set

$$\psi(\tau) = \omega_k(|\tau|), \qquad \varphi(\tau) = \omega_k(|\sin(\tau/2)|), \qquad h(\tau) = \varphi(\tau)/\psi(\tau)$$

on $[-\pi, \pi]$. Then h, $1/h \in L^\infty$ and on the basis of (2.22), (2.23)

$$0 < C_{14}(h) \leq |\pi(\varphi\,;\,z)/\pi(\psi\,;\,z)| \leq C_{15}(h) \qquad (z \in D).$$

It remains to apply Corollary 2.1 to the weight ψ.

The following result is proved with the aid of Corollary 2.2.

Theorem 2.1. *Suppose that the weight* φ *is of the form*

$$\varphi(\tau) = h(\tau) \prod_{\nu=1}^{m} F_\nu\left(\left|\sin\frac{\tau - \theta_\nu}{2}\right|\right), \tag{2.26}$$

where $-\pi < \theta_1 < \cdots < \theta_m \leq \pi$,

$$h(\tau) \geq 0, \qquad h \text{ and } 1/h \in L^\infty, \tag{2.27}$$

$$F_\nu(u) = \prod_{\mu=1}^{l_\nu} [\omega_{k(\mu,\nu)}(u)]^{\alpha(\mu,\nu)}, \qquad m, l_\nu, k(\mu,\nu) \in \mathbf{N}, \ \alpha(\mu,\nu) \in \mathbf{R}, \tag{2.28}$$

and $\omega_{k(\mu,\nu)}(u)$ $(0 \le u \le \pi)$ *is a function of modulus of continuity type. Then there exist positive constants* $C_{16}(\varphi)$ *and* $C_{17}(\varphi)$ *such that*

$$C_{16}(\varphi) \le |\pi(\varphi; \rho e^{i\theta})| \left\{ \prod_{\nu=1}^{m} F_\nu \left(\left| \sin \frac{\theta - \theta_\nu}{2} \right| + 1 - \rho \right) \right\}^{1/2} \le C_{17}(\varphi) \tag{2.29}$$

for all $\theta \in \mathbf{R}$ *and* $\rho \in [\frac{1}{2}, 1]$.

Proof. If $\theta_0 \in \mathbf{R}$, ψ is a weight and $\ln \psi \in L^1$, then by (1.9)

$$\pi(\psi^\alpha(\tau - \theta_0); z) = \alpha\pi(\psi(\tau); e^{-i\theta_0} z) \qquad (z \in D). \tag{2.30}$$

It follows from (2.26), (2.27), (2.30), and (2.22) that

$$\pi(\varphi; z) = \pi(h; z) \prod_{\nu=1}^{m} \prod_{\mu=1}^{l_\nu} \alpha(\mu, \nu)\pi(\varphi(\mu, \nu; \tau); e^{-i\theta_\nu} z),$$

where $\varphi(\mu, \nu; \tau) = \omega_{k(\mu,\nu)}(|\sin(\tau/2)|)$. In view of Corollary 2.2 this implies (2.29).

Corollary 2.3. *In the formulation of Theorem* 2.1 *in formulas* (2.26) *and* (2.29), $\sin[(\tau - \theta_\nu)/2]$ *and* $\sin[(\theta - \theta_\nu)/2]$ *may be replaced, respectively, by* $\cos\tau - \cos\theta_\nu$ *and* $\cos\theta - \cos\theta_\nu$ *for those values of* ν *for which* $\theta_\nu \ne 0$, $\theta_\nu \ne \pi$.

3. Estimates of moduli of continuity

One of our primary goals is to find classes of weights φ with a finite number of singular points for which the quantity

$$\delta_n(\varphi) = \inf_{c_0, \dots, c_n \in \mathbf{C}} \|[\pi(\varphi; e^{i\tau}) - c_0 - \cdots - c_n e^{in\tau}]\sqrt{\varphi(\tau)}\|_2 \tag{3.1}$$

is $O(n^{-1/2})$ or $o(n^{-1/2})$ as $n \to \infty$. As is known ([2], formula (2.8)), $\delta_n(\varphi)$ coincides with $\delta_{\sigma,n}$ (see (1.13)), where $d\sigma(\tau) = \varphi(\tau)\,d\tau$. It is also known ([2], formula (3.9)), that if $\ln\varphi$, $\varphi^{-1} \in L^1$, then

$$\delta_n(\varphi) \le 12\omega(\pi(\varphi; e^{i\tau}); n^{-1})_{2,\varphi}, \tag{3.2}$$

where

$$\omega(f; \delta)_{2,\varphi} = \sup_{|\lambda| \le \delta} \|[f(\tau + \lambda) - f(\tau)]/\sqrt{\varphi(\tau)}\|_2. \tag{3.3}$$

Having in mind possible use of formula (3.2), we establish a number of lemmas concerning the quantity (3.3).

Lemma 3.1. *Let* φ *be a weight such that*

$$\varphi^{-1} \in L^1. \tag{3.4}$$

Then the mean-square modulus of continuity satisfies the inequality

$$\omega(\ln\varphi; \delta)_2 \le 2\sqrt{2}\omega(\varphi^{-1/2}; \delta)_{2,\varphi} \qquad (\delta \ge 0). \tag{3.5}$$

Proof. If $0 < r < \infty$, then, partitioning $[0, 2\pi]$ into the sets e_1 and e_2 on which $\varphi(\tau) \ge 1$ and $\varphi(\tau) < 1$, respectively, we have

$$0 \le r^{-1}\ln\varphi(\tau) \le \varphi^{1/r}(\tau) - 1 \le \varphi^{1/r}(\tau) \qquad (\tau \in e_1),$$
$$0 \le r^{-1}\ln\varphi^{-1}(\tau) \le \varphi^{-1/r}(\tau) \qquad (\tau \in e_2).$$

Hence, in view of $\varphi^{\pm 1} \in L^1$, it follows that $\ln\varphi \in L^r$. Therefore, the left-hand side of (3.5) is meaningful. Partitioning $[0, 2\pi]$ into sets $e_{1,\lambda}$ and $e_{2,\lambda}$ on which

$\varphi(\tau) \geq \varphi(\tau+\lambda)$ and $\varphi(\tau) < \varphi(\tau+\lambda)$, respectively, for each $\lambda \in [-\delta, \delta]$, we conclude that

$$0 \leq \ln[\varphi(\tau)/\varphi(\tau+\lambda)]^{1/2} \leq [\varphi(\tau)/\varphi(\tau+\lambda)]^{1/2} - 1 \qquad (\tau \in e_{1,\lambda}), \tag{3.6}$$

$$0 \leq \ln[\varphi(\tau+\lambda)/\varphi(\tau)]^{1/2} \leq [\varphi(\tau+\lambda)/\varphi(\tau)]^{1/2} - 1 \qquad (\tau \in e_{2,\lambda}). \tag{3.7}$$

From (3.6), (3.7), and the equality (see (3.3))

$$\omega(\varphi^{-1/2}; \delta)_{2,\varphi} = \sup_{|\lambda|\leq\delta} \|[\varphi(\tau)/\varphi(\tau+\lambda)]^{1/2} - 1\|_2 \tag{3.8}$$

it follows that

$$\begin{aligned}\|\ln[\varphi(\tau)/\varphi(\tau+\lambda)]^{1/2}\|_2^2 &\leq \|[\varphi(\tau)/\varphi(\tau+\lambda)]^{1/2} - 1\|_2^2 + \|[\varphi(\tau+\lambda)/\varphi(\tau)]^{1/2} - 1\|_2^2 \\ &\leq 2[\omega(\varphi^{-1/2}; \delta)_{2,\varphi}]^2,\end{aligned}$$

whence (3.5) follows.

Lemma 3.2. *Under the conditions of Lemma* 3.1

$$\omega(\varphi^{-1/2}; \delta)_{2,\varphi} \leq \omega(\pi(\varphi; e^{i\tau}); \delta)_{2,\varphi} \leq (1+\sqrt{2})\omega(\varphi^{-1/2}; \delta)_{2,\varphi}. \tag{3.9}$$

Proof. Taking into account that ([2], formula (2.6))

$$\pi(\varphi; e^{i\tau}) = [\varphi(\tau)]^{-1/2} e^{i\gamma(\tau)} \tag{3.10}$$

a.e. on $[0, 2\pi]$, where $\gamma(\tau)$ is the conjugate function for $\ln[\varphi(\tau)]^{-1/2}$, we have

$$|\pi(\varphi; e^{i\tau})/\pi(\varphi; e^{i(\tau+\lambda)}) - 1| \geq |\sqrt{\varphi(\tau+\lambda)/\varphi(\tau)} - 1|,$$

whence, by (3.8), there follows the first inequality in (3.9).

Applying (3.10), we obtain the inequality

$$\left|\frac{\pi(\varphi; e^{i\tau})}{\pi(\varphi; e^{i(\tau+\lambda)})} - 1\right| \leq \left|\sqrt{\frac{\varphi(\tau+\lambda)}{\varphi(\tau)}} - 1\right| + |\gamma(\tau+\lambda) - \gamma(\tau)|, \tag{3.11}$$

which, in view of the M. Riesz theorem ([28], Chapter VII, §§1, 14), implies that for $\lambda \in [-\delta, \delta]$ the L^2-norm of the left side of (3.11) does not exceed $\omega(\varphi^{-1/2}; \delta)_{2,\varphi} + \omega(\ln\varphi^{1/2}; \delta)_2$. Now, to obtain the second inequality in (3.9) it remains to apply Lemma 3.1.

Lemma 3.3. *Let the weight* φ *be given by the formula*

$$\varphi(\tau) = \omega(|\tau|) \qquad (-\pi \leq \tau \leq \pi), \tag{3.12}$$

where $\omega(\tau)$ *is a concave modulus of continuity such that*

$$\omega^{-1} \in L^1[0, \pi], \qquad \int_0^\delta \frac{d\tau}{\omega(\tau)} = O\left(\frac{\delta}{\omega(\delta)}\right) \qquad (\delta \to +0). \tag{3.13}$$

Then

$$\omega(\varphi^{-1/2}; \delta)_{2,\varphi} = O(\delta^{1/2}) \qquad (\delta \to +0). \tag{3.14}$$

Proof. We may assume that $\delta < \varepsilon/2$, $\varepsilon < \pi/2$ in (3.8). We may also assume that $\lambda > 0$, since

$$\|[\varphi(\tau)/\varphi(\tau-\lambda)]^{1/2} - 1\|_2 = \|[\varphi(\tau)/\varphi(\tau+\lambda)]^{1/2} - 1\|_2.$$

By the concavity of $\omega(\tau)$ we have the formula ([29], p. 24)

$$\omega(\theta) - \omega(\eta) = \int_\eta^\theta g(\tau)\,d\tau \qquad (0 \leq \eta < \theta \leq \pi), \tag{3.15}$$

where $g(\tau)$ does not increase and coincides with $\omega'(\tau)$ almost everywhere on $[0, \pi]$. It follows from (3.15) that if $\varepsilon \le \delta_1 < \delta_2 \le \pi$ then

$$\omega(\delta_2) - \omega(\delta_1) \le g(\delta_1)(\delta_2 - \delta_1) \le g(\varepsilon)(\delta_2 - \delta_1). \tag{3.16}$$

Therefore, the weight φ satisfies a Lipschitz condition on the segments $[-\pi, -\varepsilon/2]$ and $[\varepsilon, \pi + \varepsilon/2]$, as a consequence of which

$$\|\sqrt{\varphi(\tau)/\varphi(\tau+\lambda)} - 1\|_2 \le J + O(\lambda), \tag{3.17}$$

where

$$J = \left\{ \frac{1}{2\pi} \int_{-\varepsilon}^{\varepsilon} \left[\sqrt{\frac{\varphi(\tau)}{\varphi(\tau+\lambda)}} - 1 \right]^2 d\tau \right\}^{1/2}.$$

Taking (3.12) into account, we have

(3.18)

$$\begin{aligned} 2\pi J^2 &= \left\{ \int_{-\varepsilon}^{-2\lambda} + \int_{-2\lambda}^{-\lambda} + \int_{-\lambda}^{0} + \int_{0}^{\lambda} + \int_{\lambda}^{\varepsilon} \right\} \left[\sqrt{\frac{\varphi(\tau)}{\varphi(\tau+\lambda)}} - 1 \right]^2 d\tau \\ &= \left\{ \int_{2\lambda}^{\varepsilon} + \int_{\lambda}^{2\lambda} \right\} \left[\sqrt{\frac{\omega(\tau)}{\omega(\tau-\lambda)}} - 1 \right]^2 d\tau + \int_0^{\lambda} \left[\sqrt{\frac{\omega(\tau)}{\omega(\lambda-\tau)}} - 1 \right]^2 d\tau \\ &\quad + \int_0^{\lambda} \left[\sqrt{\frac{\omega(\tau)}{\omega(\tau+\lambda)}} - 1 \right]^2 d\tau + \int_{\lambda}^{\varepsilon} \left[\sqrt{\frac{\omega(\tau)}{\omega(\tau+\lambda)}} - 1 \right]^2 d\tau = J_1 + \cdots + J_5. \end{aligned}$$

Using (3.13), we obtain

(3.19)

$$\begin{aligned} J_2 + J_3 + J_4 &\le \int_{\lambda}^{2\lambda} \frac{\omega(\tau) - \omega(\tau-\lambda)}{\omega(\tau-\lambda)} d\tau + \int_0^{\lambda} \frac{|\omega(\tau) - \omega(\lambda-\tau)|}{\omega(\lambda-\tau)} d\tau \\ &\quad + \int_0^{\lambda} \frac{\omega(\tau+\lambda) - \omega(\tau)}{\omega(\tau+\lambda)} d\tau \le 3\omega(\lambda) \int_0^{\lambda} \frac{du}{\omega(u)} = O(\lambda) \qquad (\lambda \to +0). \end{aligned}$$

The inequality

$$\omega(\delta) \ge g(\delta)\delta \qquad (0 < \delta \le \pi) \tag{3.20}$$

follows from (3.15) for $\theta = \delta$ and $\eta = 0$. Applying (3.16) and (3.20), we find that

$$\begin{aligned} J_1 + J_5 &\le \int_{2\lambda}^{\varepsilon} \left[\frac{\omega(\tau) - \omega(\tau-\lambda)}{\omega(\tau-\lambda)} \right]^2 d\tau + \int_{\lambda}^{\varepsilon} \left[\frac{\omega(\tau+\lambda) - \omega(\tau)}{\omega(\tau)} \right]^2 d\tau \\ &\le \lambda^2 \int_{2\lambda}^{\varepsilon} \frac{d\tau}{(\tau+\lambda)^2} + \lambda^2 \int_{\lambda}^{\varepsilon} \frac{d\tau}{\tau^2} = O(\lambda). \end{aligned} \tag{3.21}$$

Now (3.14) follows from (3.8), (3.17)–(3.19), and (3.21).

Lemma 3.4. *Let ω be a concave modulus of continuity and let E be the set of those points on the interval $(0, \pi)$ at which the derivative $\omega'(\tau)$ exists. If $\tau\omega'(\tau)/\omega(\tau) \to 0$ as $\tau \to 0$, remaining in E, and the weight φ is given by formula* (3.12), *then*

$$\omega(\varphi^{-1/2}; \delta)_{2,\varphi} = o(\delta^{1/2}) \qquad (\delta \to +0). \tag{3.22}$$

Proof. We start by showing that the first of the conditions (3.13) is fulfilled. To this end, we observe that for $\alpha > 0$ the function $G(\tau) = \tau^\alpha/\omega(\tau)$ is absolutely continuous on $[\varepsilon, \pi]$ for each $\varepsilon \in (0, \pi)$: both the numerator and the denominator of the quotient are absolutely continuous. It is clear from the equality

$$\tau G'(\tau) = [\alpha - \tau\omega'(\tau)/\omega(\tau)]G(\tau) \qquad (\tau \in E)$$

that $G'(\tau) \geq 0$, provided $\tau \in E \cap (0, \varepsilon)$ and ε is sufficiently small. Therefore

$$0 < \tau_1 < \tau_2 \leq \varepsilon \Rightarrow G(\tau_2) - G(\tau_1) = \int_{\tau_1}^{\tau_2} G'(\tau)\, d\tau \geq 0. \tag{3.23}$$

Thus, $G(\tau) \leq G(\varepsilon)$ $(0 < \tau \leq \varepsilon)$. Obviously, $G(\tau)$ is bounded on $[\varepsilon, \pi]$ as well. Therefore,

$$\forall \alpha \in (0, \infty) \qquad 1/\omega(\tau) = O(\tau^{-\alpha}) \qquad (0 < \tau \leq \pi).$$

For $0 < \alpha < r^{-1}$ it follows from here that $1/\omega \in L^r[0, \pi]$ $(0 < r < \infty)$. Thus, the norm in (3.8) is finite.

Let us show that the function g (see (3.15)) satisfies the condition

$$\lim_{\tau \to +0} \tau g(\tau)/\omega(\tau) = 0. \tag{3.24}$$

One of the conditions of the lemma states that (3.24) holds for $\tau \in E$. Therefore,

$$\forall \varepsilon > 0\ \exists \delta' > 0: \quad \tau \in (0, \delta') \cap E \Rightarrow \tau g(\tau)/\omega(\tau) < \varepsilon.$$

Let $0 < \theta < \delta'$, $\theta \notin E$. Then we can find a sequence $\{\tau_n\}_1^\infty$ such that $\tau_n \in E$, $\tau_n \uparrow \theta$. Since $g(\tau)\downarrow$, it follows that

$$\frac{\theta g(\theta)}{\omega(\theta)} \leq \frac{\theta}{\omega(\theta)} g(\tau_n) = \frac{\tau_n[1 + o(1)]}{\omega(\tau_n)[1 + o(1)]} g(\tau_n) < \varepsilon[1 + o(1)].$$

Passing here to the limit as $n \to \infty$, we obtain the inequality $\theta g(\theta)/\omega(\theta) \leq \varepsilon$. Thus, equality (3.24) is proved.

Put $\mu(\lambda) = \sup\{\tau g(\tau)/\omega(\tau) : 0 < \tau \leq \lambda\}$. It follows from (3.24) that

$$\eta \overset{\text{def}}{=} \lambda \mu(\lambda) = o(\lambda) \qquad (\lambda \to +0). \tag{3.25}$$

We shall suppose that $\eta < \lambda/2$ (this can be achieved by choosing a sufficiently small δ in (3.8)). Let us show that

$$\omega(\lambda) \int_0^\eta \frac{d\tau}{\omega(\tau)} = o(\lambda) \qquad (\lambda \to +0). \tag{3.26}$$

To this end, we note that under the conditions of the lemma

$$\int_0^\theta \frac{d\tau}{\omega(\tau)} \asymp \frac{\theta}{\omega(\theta)} \qquad (0 < \theta \leq \pi). \tag{3.27}$$

Indeed, $\omega(\tau)\uparrow$; therefore, the left-hand side of (3.27) is not smaller than the right-hand side. By (3.23) the function $\tau^{1/2}/\omega(\tau)$ does not decrease in $(0, \varepsilon]$, so that for $\theta \in (0, \varepsilon]$

$$\int_0^\theta \frac{d\tau}{\omega(\tau)} = \int_0^\theta \frac{\sqrt{\tau}\, d\tau}{\omega(\tau)\sqrt{\tau}} \leq \frac{\sqrt{\theta}}{\omega(\theta)} \int_0^\theta \frac{d\tau}{\sqrt{\tau}} = 2\frac{\theta}{\omega(\theta)}.$$

By (3.16), $\omega(\lambda) \leq \omega(\eta) + \lambda g(\eta)$. Therefore,

$$0 \leq \omega(\lambda)\eta/\omega(\eta) \leq \eta + \lambda\eta g(\eta)/\omega(\eta).$$

This and (3.24) imply (3.26).

Using (3.17), (3.18) we estimate the left-hand side of (3.17). It is not difficult to verify that

$$J_4 \leq J_2 = \int_0^\lambda \left[\sqrt{\frac{\omega(\tau + \lambda)}{\omega(\tau)}} - 1\right]^2 d\tau, \tag{3.28}$$

and that J_1 and J_5 do not exceed the integral over the segment $[\lambda, \varepsilon]$ of the integrand in (3.28). Therefore,

$$(J_2 + J_4) + (J_1 + J_5) \le 2\left(\int_0^\eta + \int_\eta^\varepsilon\right)\left[\sqrt{\frac{\omega(\tau+\lambda)}{\omega(\tau)}} - 1\right]^2 d\tau. \tag{3.29}$$

Since $(a-b)^2 \le a^2 + b^2$ $(a, b \ge 0)$, by (3.25) and (3.26) we have

$$\int_0^\eta \left[\sqrt{\frac{\omega(\tau+\lambda)}{\omega(\tau)}} - 1\right]^2 d\tau \le \int_0^\eta \frac{\omega(\tau+\lambda)}{\omega(\tau)}\, d\tau + \eta \le \eta + 2\omega(\lambda)\int_0^\eta \frac{d\tau}{\omega(\tau)} = o(\lambda). \tag{3.30}$$

At the same time, by (3.16) and (3.25)

(3.31)

$$\begin{aligned}\int_\eta^\varepsilon \left[\sqrt{\frac{\omega(\tau+\lambda)}{\omega(\tau)}} - 1\right]^2 d\tau &\le \int_\eta^\varepsilon \left[\frac{\omega(\tau+\lambda) - \omega(\tau)}{\omega(\tau)}\right]^2 d\tau \le \int_\eta^\varepsilon \left[\frac{\lambda g(\tau)}{\omega(\tau)}\right]^2 d\tau \\ &\le \lambda^2\mu^2(\lambda)\int_\eta^\lambda \frac{d\tau}{\tau^2} + \lambda^2\mu^2(\sqrt{\lambda})\int_\lambda^{\sqrt{\lambda}} \frac{d\tau}{\tau^2} + \lambda^2\mu^2(\varepsilon)\int_{\sqrt{\lambda}}^\varepsilon \frac{d\tau}{\tau^2} \\ &\le \eta + \lambda\mu^2(\sqrt{\lambda}) + \lambda^{3/2}\mu^2(\varepsilon) = o(\lambda).\end{aligned}$$

Similarly, we obtain

$$\begin{aligned}\frac{1}{2}J_3 &\le \int_0^{\lambda/2}\left[\sqrt{\frac{\omega(\lambda-\tau)}{\omega(\tau)}} - 1\right]^2 d\tau \\ &\le \int_0^\eta \frac{\omega(\lambda-\tau)}{\omega(\tau)}\, d\tau + \eta + \int_\eta^{\lambda/2}\left[\frac{\omega(\lambda-\tau)-\omega(\tau)}{\omega(\tau)}\right]^2 d\tau \\ &\le \omega(\lambda)\int_0^\eta \frac{d\tau}{\omega(\tau)} + \eta + \int_\eta^{\lambda/2}\left[\frac{\lambda g(\tau)}{\omega(\tau)}\right]^2 d\tau = o(\lambda).\end{aligned} \tag{3.32}$$

Now (3.22) follows from (3.17), (3.18), and (3.29)–(3.32).

Remark 3.1. Lemmas 3.3 and 3.4 remain in force if condition (3.12) in their formulations is replaced by the condition

$$\varphi(\tau) = \omega(|\sin(\tau/2)|) \qquad (-\pi \le \tau \le \pi). \tag{3.33}$$

Indeed, if ω is a concave modulus of continuity on $[0, \pi]$ satisfying the conditions of Lemma 3.3 or of Lemma 3.4, then the function $\Omega(\tau) = \omega(|\sin(\tau/2)|)$ has the same properties.

Lemma 3.5. *Let the weight* φ *have the form*

$$\varphi(\tau) = \omega(|\cos\tau - \cos\theta_0|) \qquad (\tau \in \mathbf{R},\ 0 < \theta_0 < \pi), \tag{3.34}$$

where ω *satisfies the conditions of Lemma* 3.3. *Then the estimate* (3.14) *holds.*

The proof is obtained by a modification of the proof of Lemma 3.3. Namely, inequality (3.17) is replaced by the inequality

$$\|\sqrt{\varphi(\tau)/\varphi(\tau+\lambda)} - 1\|_2 \le J' + J'' + O(\lambda), \tag{3.35}$$

where

$$J' = k([\theta_0 - \varepsilon, \theta_0 + \varepsilon]), \qquad J'' = k([-\theta_0 - \varepsilon, -\theta_0 + \varepsilon]),$$

$$k(T) = \left\{\frac{1}{2\pi}\int_T \left[\sqrt{\frac{\varphi(\tau)}{\varphi(\tau+\lambda)}} - 1\right]^2 d\tau\right\}^{1/2}.$$

Setting $e_1'=[\theta_0-\varepsilon\,,\ \theta_0-2\lambda]$, $e_2'=[\theta_0-2\lambda\,,\ \theta_0-\lambda]$, $e_3'=[\theta_0-\lambda\,,\ \theta_0]$, $e_4'=[\theta_0\,,\ \theta_0+\lambda]$, $e_5'=[\theta_0+\lambda\,,\ \theta_0+\varepsilon]$ and defining segments e_k'' $(k=1\,,\,\dots\,,\,5)$ by replacing θ_0 by $-\theta_0$ in the definition of e_k', we consider the integrals $J_k'=k(e_k')$, $J_k''=k(e_k'')$. In this way we obtain

$$2\pi(J')^2=J_1'+\cdots+J_5', \qquad 2\pi(J'')^2=J_1''+\cdots+J_5''.$$

The integrals J_k' and J_k'' are estimated analogously to the integrals J_k in the proof of Lemma 3.3.

The proof of the following analog of Lemma 3.4 for the weight (3.34) is more complicated than the proof of Lemma 3.5 and for that reason is carried out more fully.

Lemma 3.6. *Let the weight* φ *be defined by formula* (3.34), *with* ω *satisfying the conditions of Lemma* 3.4. *Then the estimate* (3.22) *holds.*

Proof. By (3.35), to prove (3.22) it suffices to establish that $J'=o(\lambda^{1/2})$ and $J''=o(\lambda^{1/2})$. We estimate J' only, since the arguments for J'' are absolutely analogous. Dividing the segment $[\theta_0-\varepsilon\,,\ \theta_0+\varepsilon]$ into parts, applying (3.34), and introducing the notation

$$\xi(\tau)=\cos\tau-\cos\theta_0\,, \tag{3.36}$$

we obtain

$$2\pi(J')^2=\int_{\theta_0-\varepsilon}^{\theta_0-\lambda}\left[\sqrt{\frac{\omega(\xi(\tau))}{\omega(\xi(\tau+\lambda))}}-1\right]^2 d\tau+\int_{\theta_0-\lambda}^{\theta_0}\left[\sqrt{\frac{\omega(\xi(\tau))}{\omega(-\xi(\tau+\lambda))}}-1\right]^2 d\tau$$
$$+\int_{\theta_0}^{\theta_0+\varepsilon}\left[\sqrt{\frac{\omega(-\xi(\tau))}{\omega(-\xi(\tau+\lambda))}}-1\right]^2 d\tau=A_1+A_2+A_3. \tag{3.37}$$

Using the notation (3.25) and (3.36), in analogy with the proof of the estimates (3.30) and (3.31), for sufficiently small $\varepsilon\,,\ \delta$ we have

$$A_1\le\omega(\lambda)\int_{\theta_0-\lambda-\eta}^{\theta_0-\lambda}\frac{d\tau}{\omega(\xi(\tau+\lambda))}+\lambda^2\int_{\theta_0-\varepsilon}^{\theta_0-\lambda-\eta}\left[\frac{g(\xi(\tau+\lambda))}{\omega(\xi(\tau+\lambda))}\right]^2 d\tau. \tag{3.38}$$

Considering that $\xi(\tau+\lambda)\asymp\theta_0-\tau-\lambda$ in (3.38), and setting

$$A(\tau)=\xi(\tau+\lambda)g(\xi(\tau+\lambda))/\omega(\xi(\tau+\lambda))\,, \qquad A(a\,,\,b)=\sup_{a\le\tau\le b}A(\tau)\,,$$

from (3.38), (3.26), and (3.16) we obtain

$$\begin{aligned}A_1\le C_1\omega(\lambda)\int_0^\eta\frac{d\tau}{\omega(\tau)}+\lambda^2\Bigg[&A^2(\theta_0-\varepsilon\,,\ \theta_0-\sqrt{\lambda})\int_{\theta_0-\varepsilon}^{\theta_0-\sqrt{\lambda}}\frac{d\tau}{\xi^2(\tau+\lambda)}\\&+A^2(\theta_0-\sqrt{\lambda}\,,\ \theta_0-2\lambda)\int_{\theta_0-\sqrt{\lambda}}^{\theta_0-2\lambda}\frac{d\tau}{\xi^2(\tau+\lambda)}\\&+A^2(\theta_0-2\lambda\,,\ \theta_0-\lambda-\eta)\int_{\theta_0-2\lambda}^{\theta_0-\lambda-\eta}\frac{d\tau}{\xi^2(\tau+\lambda)}\Bigg]\\\le o(\lambda)+C_2\lambda^2&\left\{\mu^2(\varepsilon)\lambda^{-1/2}+\mu^2(\sqrt{\lambda})\lambda^{-1}+\mu^2(\lambda)\eta^{-1}\right\}=o(\lambda).\end{aligned} \tag{3.39}$$

Analogously, for A_3 we have the estimates

(3.40)

$$A_3 = \left\{\int_{\theta_0}^{\theta_0+\eta} + \int_{\theta_0+\eta}^{\theta_0+\lambda} + \int_{\theta_0+\lambda}^{\theta_0+\sqrt{\lambda}} + \int_{\theta_0+\sqrt{\lambda}}^{\theta_0+\varepsilon}\right\} \left[\sqrt{\frac{\omega(-\xi(\tau))}{\omega(-\xi(\tau+\lambda))}} - 1\right]^2 d\tau$$
$$\leq C_3\omega(\lambda)\int_0^\eta \frac{d\tau}{\omega(\tau)}$$
$$+ \lambda^2\left[\mu^2(\lambda)\int_{\theta_0+\eta}^{\theta_0+\lambda}\frac{d\tau}{\xi^2(\tau)} + \mu^2(\sqrt{\lambda})\int_{\theta_0+\lambda}^{\theta_0+\sqrt{\lambda}}\frac{d\tau}{\xi^2(\tau)} + \mu^2(\varepsilon)\int_{\theta_0+\sqrt{\lambda}}^{\theta_0+\varepsilon}\frac{d\tau}{\xi^2(\tau)}\right] = o(\lambda).$$

When applying (3.16) to obtain the estimates (3.39) and (3.40), we could easily determine which of the quantities $|\xi(t)|$ and $|\xi(\tau+\lambda)|$ was smaller than the other. In the estimation of the integral A_2 this can be done only for the segment $[\theta_0-\lambda/2, \theta_0]$. For the segment $[\theta_0-\lambda, \theta_0-\lambda/2]$ we can only establish that

$$\xi_1(\tau) \overset{\text{def}}{=} \min\{-\xi(\tau), -\xi(\tau+\lambda)\} \asymp \theta_0 - \tau,$$

which, nevertheless, is sufficient for obtaining the estimates

$$A_2 \leq \left\{\int_{\theta_0-\eta}^{\theta_0} + \int_{\theta_0-\lambda/2}^{\theta_0-\eta} + \int_{\theta_0-\lambda+\eta}^{\theta_0-\lambda/2} + \int_{\theta_0-\lambda}^{\theta_0-\lambda+\eta}\right\} \left[\sqrt{\frac{\omega(\xi(\tau))}{\omega(-\xi(\tau+\lambda))}} - 1\right]^2 d\tau$$
$$\leq \int_{\theta_0-\eta}^{\theta_0}\left[\frac{\omega(\xi(\tau))}{\omega(-\xi(\tau+\lambda))} + 1\right] d\tau + \int_{\theta_0-\lambda/2}^{\theta_0-\eta}\left[\frac{g(\xi(\tau)\lambda)}{\omega(\xi(\tau))}\right]^2 d\tau$$
$$(3.41)\qquad + \int_{\theta_0-\lambda+\eta}^{\theta_0-\lambda/2}\left[\frac{g(\xi_1(\tau))\lambda}{\omega(\xi_1(\tau))}\right]^2 d\tau + \int_{\theta_0-\lambda}^{\theta_0+\eta-\lambda}\frac{\omega(\lambda)\,d\tau}{\omega(-\xi(\tau+\lambda))}$$
$$\leq 2\eta + C_4\lambda^2\mu^2(\lambda)\int_{\theta_0-\lambda/2}^{\theta_0-\eta}\frac{d\tau}{(\theta_0-\tau)^2}$$
$$+ C_5\lambda^2\mu^2(C_6\lambda)\int_{\theta_0-\lambda+\eta}^{\theta_0-\lambda/2}\frac{d\tau}{(\theta_0-\tau)^2} + C_7\omega(\lambda)\int_0^\eta\frac{d\tau}{\omega(\tau)} = o(\lambda).$$

It follows from (3.37) and (3.39)–(3.41) that $J' = o(\sqrt{\lambda})$; from this and the analogous estimate for J'' in view of (3.35) there follows the lemma.

Remark 3.2. An analysis of the proofs of Lemmas 3.3–3.6 shows (see (3.17) and (3.35)) that these lemmas remain valid also in the case when the corresponding restrictions on $\omega(\tau)$ are fulfilled not on the entire segment $[0, \pi]$, but only on a part $[0, a]$ of it. On $[a, \pi]$ these restrictions may be weakened, requiring only that the modulus of continuity in $L^2[a, \pi]$ of the function $\omega(\tau)$ be $O(\sqrt{\delta})$ in Lemmas 3.3 and 3.5, and $o(\sqrt{\delta})$ in Lemmas 3.4 and 3.6.

The upcoming Lemmas 3.7–3.10 provide examples of concave moduli of continuity needed in the sequel.

Lemma 3.7. *Let*

$$(3.42)\qquad F(t) = \exp\{\alpha_l F_l^{\beta_l}(t) + \cdots + \alpha_m F_m^{\beta_m}(t)\} \qquad (0 \leq t \leq 1),$$

where $l, m \in \mathbf{N}$, $l \leq m$; $\alpha_l, \ldots, \alpha_m \in \mathbf{R}$; $\beta_l, \ldots, \beta_m \geq 0$; $\alpha_l < 0$, $\beta_l > 0$ *(for* $l = 1$ *we require* $\beta_1 < 1$, *in addition)*;

$$(3.43)\qquad F_0(t) = 1/t, \qquad F_k(t) = 1 + \ln F_{k-1}(t) \qquad (0 < t \leq 1,\ k \in \mathbf{N}).$$

Then there exists a segment $[0, \delta_F]$ $(0 < \delta_F < 1)$ *on which* $F(t)$ *is a concave modulus of continuity.*

Proof. By induction, we verify that

$$F_k'(t)/F_k(t) = -1/[tF_1(t)\cdots F_k(t)] \qquad (k \in \mathbf{N}), \tag{3.44}$$

whence it follows that

$$\begin{gathered} F_k'(t) \le 0 \qquad (0 < t \le 1,\ k \in \mathbf{N}), \\ \ln[-F_k'(t)] = \ln F_k(t) + \ln t - \sum_{\nu=1}^{k} \ln F_\nu(t). \end{gathered} \tag{3.45}$$

Differentiating the last equality and taking account of (3.44), (3.45), we have

$$\frac{F_k''}{F_k'} = -\frac{F_k'}{F_k} + \frac{1}{t} + \sum_{\nu=1}^{k} \frac{F_\nu'}{F_\nu} = \frac{1}{t}[1 + o(1)] \qquad (t \to +0). \tag{3.46}$$

Taking logarithms and differentiating (3.42), we obtain

$$\frac{F'}{F} = \sum_{k=l}^{m} \alpha_k \beta_k F_k^{\beta_k - 1} F_k' = \alpha_l \beta_l F_l^{\beta_l - 1} F_l'[1 + o(1)]. \tag{3.47}$$

Since $\alpha_l < 0$, it follows from (3.47) and (3.45) that there is a half-open interval $(0, \delta']$ on which $F'(t) \ge 0$. Taking logarithms and differentiating on $(0, \delta']$ the first of the equalities (3.47), by (3.44)–(3.46) we have

$$\begin{aligned} \frac{F''}{F'} &= \frac{F}{F'}\left\{\sum_{k=l}^{m} \alpha_k \beta_k(\beta_k - 1)F_k^{\beta_k}\left(\frac{F_k'}{F_k}\right)^2 + \left(\frac{F'}{F}\right)^2 + \sum_{k=l}^{m} \alpha_k \beta_k F_k^{\beta_k} \frac{F_k''}{F_k'} \frac{F_k'}{F_k}\right\} \\ &= \frac{F}{F'} \alpha_l \beta_l F_l^{\beta_l} \frac{1 + o(1)}{t^2 F_1 \cdots F_l} \qquad (t \to +0). \end{aligned} \tag{3.48}$$

This implies the existence of a half-open interval $(0, \delta'']$ on which $F''(t) \le 0$. Thus, on some $[0, \delta_F]$, $F(t)$ is a nondecreasing concave function with $F(0) = 0$, and, consequently ([25], §3.2.3), a concave modulus of continuity.

Lemma 3.8. *Let* $\alpha, \beta > 0$, $m \in \mathbf{N}$, $m \ge 2$, *and*

$$F(t) = \exp\{-\alpha F_1(t) F_m^{-\beta}(t)\} \qquad (0 \le t \le 1), \tag{3.49}$$

where F_1 *and* F_m *are defined by formulas* (3.43). *Then the claim of Lemma* 3.7 *is valid.*

Proof. Using (3.44) and (3.46), in analogy with (3.47) and (3.48) we derive the following relations (as $t \to +0$):

$$\begin{aligned} F' &= \alpha t^{-1} F F_m^{-\beta}[1 + o(1)], \\ F'' &= -\alpha t^{-2} F F_m^{-\beta}[1 + o(1)], \end{aligned}$$

which imply Lemma 3.8.

Lemma 3.9. *Let*

$$F(\tau) = |\exp\{\alpha_l f_l^{\beta_l}(e^{i\tau}) + \cdots + \alpha_m f_m^{\beta_m}(e^{i\tau})\}| \qquad (\tau \in \mathbf{R}), \tag{3.50}$$

where $\alpha_l, \ldots, \alpha_m$, $\beta_l, \ldots, \beta_m$ *satisfy the conditions of Lemma* 3.7,

$$f_0(z) = 2/(1 - z), \qquad f_k(z) = 1 + \ln f_{k-1}(z) \qquad (k \in \mathbf{N}), \tag{3.51}$$

and

$$\ln z = \ln|z| + i\arg z, \qquad -\pi < \arg z \le \pi.$$

Then there exists a $\delta_F \in (0, \pi)$ *such that* $F(\tau)$ *is a concave modulus of continuity on* $[0, \delta_F]$.

Proof. Introducing the notation

(3.52) $$a_k = |f_k(e^{i\tau})|^{-1}, \qquad \gamma_k = \arg f_k(e^{i\tau}),$$

in view of (3.51) we obtain for $k \in \mathbf{Z}_+$

(3.53) $$a_{k+1}^{-2} = \ln^2(e/a_k) + \gamma_k^2,$$

(3.54) $$\cos\gamma_{k+1} = a_{k+1}\ln(e/a_k), \qquad \sin\gamma_{k+1} = \gamma_k a_{k+1},$$

(3.55) $$\gamma_0 = (\pi - \tau)/2, \qquad a_0 = \sin(\tau/2) \qquad (0 < \tau < 2\pi).$$

Applying (3.54) and (3.55), we establish by induction that

(3.56) $$\gamma_k = (\pi/2)a_1\cdots a_k[1 + O(a_1^2)] \qquad (\tau \to +0,\ k \in \mathbf{N}).$$

Using (3.56) and the first of the formulas (3.54), we find that

(3.57) $$\ln(e/a_k) = a_{k+1}^{-1}[1 + O(a_1^2\cdots a_{k+1}^2)] \qquad (\tau \to +0,\ k \in \mathbf{Z}_+).$$

With the aid of (3.53)–(3.57) and induction we verify the relations

(3.58) $$a_k' a_k^{-1} = a_1\cdots a_k\tau^{-1}[1 + O(a_1^2)] \qquad (\tau \to +0,\ k \in \mathbf{N}),$$

(3.59) $$\gamma_k' = \pi a_1(2\tau)^{-1}a_1\cdots a_k[1 + O(a_1^2)] \qquad (\tau \to +0,\ k \in \mathbf{N}).$$

Indeed, differentiating (3.53) for $k = 0$ and taking account of (3.55)–(3.57), we have (as $\tau \to +0$)

$$\frac{a_1'}{a_1^3} = \frac{1}{2}\cot\frac{\tau}{2}\ln\frac{e}{a_0} + \frac{\pi - \tau}{4} = \tau^{-1}[1 + O(\tau^2)]a_1^{-1}[1 + O(a_1^2)] + (\pi/2)[1 + O(\tau)],$$

whence (3.58) follows for $k = 1$. Differentiating the second formula in (3.54) for $k = 0$ and taking account of (3.55), (3.56), and the relation just derived, we obtain that, as $\tau \to +0$,

$$\gamma_1'\cos\gamma_1 = a_1'\cos\gamma_0 + a_1\gamma_0' = \pi(2\tau)^{-1}a_1^2[1 + O(a_1^2)],$$

whence, in view of $\cos\gamma_1 = 1 + O(a_1^2)$, there follows (3.59) for $k = 1$. Thus, formulas (3.58) and (3.59) are valid for $k = 1$. Suppose they are valid for some k. Then, differentiating (3.53), with (3.56), (3.57) taken into account, we have $(\tau \to +0)$

(3.60) $$\frac{a_{k+1}'}{a_{k+1}^3} = \frac{a_k'}{a_k}\ln\frac{e}{a_k} - \gamma_k\gamma_k' = \frac{a_1\cdots a_k}{\tau a_{k+1}}[1 + O(a_1^2)],$$

i.e., (3.58) is valid with k replaced by $k + 1$ as well. With this in mind, we differentiate the second formula in (3.54) and apply (3.56), (3.59), and the relation $\cos\gamma_{k+1} = 1 + O(a_1^2)$ $(\tau \to +0)$. As a result, we obtain

$$\gamma_{k+1}' = (a_{k+1}'\gamma_k + a_{k+1}\gamma_k')/\cos\gamma_{k+1} = \pi a_1(2\tau)^{-1}a_1\cdots a_{k+1}[1 + O(a_1^2)] \qquad (\tau \to +0).$$

Thus, formulas (3.58), (3.59) are valid for all $k \in \mathbf{N}$.

Further, taking logarithms and differentiating the first of the equalities (3.60), and also differentiating twice the second formula in (3.54), we have

$$\frac{a_{k+1}''}{a_{k+1}'} = 3\frac{a_{k+1}'}{a_{k+1}} + \frac{a_{k+1}^3}{a_{k+1}'}\left\{\frac{a_k''}{a_k'}\frac{a_k'}{a_k}\ln\frac{e}{a_k} - \left(\frac{a_k'}{a_k}\right)^2\ln\frac{e^2}{a_k} - (\gamma_k')^2 - \gamma_k\gamma_k''\right\},$$

$$\gamma_k''\cos\gamma_{k+1} = (\gamma_{k+1}')^2\sin\gamma_{k+1} + a_{k+1}''\gamma_k + 2a_{k+1}'\gamma_k + a_{k+1}\gamma_k''.$$

Using these equalities and formulas (3.55)–(3.59) we prove by induction, similarly to (3.58), (3.59), that

$$a''_{k+1}/a'_{k+1} = -\tau^{-1}[1 + O(a_1)] \qquad (\tau \to +0,\ k \in \mathbf{Z}_+), \tag{3.61}$$

$$\gamma''_k = -\pi a_1(2\tau)^{-2}a_1 \cdots a_k[1 + O(a_1)] \qquad (\tau \to +0,\ k \in \mathbf{N}). \tag{3.62}$$

Taking logarithms and differentiating the equality

$$F = \exp\{\alpha_l a_l^{-\beta_l}\cos(\beta_l\gamma_l) + \cdots + \alpha_m a_m^{-\beta_m}\cos(\beta_m\gamma_m)\},$$

which follows from (3.50) and (3.52), we obtain

$$\frac{F'}{F} = -\sum_{k=l}^{m}\alpha_k\beta_k a_k^{-\beta_k}\left\{\frac{a'_k}{a_k}\cos(\beta_k\gamma_k) + \gamma'_k\sin(\beta_k\gamma_k)\right\}. \tag{3.63}$$

In view of (3.56), (3.58), and (3.59), it follows from (3.63) that

$$F'F^{-1} = -\alpha_l\beta_l a_l^{-\beta_l}a_1\cdots a_l\tau^{-1}[1 + O(a_1^2)] \qquad (\tau \to +0). \tag{3.64}$$

Differentiating (3.63), we have

$$\begin{aligned}\frac{F''}{F} = \left(\frac{F'}{F}\right)^2 + \sum_{k=l}^{m}\alpha_k\beta_k a_k^{-\beta_k}\frac{a'_k}{a'_k}\bigg\{&(1+\beta_k)\frac{a'_k}{a_k}\cos(\beta_k\gamma_k)\\ &+ 2\beta_k\gamma'_k\sin(\beta_k\gamma_k) - \frac{a''_k}{a'_k}\cos(\beta_k\gamma_k)\\ &- \frac{a_k}{a'_k}\gamma''_k\sin(\beta_k\gamma_k) - \beta_k\frac{a_k}{a'_k}\gamma'_k\cos(\beta_k\gamma_k)\bigg\}.\end{aligned} \tag{3.65}$$

From (3.65), (3.64), (3.58), (3.59), (3.61), and (3.62) we find that

$$F''/F = \alpha_l\beta_l a_l^{-\beta_l}a_1\cdots a_l\tau^{-2}[1 + o(1)] \qquad (\tau \to +0). \tag{3.66}$$

By (3.64) and (3.66) a half-open interval $(0, \delta_F]$ $(\delta_F > 0)$ can be found on which $F'(\tau) > 0$, $F''(\tau) < 0$. Thus, F is a nondecreasing concave function on $[0, \delta_F]$ with $F(0) = 0$, and, consequently, the lemma is valid.

Lemma 3.10. *Let* $\alpha, \beta > 0$, $m \in \mathbf{N}$, $m \geq 2$, *and*

$$F(\tau) = |\exp\{-\alpha f_1(e^{i\tau})f_m^{-\beta}(e^{i\tau})\}| \qquad (\tau \in \mathbf{R}), \tag{3.67}$$

where f_1 *and* f_m *are defined by formulas* (3.51). *Then the claim of Lemma* 3.7 *holds.*

Proof. It follows from (3.67) and (3.52) that

$$F(\tau) = \exp\{-\alpha a_1^{-1}a_m^{\beta}\cos(\gamma_1 - \beta\gamma_m)\}. \tag{3.68}$$

Taking logarithms and differentiating (3.68), with account taken of (3.58), (3.56), and (3.59), we have

$$\begin{aligned}\frac{F'}{F} &= \alpha\frac{a_m^{\beta}}{a_1}\left[\left(\frac{a'_1}{a_1} - \beta\frac{a'_m}{a_m}\right)\cos(\gamma_1 - \beta\gamma_m) + (\gamma'_1 - \beta\gamma'_m)\sin(\gamma_1 - \beta\gamma_m)\right]\\ &= \alpha a_1^{-1}a_m^{\beta}\tau^{-1}[1 + o(1)] \qquad (\tau \to +0).\end{aligned} \tag{3.69}$$

Differentiating the first of the equalities (3.69), we get

$$\frac{F''}{F} = \left(\frac{F'}{F}\right)^2 - \alpha\frac{a_m^{\beta}}{a_1}\{A_1\cos(\gamma_1 - \beta\gamma_m) + A_2\sin(\gamma_1 - \beta\gamma_m)\},$$

where

$$A_1 = \left(\frac{a_1'}{a_1} - \beta\frac{a_m'}{a_m}\right)^2 - \frac{a_1''}{a_1} + \left(\frac{a_1'}{a_1}\right)^2 + \beta\frac{a_m''}{a_m} - \beta\left(\frac{a_m'}{a_m}\right)^2 + (\gamma_1' - \beta\gamma_m')^2,$$

$$A_2 = (\gamma_1' - \beta\gamma_m')\left(\frac{a_1'}{a_1} - \beta\frac{a_m'}{a_m} - 1\right) + (\gamma_1'' - \beta\gamma_m'').$$

From this, with (3.69), (3.58), (3.62), (3.59), and (3.56), we find that

$$F'' = -\alpha F a_m^{\beta}\tau^{-2}[1 + o(1)] \qquad (\tau \to +0). \tag{3.70}$$

Lemma 3.10 follows from (3.69) and (3.70).

4. Estimates of best approximations in $L^2_\varphi(\Gamma_1)$ of the Szegő function

In §3 we gave examples of continuous weights φ with one or two zeros on a period, satisfying one of the conditions (3.14) or (3.22). In that case, by (3.2), the corresponding relation $\delta_n(\varphi) = O(n^{-1/2})$ or $\delta_n(\varphi) = o(n^{-1/2})$ $(n \to \infty)$ held. In this section we give examples of more general weights with the same behavior of $\delta_n(\varphi)$. Along with the propositions of §3 we also need the following two lemmas.

Lemma 4.1. *Let ξ, η be weights with integrable logarithms, and let $\{\xi_n(z)\}_{n=0}^{\infty}$ be a system of polynomials orthonormal on Γ_1 with weight ξ. Assume that $\xi\eta \in L$ and*

$$\sup_{n\in\mathbf{Z}_+} \|\xi_n(e^{i\theta})\sqrt{\xi(\theta)}\|_\infty < \infty. \tag{4.1}$$

Then

$$\delta_n(\xi\eta) \le \delta_{[n/2]}(\xi) + O(\delta_{[n/2]}(\eta)) \qquad (n \to \infty). \tag{4.2}$$

This lemma was proved in [18], and implicitly even earlier in [15].

Lemma 4.2. *Let $0 < c < \infty$, $R_n = 1 - cn^{-1}$, and let condition (1.7) be satisfied. Then a constant $C_3(\sigma, c)$ can be found such that in the disc $|z| < 1$ and for all $n > c$*

$$|\varphi^*_{\sigma,\eta}(z)| \le C_3(\sigma, c)|\pi(\sigma'; R_n z)|(1 + \delta_{\sigma,n}\sqrt{n}), \tag{4.3}$$

where $\delta_{\sigma,n}$ is defined by the equality (1.13).

This lemma is a simple corollary of Lemma 3 in [9].

We prove two theorems providing needed estimates for $\delta_n(\varphi)$. In the formulation of the first theorem we waive the restriction $\varphi \in L^1$.

Theorem 4.1. *Suppose φ has the form*

$$\varphi(\tau) = h(\tau)\prod_{\nu=1}^{m} w_\nu\left(\left|\sin\frac{\tau - \theta_\nu}{2}\right|\right), \tag{4.4}$$

where $-\pi < \theta_1 < \cdots < \theta_m \le \pi$,

$$h,\, 1/h \in L^\infty, \qquad h(\tau) \ge 0, \qquad \omega(h;\delta)_2 = O(\sqrt{\delta}), \tag{4.5}$$

$$w_\nu(u) = \prod_{\mu=1}^{l_\nu}[g_{\mu,\nu}(u)]^{\alpha(\mu,\nu)}, \tag{4.6}$$

$m, l_\nu \in \mathbf{N}$, $\alpha(\mu,\nu) \in \mathbf{R}$, $g_{\mu,\nu}(u)$ are concave moduli of continuity ($\mu = 1, \dots, l_\nu$; $\nu = 1, \dots, m$). Then

$$\delta_n(\varphi) = O(n^{-1/2}) \qquad (n \to \infty). \tag{4.7}$$

Proof. For $w_\nu(u) = u^{\gamma_\nu}$ $(\gamma_\nu \in \mathbf{R};\ \nu = 1, \ldots, m)$ this theorem is known [15], [17]. Let $0 < \alpha < 1$, and let $\omega(u)$ be a concave modulus of continuity. Then $g(u) = \omega^\alpha(u)$ is also a concave modulus of continuity and, by (2.2),

$$\int_0^\delta \frac{du}{g(u)} \le \int_0^\delta \left[\frac{2\delta}{\omega(\delta)u}\right]^\alpha du = \frac{2^\alpha}{1-\alpha}\frac{\delta}{g(\delta)},$$

i.e., condition (3.3) is satisfied for g. Therefore, by (3.2) and Remark 3.1 we conclude that

$$\varphi(t) = \omega^\alpha(|\sin(\tau/2)|) \qquad (\tau \in \mathbf{R}) \tag{4.8}$$

satisfies (4.7) for $0 < \alpha < 1$. Note that if φ is a weight, $\varphi^{\pm 1} \in L^1$, $\theta_0 \in \mathbf{R}$, and $\psi(\tau) = \varphi(\tau - \theta_0)$, then by (3.8)

$$\omega(\psi^{1/2}; \delta)_{1/\psi, 2} = \omega(\psi^{-1/2}; \delta)_{\psi, 2} = \omega(\varphi^{-1/2}; \delta)_{\varphi, 2}. \tag{4.9}$$

From this we conclude that for a weight (4.8) condition (4.7) is satisfied also for $\alpha \in (-1, 0)$. Consider the weights

$$\xi(\tau) = |\sin[(\tau - \theta_\nu)/2]|^{\gamma_\nu}, \qquad \eta(\tau) = g_{\mu,\nu}^{\pm\alpha}(|\sin[(\tau - \theta_\nu)/2]|)$$

for $\gamma_\nu \ge 1$, $\alpha \in (0, 1)$. By (1.3), Lemma 4.2, Theorem 2.1, and the remark at the beginning of the proof, $\delta_n(\xi) = O(n^{-1/2})$, and (4.1) holds. From this and the estimate $\delta_n(\eta) = O(n^{-1/2})$, just proved, by Lemma 4.1 there follows the estimate (4.7) for the weight

$$\varphi(\tau) = |\sin[(\tau - \theta_\nu)/2]|^{\gamma_\nu} g_{\mu,\nu}^{\pm\alpha}(|\sin[(\tau - \theta_\nu)/2]|) \tag{4.10}$$

for $\gamma_\nu \ge 1$, $\alpha \in (0, 1)$. The weight (4.10) is bounded and satisfies (by Lemma 4.2 and Theorem 2.1) condition (4.1). Therefore, setting $\xi = \eta = \varphi$, in view of Lemma 4.1 we see that the estimate (4.7) is valid for the weight φ^2. Then setting $\xi = \varphi$, $\eta = \varphi^2$, in the same manner we arrive at the estimate $\delta_n(\varphi^3) = O(n^{-1/2})$. We prove by induction that, in general, for $\gamma_\nu \ge 1$, $0 < \alpha < 1$, the estimate $\delta_n(\varphi^k) = O(n^{-1/2})$ holds for the weight (4.10) for any $k \in \mathbf{N}$. This means that condition (4.7) is fulfilled for the weight φ for $\gamma_\nu \ge k$, $0 < \alpha < k$. With the aid of Lemma 4.1 we extend this result to the weight

$$\varphi(\tau) = h(\tau) \prod_{\nu=1}^m |\sin[(\tau - \theta_\nu)/2]|^{\gamma_\nu} w_\nu(|\sin[(\tau - \theta_\nu)/2]|), \tag{4.11}$$

where h satisfies (4.5) and in the expression (4.6) $\alpha^{\pm 1}(\mu, \nu) \in (0, k)$, $\gamma_\nu \ge k l_\nu$, $k \in \mathbf{N}$. Here we use the fact ([2], §3.7) that for a weight h satisfying (4.5) conditions (4.7) and (4.1) are fulfilled.

We now show that the restriction $\gamma_\nu \ge k l_\nu$ in the result obtained can be dropped. Indeed, let

$$v(\tau) = |Q_N(e^{i\tau})|^{-2}, \qquad Q_N(z) = (1 - e^{-i\theta_1} z)^{\Gamma_1} \cdots (1 - e^{-i\theta_m} z)^{\Gamma_m},$$

where $\Gamma_1, \ldots, \Gamma_m \in \mathbf{Z}_+$, $N = \Gamma_1 + \cdots + \Gamma_m$. Then $\pi(v; z) = Q_N(z)$. Suppose that for a polynomial q_n of degree at most n

$$\delta_n(\varphi) = \|q_n(e^{i\tau})/\pi(\varphi; e^{i\tau}) - 1\|_2.$$

Then by (3.1), (1.10), and (2.22) we obtain

$$\delta_{n+N}(v\varphi) \le \|q_n(e^{i\tau}) Q_N(e^{i\tau})/\pi(v\varphi; e^{i\tau}) - 1\|_2 = \delta_n(\varphi). \tag{4.12}$$

This inequality tells us that for the weight (4.11) condition (4.7) is satisfied, provided $\alpha^{\pm}(\mu, \nu) \in (0, k)$, $\gamma_\nu \ge k l_\nu - 2\Gamma_\nu$, and since Γ_ν and k may be taken as large as desired, it follows that (4.7) holds for the weight (4.11) for all $\alpha(\mu, \nu)$ and $\gamma_\nu \in \mathbf{R}$. Setting $\gamma_\nu = 0$, we get the conclusion of Theorem 4.1.

Theorem 4.2. *Let* $\varphi(\tau) = h(\tau)G(\tau)$, *with*

$$h,\, 1/h \in L^\infty, \qquad h(\tau) \geq 0, \qquad \omega(h;\delta)_2 = o(\sqrt{\delta}) \qquad (\delta \to +0), \tag{4.13}$$

$$G(\tau) = \prod_{\nu=1}^{m} g_\nu^{\varepsilon_\nu}(|\sin[(\tau - \theta_\nu)/2]|), \tag{4.14}$$

where $-\pi < \theta_1 < \cdots < \theta_m \leq \pi$, $\varepsilon_\nu = \pm 1$, *the* $g_\nu(\tau)$ *are concave moduli of continuity;* $\tau g_\nu'(\tau)/g_\nu(\tau) \to 0$ *as* $\tau \to 0$ *remaining on the set where the derivative* $g_\nu'(\tau)$ *exists* $(\nu = 1, \ldots, m)$. *Then*

$$\delta_n(\varphi) = o(n^{-1/2}) \qquad (n \to \infty). \tag{4.15}$$

Proof. First we consider the weight (4.8) for $\alpha = 1$, assuming that ω satisfies the conditions of Lemma 3.4. On the basis of Lemma 3.4, Remark 3.1, and the equalities (4.9) we conclude that for each $\nu = 1, \ldots, m$

$$\psi(\tau) = g_\nu^{-1}(|\sin[(\tau - \theta_\nu)/2]|) \Rightarrow \delta_n(\psi) = o(n^{-1/2}). \tag{4.16}$$

Assume now that $\nu \geq 2$, that φ is the weight from (4.16), and that the relation $\delta_n(\chi) = o(n^{-1/2})$ holds for the weight

$$\chi(\tau) = \prod_{\mu=1}^{\nu-1} g_\mu^{-1}(|\sin[(\tau - \theta_\mu)/2]|).$$

Then it follows from (3.1), (1.10), and (2.22) that

$$\begin{aligned} \delta_n(\chi\psi) &\leq \left\| \frac{\chi^*_{[n/2]}(e^{i\tau})\psi^*_{[n/2]}(e^{i\tau})}{\pi(\chi\psi; e^{i\tau})} - 1 \right\|_2 \\ &\leq I_1 + I_2 = \left\| \frac{\psi^*_{[n/2]}(e^{i\tau})}{\pi(\psi; e^{i\tau})} \left[\frac{\chi^*_{[n/2]}(e^{i\tau})}{\pi(\chi; e^{i\tau})} - 1 \right] \right\|_2 + \left\| \frac{\psi^*_{[n/2]}(e^{i\tau})}{\pi(\psi; e^{i\tau})} - 1 \right\|_2 . \end{aligned} \tag{4.17}$$

It is known ([2], §5.7) that $I_2 = O(\delta_{[n/2]}(\psi))$, and, hence, $I_2 = o(n^{-1/2})$ $(n \to \infty)$. Setting

$$e_1 = [\theta_\nu - n^{-1}, \theta_\nu + n^{-1}], \qquad e_2 = [\theta_\nu - \pi, \theta_\nu + \pi] \setminus e_1,$$

by Lemma 4.2 and Theorem 2.1 we have the following estimates uniformly with respect to τ:

$$\psi^*_{[n/2]}(e^{i\tau}) = O(g_\nu^{1/2}(n^{-1})) \qquad (\tau \in e_1), \tag{4.18}$$

$$\psi^*_{[n/2]}(e^{i\tau}) = O(\psi^{-1/2}(\tau)) \qquad (\tau \in e_2). \tag{4.19}$$

Since on $[\theta_\nu - \varepsilon, \theta_\nu + \varepsilon]$ for a sufficiently small $\varepsilon > 0$ the weight χ is bounded away from zero and satisfies the Dini (even Lipschitz) condition, it follows (see, for example, [6]) that

$$\chi^*_{[n/2]}(e^{i\tau}) - \pi(\chi; e^{i\tau}) \to 0 \qquad (n \to \infty) \tag{4.20}$$

uniformly in $\tau \in e_1$.

It follows from (4.18)–(4.20) and (3.27) that, as $n \to \infty$,

$$I_1 \le g_\nu^{1/2}(n^{-1})o(1)\left\{\int_{e_1} g_\nu^{-1}(|\sin[(\tau-\theta_\nu)/2]|)d\tau\right\}^{1/2}$$

$$+ O(1)\left\{\frac{1}{2\pi}\int_{e_2}\left|\frac{\chi^*_{[n/2]}(e^{i\tau})}{\pi(\chi\,;\,e^{i\tau})}-1\right|^2 d\tau\right\}^{1/2} = o(n^{-1/2}).$$

It is thereby established that if $\eta(\tau)$ is the product of the factors $g_\nu^{\varepsilon_\nu}(|\sin[(\tau-\theta_\nu)/2]|)$ appearing in (4.14) with exponents $\varepsilon_\nu = -1$, then $\delta_n(\eta) = o(n^{-1/2})$.

It follows from Lemma 4.1 that the condition $\delta_n(\xi) = o(n^{-1/2})$ holds for the weight ξ which is the product of h and the factors $g_\nu^{\varepsilon_\nu}(|\sin[(\tau-\theta_\nu)/2]|)$ in (4.14) with exponents $\varepsilon_\nu = 1$. The weights ξ and ν satisfy Lemma 4.1. Therefore $\delta_n(\xi\eta) = o(n^{-1/2})$, and Theorem 4.2 is proved.

Remark 4.1. Completely analogously, on the basis of Lemmas 3.5 and 3.6, one can prove two more theorems, the formulations of which are obtained from those of Theorems 4.1 and 4.2 by replacing $|\sin[(\tau-\theta_\nu)/2]|$ and $|\sin[(\theta-\theta_\nu)/2]|$ in (4.4) and (4.14), respectively, by $|\cos\tau - \cos\theta_\nu|$ and $|\cos\theta - \cos\theta_\nu|$ for those ν for which $\theta_\nu \ne 0$ and $\theta_\nu \ne \pi$.

5. Inequalities for integral means of an analytic function and its derivatives

We consider some refinements and generalizations of known inequalities for the means $M_r(f^{(j)}\,;\,\rho)$ and $M_q(f\,;\,R)$ (see (1.2)).

Lemma 5.1. *If* $\delta > 0$, $0 < \rho < 1$, *and* $f \in H^\delta$, *then*

$$M_\infty(f\,;\,\rho) \le (1-\rho^2)^{-1/\delta} M_\delta(f\,;\,1). \tag{5.1}$$

Proof. We argue as in [29], p. 284 (where the Hardy-Littlewood inequality is proved; the latter is obtained from (5.1) by replacing $(1-\rho^2)^{-1/\delta}$ with $(1-\rho)^{-1/\delta}$ on the right-hand side), with the distinction that we first consider the case $\delta = 2$, and not $\delta = 1$.

If $f \in H^2$ and $f(z) = a_0 + a_1 z + \cdots$ in D, then by the Cauchy inequality

$$|f(\rho e^{i\theta})| \le (|a_0|^2 + |a_1|^2 + \cdots)^{1/2}(1+\rho^2+\rho^4+\cdots)^{1/2},$$

whence in view of the Parseval equality we obtain (5.1) for $\delta = 2$. For $\delta \ne 2$ we set $f(z) = G(z)B(z)$, where $G \ne 0$ in D, and B is a Blaschke product, $|B(z)| \le 1$ in D, $|B(e^{i\theta})| = 1$ a.e. (see [29], p. 274). Then $G^{\delta/2} \in H^2$ and, since (5.1) holds for $\delta = 2$,

$$|G(\rho e^{i\theta})|^{\delta/2} \le M_2(G^{\delta/2}\,;\,1)(1-\rho^2)^{-1/2} = M_\delta^{\delta/2}(G\,;\,1)(1-\rho^2)^{-1/2}. \tag{5.2}$$

Since $|f(z)| \le |G(z)|$ in D and $M_\delta(f\,;\,1) = M_\delta(G\,;\,1)$, (5.2) implies (5.1).

Considering $f(Rz)$ and ρ/R instead of $f(z)$ and ρ, from Lemma 5.1 we obtain

Corollary 5.1. *If* $0 < \delta \le \infty$, $0 < \rho < R < 1$, *and* f *is analytic in* D, *then*

$$M_\infty(f\,;\,\rho) \le [R^2/(R^2-\rho^2)]^{1/\delta} M_\delta(f\,;\,R). \tag{5.3}$$

Remark 5.1. It is easily verified by direct calculation that for $f(z) = (1-\rho z/R^2)^{-2/\delta}$ (5.3) becomes an equality.

Lemma 5.2. *If* $0 < q < r \le \infty$, $0 < \rho < R < 1$, *and* f *is analytic in* D, *then*

$$M_r(f;\rho) \le [R^2/(R^2-\rho^2)]^{1/q-1/r} M_q(f;R). \tag{5.4}$$

Proof. For $r = \infty$, $q = \delta$, (5.4) coincides with (5.3). For $r < \infty$, applying the known method of extracting and estimating the factor $|f(\rho e^{i\theta})|^{r-q}$ in the integrand in (1.2), we obtain

$$M_r(f;\rho) \le M_\infty^{1-q/r}(f;\rho) M_q^{q/r}(f;\rho).$$

In view of (5.3) and the fact that $M_q(f;\rho)$ does not decrease with respect to ρ ([29], p. 273), this implies (5.4).

Lemma 5.3. *Let* $j \in \mathbf{N}$, $1 \le q \le r \le \infty$, $0 < \rho < R < 1$, *and let* f *be analytic in* D. *Then*

$$M_r(f^{(j)};\rho) \le j!R(R+\rho)^{1/q-1/r-1}(R-\rho)^{1/q-1/r-j} M_q(f;R). \tag{5.5}$$

Proof. Put $s = (1+1/r-1/q)^{-1}$. It is clear that $1 \le s \le \infty$. Applying the theorem on the norm of a convolution ([29], p. 37), we have

$$\begin{aligned} M_r(f^{(j)};\rho) &\left\| \frac{j!}{2\pi} \int_{|\zeta|=R} f(\zeta)(\zeta-\rho e^{i\theta})^{-j-1} \right\|_r \\ &= j!R \left\| (2\pi)^{-1} \int_0^{2\pi} [f(Re^{i\tau})e^{-ij\tau}](R-\rho e^{i(\theta-\tau)})^{-j-1} d\tau \right\|_r \\ &\le j!RA_{\rho,R}^{j,s} M_q(f;R), \qquad A_{\rho,R}^{j,s} = \|(R-\rho e^{-i\tau})^{-j-1}\|_s. \end{aligned} \tag{5.6}$$

Introducing the notation $\alpha = (j+1)s/2$, for $s < \infty$ we obtain

$$A_{\rho,R}^{j,s} = \left\{ (2/\pi) \int_0^{\pi/2} [(R-\rho)^2 + 4R\rho \sin^2 u]^{-\alpha} du \right\}^{1/s}. \tag{5.7}$$

Extracting the factor $(\sin^2 u + \cos^2 u)^{\alpha-1} \equiv 1$ in the integrand in (5.7) and applying the identity

$$(R-\rho)^2 + 4R\rho\sin^2 u = (R+\rho)^2 \sin^2 u + (R-\rho)^2 \cos^2 u,$$

after the change of variable $t = \tan u$ we obtain

$$A_{\rho,R}^{j,s} = \left\{ (2/\pi) \int_0^\infty (1+t^2)^{\alpha-1} [(R+\rho)^2 t^2 + (R-\rho)^2]^{-\alpha} dt \right\}^{1/s}. \tag{5.8}$$

The change $(R+\rho)t = (R-\rho)v$ reduces (5.8) to

$$A_{\rho,R}^{j,s} = \left\{ \frac{2}{\pi} \frac{R-\rho}{R+\rho} (R-\rho)^{-2\alpha} \int_0^\infty \left[1 + \left(\frac{R-\rho}{R+\rho} v \right)^2 \right]^{\alpha-1} \frac{dv}{(1+v^2)^\alpha} \right\}^{1/s}. \tag{5.9}$$

The integrand in (5.9) is majorized by the function $(1+v^2)^{-1}$, the integral of which over $[0,\infty)$ is equal to $\pi/2$. Hence, for $s < \infty$,

$$A_{\rho,R}^{j,s} \le [(R-\rho)/(R+\rho)]^{1/s} (R-\rho)^{-j-1}. \tag{5.10}$$

Since $A_{\rho,R}^{j,\infty} = (R-\rho)^{-j-1}$, (5.10) holds also for $s = \infty$. Now (5.5) follows from (5.6) and (5.10).

Corollary 5.2. *Under the conditions of Lemma* 5.3

$$M_r(f^{(j)}; \rho) \le j!(R-\rho)^{1/r-1/q-j} M_q(f; R). \tag{5.11}$$

Lemma 5.4. *If* $0 < q \le r \le \infty$, $j \in \mathbf{Z}_+$, $0 < \rho < R < 1$, *then there exists a constant* $C_1 = C_1(j, q, r)$ *such that*

$$M_r(f^{(j)}; \rho) \le C_1(R-\rho)^{1/r-1/q-j} M_q(f; R) \tag{5.12}$$

for any function f *analytic in* D.

Proof. Lemmas 5.1–5.3 and Corollaries 5.1 and 5.2 are special cases of this proposition and contain more precise information on C_1. For $r = q$ Lemma 5.4 and Corollary 5.2 are known (see, for example, [30]). Thus, for $0 < r \le \infty$, $0 < \rho < R_1 < 1$,

$$M_r(f^{(j)}; \rho) \le C_2(j, r)(R_1-\rho)^{-j} M_r(f; R_1).$$

Putting here $R_1 = (R+\rho)/2$ and estimating $M_r(f; R_1)$ with the aid of (5.4), we arrive at (5.12).

6. Estimates for the maximum of the modulus of the logarithmic derivative of the Szegő function

In what follows an important role is played by the estimate

$$\max_{|z|=\rho} |\pi'(\varphi; z)/\pi(\varphi; z)| = O((1-\rho)^{-1}) \qquad (\rho \to 1-0), \tag{6.1}$$

and sometimes by the stronger one

$$\max_{|z|=\rho} |\pi'(\varphi; z)/\pi(\varphi; z)| = o((1-\rho)^{-1}) \qquad (\rho \to 1-0). \tag{6.2}$$

We indicate certain conditions on φ that are sufficient for (6.1) or (6.2).

Lemma 6.1. *Let a weight* φ *satisfy the conditions of Theorem* 2.1. *Then* (6.1) *holds.*

Proof. Consider first the case of the weight

$$\varphi(\tau) = \omega_k(|\tau|) \qquad (-\pi \le \tau \le \pi), \tag{6.3}$$

where $\omega_k(\delta)$ is a function of the kth ($k \in \mathbf{N}$) modulus of continuity type. By (2.2)

$$\frac{\pi'(uv; z)}{\pi(uv; z)} = \frac{d}{dz} \ln \pi(uv; z) = \frac{\pi'(u; z)}{\pi(u; z)} + \frac{\pi'(v; z)}{\pi(v; z)} \tag{6.4}$$

for all $z \in D$. Since $g(\delta) = \{\omega_k(\delta)\}^{1/k}$ is a function of the first modulus of continuity type, in view of (6.4) we may assume that $k = 1$. As in the proof of Corollary 2.1, we refer to the Bari-Stechkin theorem and consider a modulus of continuity ω such that $\omega(\tau) \asymp g(\tau)$ $(0 \le \tau \le \pi)$ and define weights ψ, ξ, and h by the formulas (2.21). In view of Stechkin's lemma ([27], p. 154), the function ω in (2.21) may be considered to be concave. But then, by M. Riesz' theorem on conjugate series ([29], Chapter III, Theorem 5.1), and also by Lemmas 3.1 and 3.2,

$$\omega(\ln \pi(\psi; e^{i\tau}); \delta)_2 \asymp \omega(\ln \psi; \delta)_2 = O(\delta^{1/2}). \tag{6.5}$$

For analytic functions f in D and $r \in (0, \infty)$ Storozhenko [30] established the inequality

$$M_\infty(f'; \rho) \le C_1(r)(1-\rho)^{-1-1/r} \omega(f(e^{i\tau}); 1-\rho)_r \qquad (0 < \rho < 1). \tag{6.6}$$

Applying (6.6) for $r = 2$ to the function $\ln \pi(\psi; z)$ and taking account of (6.5), we arrive at the estimate

$$\max_{|z|=\rho} |\pi'(\psi; z)/\pi(\psi; z)| = O((1-\rho)^{-1}) \qquad (\rho \to 1-0). \tag{6.7}$$

By (1.10) and Lemma 5.3

$$\|\pi'(h;\rho e^{i\theta})/\pi(h;\rho e^{i\theta})\|_\infty \le \|h\|_\infty^{1/2}\|\pi'(h;\rho e^{i\theta})\|_\infty,$$
$$\|\pi'(h;\rho e^{i\theta})\|_\infty \le (1-\rho)^{-1}\|\pi(h;e^{i\theta})\|_\infty = (1-\rho)^{-1}\|1/h\|_\infty^{1/2}.$$

Consequently, condition (6.7) still holds if in it ψ is replaced by h. But then, by (6.4), the estimate (6.7) remains in force if ψ is replaced by ξ as well. Thus Lemma 6.1 is proved for the weight (6.3). If we argue as in the proof of Corollary 2.2, this and (2.22) imply that the lemma is also valid for the weight (2.24).

To extend this result to the case of a weight φ satisfying the conditions of Theorem 2.1, it suffices to use formulas (2.30), (2.22) and the following consequence of (1.9):

$$\pi(u^\alpha;z) = [\pi(u;z)]^\alpha \qquad (z\in D,\ \alpha\in \mathbf{R}). \tag{6.8}$$

Lemma 6.2. *Let the weight φ be of the form* (2.26), *with* (2.27) *and also the conditions*

$$\exists r\in(1,\infty):\quad \omega(h;\delta)_r = o(\delta^{1/r}) \qquad (\delta\to+0), \tag{6.9}$$

$$F_\nu(u) = \prod_{\mu=1}^{l_\nu}[\omega(\mu,\nu;u)]^{\alpha(\mu,\nu)} \qquad (\nu=1,\dots,m) \tag{6.10}$$

fulfilled, where $m, l_\nu\in\mathbf{N}$, $\alpha(\mu,\nu)\in\mathbf{R}$, and $\omega(\mu,\nu;\delta)$ are concave moduli of continuity such that $\tau\omega'(\mu,\nu;\tau)/\omega(\mu,\nu;\tau)\to 0$ as $\tau\to 0$ remaining on the set where the derivative $\omega'(\mu,\nu;\tau)$ exists. Then the estimate (6.2) *holds.*

Proof. By (2.26), (6.10), (2.22), (6.8), and (6.4)

$$\frac{\pi'(\varphi;z)}{\pi(\varphi;z)} = \frac{\pi'(h;z)}{\pi(h;z)} + \sum_{\nu=1}^{m}\sum_{\mu=1}^{l_\nu}\alpha(\mu,\nu)\frac{\pi'_{\mu,\nu}(z)}{\pi_{\mu,\nu}(z)},$$

where

$$\pi_{\mu,\nu}(z) = \pi(\omega(\mu,\nu;|\sin[(\tau-\theta_\nu)/2]|);z).$$

As shown in [9], p. 52, (6.9) implies (6.2) for $\varphi=h$. Applying the inequality (6.6) for $r=2$ to $f(z)=\pi_{\mu,\nu}(z)$, we obtain that for all $\rho\in(0,1)$

$$\|\pi'_{\mu,\nu}(\rho e^{i\tau})/\pi_{\mu,\nu}(\rho e^{i\tau})\|_\infty \le C_1(1-\rho)^{-3/2}\omega(\ln\pi_{\mu,\nu}(e^{i\tau});1-\rho)_2.$$

It remains to take into account that by (2.30), Lemmas 3.1 and 3.4, Remark 3.1, and M. Riesz' theorem on conjugate functions

$$\omega(\ln\pi_{\mu,\nu}(e^{i\tau});1-\rho)_2 = o((1-\rho)^{1/2}) \qquad (\rho\to 1-0).$$

Remark 6.1. Completely analogously, on the basis of Lemmas 3.5 and 3.6, one can prove two more lemmas, the formulations of which are obtained from those of Lemmas 6.1 and 6.2 by replacing $|\sin[(\tau-\theta_\nu)/2]|$ and $|\sin[(\theta-\theta_\nu)/2]|$ by $|\cos\tau-\cos\theta_\nu|$ and $|\cos\theta-\cos\theta_\nu|$, respectively, for those values of ν for which $\theta_\nu\ne 0$, $\theta_\nu\ne\pi$.

7. Inequalities for polynomials and their derivatives

In this section we establish certain weight analogs of A. A. Markov, S. N. Bernstein, and S. M. Nikol'skiĭ's inequalities. The method we propose is based on the application of the inequalities (5.1) and (5.12), Theorem 2.1, and Lemma 6.1.

Theorem 7.1. *Let* $0 < q < r \le \infty$, $0 < c < \infty$, $0 < \lambda < 1$, *and let the weight* φ *satisfy the conditions*:

a)

$$\varphi \in \bigcup_{\alpha>0} L^\alpha, \qquad \varphi(\tau) \ge 0 \text{ a.e.}, \qquad \ln\varphi \in L^1; \tag{7.1}$$

b) *there is a constant* $C_1 = C_1(\varphi)$ *such that* $\forall\rho \in (0, 1)$

$$\|\pi'(\varphi; \rho e^{i\theta})/\pi(\varphi; \rho e^{i\theta})\|_\infty \le C_1/(1-\rho), \tag{7.2}$$

where $\pi(\varphi; z)$ *is the Szegő function* (1.7).

Then for all $n > c$ *and for any polynomial* $Q_n(z)$ *of degree* $\le n$

$$\left\|\frac{Q_n(e^{i\theta})}{\pi(\varphi; R_n e^{i\theta})}\right\|_r \le A^{q,r,\varphi}_{n,c,\lambda} n^{1/q-1/r} \left\|\frac{Q_n(e^{i\theta})}{\pi(\varphi; R_n e^{i\theta})}\right\|_q, \tag{7.3}$$

where $R_n = 1 - cn^{-1}$,

$$\begin{aligned} A^{q,r,\varphi}_{n,c,\lambda} &\overset{\text{def}}{=} (1+2\lambda)^{C_1}\left[\left(2-\frac{\lambda c}{n}\right)\lambda c\right]^{1/q-1/r}\left(1-\frac{\lambda c}{n}\right)^{-n} \\ &\le (1+2\lambda)^{C_1}(\lambda c)^{1/q-1/r}([\lambda c]+1)/([\lambda c]+1-\lambda c). \end{aligned} \tag{7.4}$$

Proof. If $0 < \rho_1 < \rho_2 < 1$, $z_1 = \rho_1 e^{i\theta}$, $z_2 = \rho_2 e^{i\theta}$, then by (7.2)

$$\left|\ln\left|\frac{\pi(\varphi; z_1)}{\pi(\varphi; z_2)}\right|\right| \le \left|\int_{z_1}^{z_2}\frac{\pi'(\varphi; z)dz}{\pi(\varphi; z)}\right| \le C_1 \ln\frac{1-\rho_1}{1-\rho_2}$$

and, consequently,

$$\left(\frac{1-\rho_1}{1-\rho_2}\right)^{-C_1} \le \left|\frac{\pi(\varphi; \rho_2 e^{i\theta})}{\pi(\varphi; \rho_1 e^{i\theta})}\right| \le \left(\frac{1-\rho_1}{1-\rho_2}\right)^{C_1}. \tag{7.5}$$

Let, in particular,

$$\rho_1 = R_n\rho_0^2, \qquad \rho_0 = 1 - \lambda c n^{-1}, \qquad \rho_2 = R_n. \tag{7.6}$$

Then for $n > c$ the inequalities $0 < \rho_1 < \rho_2 < 1$ hold and

$$\frac{1-\rho_1}{1-\rho_2} = 1 + \rho_2\frac{1-\rho_0^2}{1-\rho_2} = 1 + \left(1-\frac{c}{n}\right)\left(2-\frac{\lambda c}{n}\right) < 1 + 2\lambda.$$

Therefore, it follows from (7.5) that for all $n > c$ and $\theta \in \mathbf{R}$

$$(1+2\lambda)^{-C_1} \le |\pi(\varphi; \rho_0^2 R_n e^{i\theta})/\pi(\varphi; R_n e^{i\theta})| \le (1+2\lambda)^{C_1}. \tag{7.7}$$

By (1.3),

$$|Q_n(e^{i\theta})| = |Q_n^*(e^{i\theta})|, \qquad |Q_n^*(\rho_0^{-1}e^{i\theta})| \le \rho_0^{-n}|Q_n(\rho_0 e^{i\theta})|. \tag{7.8}$$

Using (7.5), (7.6), the nondecrease of $M_r(f; \rho)$ with respect to ρ and Lemma 5.2, we obtain the inequalities

$$\begin{aligned} \|Q_n(e^{i\theta})/\pi(\varphi; R_n e^{i\theta})\|_r &\le (1+2\lambda)^{C_1}\|Q_n^*(e^{i\theta})/\pi(\varphi; \rho_0^2 R_n e^{i\theta})\|_r \\ &\le (1+2\lambda)^{C_1}\|Q_n^*(\rho_0^{-1}e^{i\theta})/\pi(\varphi; \rho_0 R_n e^{i\theta})\|_r \\ &\le \rho_0^{-n}(1+2\lambda)^{C_1}\|Q_n(\rho_0 e^{i\theta})/\pi(\varphi; \rho_0 R_n e^{i\theta})\|_r \\ &\le \rho_0^{-n}(1+2\lambda)^{C_1}(1-\rho_0^2)^{1/r-1/q}\|Q_n(e^{i\theta})/\pi(\varphi; R_n e^{i\theta})\|_q. \end{aligned}$$

Thus the estimate (7.3) is proved. Using the fact that

$$n > \lambda c, \qquad x\ln\{x/(x-x_0)\}\downarrow \qquad (x > x_0), \tag{7.9}$$

we easily arrive at the estimate (7.4).

Remark 7.1. By Lemma 6.1 the conditions of Theorem 7.1 are satisfied for a weight φ that satisfies the conditions of Theorem 2.1.

Corollary 7.1. *Under the conditions of Theorem 7.1, for every $n > c/2$ any trigonometric polynomial $t_n(\tau)$ of order at most n satisfies the inequality*

$$\left\| \frac{t_n(\theta)}{\pi(\varphi;\, R_{2n}e^{i\theta})} \right\|_r \le A^{q,r,\varphi}_{2n,c,\lambda}(2n)^{1/q-1/r} \left\| \frac{t_n(\theta)}{\pi(\varphi;\, R_{2n}e^{i\theta})} \right\|_q . \tag{7.10}$$

The proof reduces to the application of Theorem 7.1 to the polynomial Q_{2n} in the obvious representation $t_n(\theta) = e^{-in\theta}Q_{2n}(e^{i\theta})$.

If $\varphi(\tau) \equiv 1$ $(\tau \in \mathbf{R})$, then $\pi(\varphi;\, z) \equiv 1$ $(z \in D)$ and $C_1 = 0$ in (7.2). In this case for $\lambda = 1$ relation (7.10) takes the form

$$\|t_n\|_r \le B^{q,r}_{n,c} n^{1/q-1/r}\|t_n\|_q \qquad (0 < q \le r \le \infty,\ n > c/2), \tag{7.11}$$

where

$$B^{q,r}_{n,c} = \{c(1 - c/(4n))\}^{1/q-1/r}(1 - c/(2n))^{-2n}. \tag{7.12}$$

Similar inequalities have been studied by many authors. Thus, Jackson [31] established that

$$\|t_n\|_\infty \le 2(2\pi n)^{1/q}\|t_n\|_q \qquad (q > 0,\ n \in \mathbf{N}); \tag{7.13}$$

Nikol′skiĭ [24] proved that

$$\|t_n\|_r \le 2(2\pi n)^{1/q-1/r}\|t_n\|_q \qquad (1 \le q \le r \le \infty,\ n \in \mathbf{N}). \tag{7.14}$$

As Stechkin observed (see [32], p. 172, and [33], Russian p. 539, English p. 265), the following inequality, refining (7.14),

$$\|t_n\| \le 2^{(r-q)/r}(2\pi n)^{1/q-1/r}\|t_n\|_q \qquad (0 < q \le r \le \infty,\ n \in \mathbf{N}) \tag{7.15}$$

easily follows from (7.13). Arestov [33] found the best value of the constant $\lambda(r, n)$ in the inequality

$$\|t_n\|_r \le \lambda(r, n)\|t_n\|_0 \qquad (0 < r \le \infty,\ n \in \mathbf{N}). \tag{7.16}$$

The following theorem makes (7.13)–(7.15) more precise for $q \le (\ln 2)^{-1}\ln(2\pi)$.

Theorem 7.2. *Let $0 < q \le r \le \infty$. Then any trigonometric polynomial t_n of order at most $n \in \mathbf{N}$ satisfies the inequality*

$$\|t_n\|_r \le 4^{(r-q)/r} n^{1/q-1/r}\|t_n\|_q . \tag{7.17}$$

Proof. Put $c = 2(n - \sqrt{n^2 - n})$ $(n \ge 2)$. Then $0 < c < n$, and $\{1 - c/(4n)\}c = 1$ in (7.12). Therefore and by (7.9)

$$B^{q,r}_{n,c} = \{1 - c/(2n)\}^{-2n} = \{n/(n-1)\}^n \le 4 \qquad (n \ge 2). \tag{7.18}$$

It follows from (7.11), (7.12), and (7.18) that for $n \ge 2$

$$\|t_n\|_r \le 4n^{1/q-1/r}\|t_n\|_q \qquad (0 < q \le r \le \infty). \tag{7.19}$$

Using the fact that $\|F\|_q$ does not decrease with respect to q for $0 \le q \le \infty$ and the equality $\sup_{t_1}(\|t_1\|_\infty/\|t_1\|_0) = 4$ [33], we conclude that (7.19) remains in force also for $n = 1$, and that $c = 4$ is the smallest among the constants in the inequality (1.25) independent of q, r $(0 < q \le r \le \infty)$ and $n \in \mathbf{N}$. Now, repeating Stechkin's argument ([33], Russian p. 539, English p. 265), we derive (7.17) from (7.19) with $r = \infty$:

$$\|t_n\|_r \le \|t_n\|_\infty^{1-q/r}\|t_n\|_q^{q/r} \le 4^{(r-q)/r} n^{1/q-1/r}\|t_n\|_q .$$

It is easy to see that for $0 < q < (\ln 2)^{-1}\ln(2\pi)$ the right-hand side of (7.17) does not exceed the right-hand side of (7.15), and, consequently, (7.17) is a refinement of (7.13)–(7.15).

Setting $\varphi \equiv 1$, $\lambda = \sqrt{n}/(\sqrt{n}+\sqrt{n-c/2})$ $(n > c/2)$ in (7.10), we have $0 < \lambda < 1$, $\{1-\lambda c/(4n)\}\lambda c = c/2$. From this and (7.4) we find that

$$2^{1/q-1/r} A^{q,r,\varphi}_{2n,c,\lambda} = (2/c)^{1/q-1/r}\{n/(n-c/2)\}^n \qquad (n > c/2),$$

and, since $\{n/(n-c/2)\}^n \downarrow$ $(n > c/2)$, (7.10) implies the inequality

$$\|t_n\|_r \le (2/c)^{1/q-1/r}(1+c/2)^{1+c/2} n^{1/q-1/r}\|t_n\|_q \qquad (n \ge 1+c/2),$$

in which for fixed $c > 2$ the constant independent of n tends to zero as $1/q - 1/r \to \infty$.

We now establish an analog of the inequality (7.3) for the case when in its left-hand side the derivative $Q_n^{(j)}(e^{i\theta})$ appears instead of $Q_n(e^{i\theta})$; we are not interested here in how the constant depends on the parameters.

Theorem 7.3. *Let $j \in \mathbf{Z}_+$, $0 < q \le r \le \infty$, $0 < c < \infty$, $R_n = 1 - cn^{-1}$, and let the weight φ satisfy the conditions (7.1)–(7.2). Then there is a constant C_2 (depending on φ, c, j, q, and r) such that for all $n > c$ any polynomial $Q_n(z)$ of degree at most n satisfies the inequality*

$$\|Q_n^{(j)}(e^{i\theta})/\pi(\varphi; R_n e^{i\theta})\|_r \le C_2 n^{j+1/q-1/r}\|Q_n(e^{i\theta})/\pi(\varphi; R_n e^{i\theta})\|_q. \tag{7.20}$$

Proof. The case $j = 0$ was already considered (see Theorem 7.1), and the case $j \in \mathbf{N}$ reduces to the case $j = 1$. Using the relations (7.6)–(7.8) and arguing as in the proof of Theorem 7.1, we obtain the inequalities

$$\begin{aligned} \left\|\frac{Q_n'(e^{i\theta})}{\pi(\varphi; R_n e^{i\theta})}\right\|_r &\le (1+2\lambda)^{C_1}\left\|\frac{Q_n'^*(e^{i\theta})}{\pi(\varphi; \rho_0^2 R_n e^{i\theta})}\right\|_r \\ &\le (1+2\lambda)^{C_1}\left\|\frac{Q_n'^*(\rho_0^{-1}e^{i\theta})}{\pi(\varphi; \rho_0 R_n e^{i\theta})}\right\|_r \le \rho_0^{-n}(1+2\lambda)^{C_1}\left\|\frac{Q_n'(\rho_0 e^{i\theta})}{\pi(\varphi; \rho_0 R_n e^{i\theta})}\right\|_r. \end{aligned} \tag{7.21}$$

Differentiating the formula $F_n(z) \overset{\text{def}}{=} Q_n(z)/\pi(\varphi; R_n z)$, we have

$$F_n'(z) = \frac{Q_n'(z)}{\pi(\varphi; R_n z)} - R_n \frac{\pi'(\varphi; R_n z)}{\pi(\varphi; R_n z)} F_n(z).$$

From this, using (7.2), we get

$$\left|\frac{Q_n'(\rho_0 e^{i\theta})}{\pi(\varphi; \rho_0 R_n e^{i\theta})}\right| \le \frac{C_1 R_n}{1-\rho_0 R_n}|F_n(\rho_0 e^{i\theta})| + |F_n'(\rho_0 e^{i\theta})|.$$

Consequently,

$$\left\|\frac{Q_n'(\rho_0 e^{i\theta})}{\pi(\varphi; \rho_0 R_n e^{i\theta})}\right\|_r \le C_3(C_1, \lambda, c, r)\{n\|F_n(e^{i\theta})\|_r + M_r(F_n'; \rho_0)\}. \tag{7.22}$$

By Theorem 7.1 and Corollary 5.2, (7.21) and (7.22) imply (7.20) for $j = 1$.

Arguing as in the proof of Corollary 7.1, from Theorem 7.3 we obtain

Corollary 7.2. *Let the conditions of Theorem 7.3 be satisfied. Then there is a constant $C_4 = C_4(\varphi, c, j, q, r)$ such that*

$$\left\|\frac{t_n^{(j)}(\theta)}{\pi(\varphi; R_{2n}e^{i\theta})}\right\|_r \le C_4 n^{j+1/q-1/r}\left\|\frac{t_n(\theta)}{\pi(\varphi; R_{2n}e^{i\theta})}\right\|_q \tag{7.23}$$

for all $n > c/2$.

We note that, in the case $\pi(\varphi; z) \equiv 1$ $(z \in D)$, $j \in \mathbf{N}$, many papers (see, for example, [34]–[45]) have been devoted to the inequality (7.23). Among them we single out those of Bernstein [34], Zygmund [35], [45], p. 11, and Arestov [41], [44], where it was established that if $j = 1$, $0 < q = r \le \infty$, then $C_4 = 1$ is the best constant in (7.23).

Using Theorem 7.3, we shall generalize it a bit more by considering, along with the variable weight $|\pi(\varphi; R_n e^{i\theta})|^{-1}$ without singularities, a weight that may become zero or infinity at separate points of $[-\pi, \pi]$.

Theorem 7.4. *Let $j \in \mathbf{Z}_+$, $0 < q \le r \le \infty$, $0 < c < \infty$, and let the weights φ and ψ satisfy the conditions* a) *and* b) *of Theorem* 7.1 *and, in addition, let there be a constant $C_5 = C_5(\psi)$ such that, for all $\rho \in (0, 1)$,*

$$\|\pi(\psi; \rho e^{i\theta})\sqrt{\psi(\theta)}\|_\infty \le C_5. \tag{7.24}$$

We define a weight g by the relations

$$g(\tau) = \prod_{\nu=1}^{m} g_\nu\left(\left|\sin\frac{\tau-\theta_\nu}{2}\right|\right), \qquad g_\nu(t) = \prod_{\mu=1}^{l_\nu}[g_{\mu,\nu}(t)]^{\alpha(\mu,\nu)}, \tag{7.25}$$

where $m, l_\nu \in \mathbf{N}$, $-\pi < \theta_1 < \cdots < \theta_m \le \pi$, $\alpha(\mu, \nu) \ge 0$; the $g_{\mu,\nu}(t)$ are moduli of continuity $(\mu = 1, \dots, l_\nu; \nu = 1, \dots, m)$, where for $r = \infty$ all $\alpha(\mu, \nu) = 0$, and for $r < \infty$

$$\int_0^\theta g_\nu^{-r}(t)\,dt = O(\theta g_\nu^{-r}(\theta)) \qquad (\theta \to +0;\ \nu = 1, \dots, m). \tag{7.26}$$

Then there is a constant $C_6 = C_6(\varphi, \psi, c, j, q, r)$ such that if $n > c$, $R_n = 1 - cn^{-1}$, and Q_n is a polynomial of degree $\le n$, then

$$\left\|\frac{Q_n^{(j)}(e^{i\theta})\psi(\theta)}{\pi(\varphi; R_n e^{i\theta})g(\theta)}\right\|_r \le C_6 n^{j+1/q-1/r}\left\|\frac{Q_n(e^{i\theta})\psi(\theta)}{\pi(\varphi; R_{2n}e^{i\theta})g(\theta)}\right\|_q. \tag{7.27}$$

Proof. First let $g(\tau) \equiv 1$. Applying (7.24), (2.22), and Theorem 7.3, we obtain the inequalities

$$\begin{aligned}\left\|\frac{Q_n^{(j)}(e^{i\theta})\psi(\theta)}{\pi(\varphi; R_n e^{i\theta})}\right\|_r &\le C_4^2\left\|\frac{Q_n^{(j)}e^{i\theta}}{\pi(\varphi\psi^2; R_n e^{i\theta})}\right\|_r \\ &\le C_6 n^{j+1/q-1/r}\|Q_n(e^{i\theta})/\pi(\varphi; R_n e^{i\theta})\pi(\psi^2; R_n e^{i\theta})\|_q.\end{aligned} \tag{7.28}$$

Since φ satisfies the conditions a)–b) of Theorem 7.2, the inequalities (7.5) are fulfilled. Using them for $\rho_1 = R_n^3$ and $\rho_2 = R_n$, we find that

$$3^{-C_1(\varphi)} \le |\pi(\varphi; R_n^3 e^{i\theta})/\pi(\varphi; R_n e^{i\theta})| \le 3^{C_1(\varphi)}. \tag{7.29}$$

Similarly, putting $\rho_1 = R_n^2$ and $\rho_2 = R_n$, we have

$$4^{-C_1(\psi)} \le |\pi(\psi^2; R_n^2 e^{i\theta})/\pi(\psi^2; R_n e^{i\theta})| \le 4^{C_1(\psi)}. \tag{7.30}$$

By (7.29), (7.30), (1.3), (1.10), and the fact that $M_q(f; \rho)$ is nondecreasing with respect to ρ,

$$\begin{aligned}&\|Q_n(e^{i\theta})/[\pi(\varphi; R_n e^{i\theta})\pi(\psi^2; R_n e^{i\theta})]\|_q \\ &\quad\le 3^{C_1(\varphi)}4^{C_1(\psi)}\|Q_n^*(e^{i\theta})/[\pi(\varphi; R_n^3 e^{i\theta})\pi(\psi^2; R_n^2 e^{i\theta})]\|_q \\ &\quad\le 3^{C_1(\varphi)}4^{C_1(\psi)}R_n^{-n}\|Q_n(R_n e^{i\theta})/[\pi(\varphi; R_n^2 e^{i\theta})\pi(\psi^2; R_n e^{i\theta})]\|_q \\ &\quad\le 3^{C_1(\varphi)}4^{C_1(\psi)}R_n^{-n}\|Q_n(e^{i\theta})\psi(\theta)/\pi(\varphi; R_n e^{i\theta})\|_q.\end{aligned} \tag{7.31}$$

From (7.28) and (7.31) we obtain (7.27) for $g(\tau) \equiv 1$.

Now we consider the general case, waiving the restriction $g(\tau) \equiv 1$. We introduce the function $g_n(\tau) = |\pi(g\,;\, R_n e^{i\tau})|^{-2}$. By Theorem 2.1

$$g_n(\tau) \asymp \prod_{\nu=1}^{m} g_\nu\left(\left|\sin\frac{\tau-\theta_\nu}{2}\right| + \frac{1}{n}\right).$$

Let $\varepsilon > 0$ be so small that $(\varepsilon-\pi\,,\,\varepsilon+\pi)$ contains each one of the points $\theta_1\,,\,\ldots\,,\,\theta_m$. Choose $\delta > 0$ such that the segments $e_\nu = [\theta_\nu - \delta/n\,,\,\theta_\nu + \delta/n]$ $(\nu = 1\,,\,\ldots\,,\,m)$ are contained in $(\varepsilon - \pi\,,\,\varepsilon + \pi)$. Using the relations

$$\begin{aligned} &\|F_1 + F_2\|_r \le (\|F_1\|_r + \|F_2\|_r)\min(1\,,\,2^{1/r}) && (0 < r \le \infty),\\ &g_n(\tau) \le C_7(g\,,\,\delta)g(\tau) && \left(\tau \in [\varepsilon-\pi\,,\,\varepsilon+\pi]\setminus \bigcup_{\nu=1}^{m} e_\nu\right),\\ &g_n(\tau) \le C_8(g\,,\,\delta)g_n(\delta/n) && (\tau \in e_\nu\,;\ \nu = 1\,,\,\ldots\,,\,m), \end{aligned}$$

and also condition (7.26), we obtain the estimate

$$(7.32)\quad \left\|\frac{Q_n^{(j)}(e^{i\theta})\psi(\theta)}{\pi(\varphi\,;\,R_n e^{i\theta})g(\theta)}\right\|_r \le C_9\left\|\frac{Q_n^{(j)}(e^{i\theta})\psi(\theta)}{\pi(\varphi\,;\,R_n e^{i\theta})g_n(\theta)}\right\|_r + C_{10}n^{-1/r}\|Q_n^{(j)}(e^{i\theta})\psi(\theta)/[\pi(\varphi\,;\,R_n e^{i\theta})g_n(\theta)]\|_\infty.$$

Using the validity of the theorem for $\varphi(\tau) \equiv 1$ and the condition (7.24), we conclude that the right-hand side of (7.32) does not exceed

$$\begin{aligned} &C_{11}n^{j+1/q-1/r}\left\|\frac{Q_n(e^{i\theta})\psi(\theta)}{\pi(\varphi\,;\,R_n e^{i\theta})\pi(g^{-2}\,;\,R_n e^{i\theta})}\right\|_q\\ &\le C_{12}n^{j+1/q-1/r}\left\|\frac{Q_n(e^{i\theta})}{\pi(\varphi\,;\,R_n e^{i\theta})\pi(\psi^2 g^{-2}\,;\,R_n e^{i\theta})}\right\|_q. \end{aligned}$$

To obtain (7.27) it remains now to argue as in the proof of (7.31).

The inequalities (7.3), (7.20), and (7.27) contain the same weights under the "norms" in the left- and right-hand sides. In the next theorem we consider the case of different weights.

Theorem 7.5. *Suppose that* $0 < r \le \infty$, $0 < c < \infty$, *the weights* φ, ψ, Φ, Ψ *satisfy the conditions* a)–b) *of Theorem* 7.1, *and the weights* ψ *and* Ψ *satisfy the condition* (7.24). *Let the weight* $g(\tau)$ *be* $\equiv 1$ *for* $r = \infty$ *and satisfy* (7.25)–(7.26) *for* $1 < r < \infty$. *Let the weight* G *be defined by the relations*

$$G(\tau) = \prod_{\nu=1}^{m} G_\nu\left(\left|\sin\frac{\tau-\theta_\nu}{2}\right|\right), \qquad G_\nu(t) = \prod_{\mu=1}^{l_\nu}[G_{\mu,\nu}(t)]^{A(\mu,\nu)},$$

where $m\,,\,l_\nu \in \mathbf{N}$, $-\pi < \theta_1 < \cdots < \theta_m \le \pi$, $A(\mu\,,\,\nu) \in \mathbf{R}$, *and* $G_{\mu,\nu}(t)$ *are moduli of continuity* $(\mu = 1\,,\,\ldots\,,\,l_\nu\,;\ \nu = 1\,,\,\ldots\,,\,m)$. *Then there is a constant* C_{13}, *depending on* c, r, *and the weights considered, such that if* $n > c$, $R_n = 1 - cn^{-1}$ *and* $Q_n(z)$ *is a polynomial of degree at most* n, *then*

$$(7.33)\quad \begin{aligned} &\|Q_n(e^{i\theta})\psi(\theta)/[\pi(\varphi\,;\,R_n e^{i\theta})g(\theta)]\|_r\\ &\le C_{13}\left\|\pi\left(\frac{\Phi\Psi^2 g^2}{\varphi\psi^2 G^2}\,;\,R_n e^{i\theta}\right)\right\|_\infty \left\|\frac{Q_n(e^{i\theta})\Psi(\theta)}{\pi(\Phi\,;\,R_n e^{i\theta})G(\theta)}\right\|_r. \end{aligned}$$

Proof. Set $j=0$ in the inequality (7.32) and then apply Theorem 7.4 to the second term on the right. We get

$$\|Q_n(e^{i\theta})\psi(\theta)/[\pi(\varphi;R_ne^{i\theta})g(\theta)]\|_r \le C_{14}J_1, \tag{7.34}$$

where

$$J_1 = \|Q_n(e^{i\theta})\psi(\theta)/[\pi(\varphi;R_ne^{i\theta})g_n(\theta)]\|_r.$$

By (7.24) and (2.22)

$$\begin{aligned} J_1 &\le C_{15}\|Q_n(e^{i\theta})/[\pi(\varphi;R_ne^{i\theta})\pi(\psi^2g^{-2};R_ne^{i\theta})]\|_r \\ &\le C_{15}\|\pi(\Phi g^2\Psi^2\varphi^{-1}G^{-2}\psi^{-2};R_ne^{i\theta})\|_\infty J_2, \\ J_2 &= \|Q_n(e^{i\theta})/[\pi(\Phi;R_ne^{i\theta})\pi(\Psi^2G^{-2};R_ne^{i\theta})]\|_r. \end{aligned} \tag{7.35}$$

The weights Φ, Ψ^2, and G^{-2} satisfy the conditions a)–b) of Theorem 7.1 (the last one by Lemma 6.1). Therefore, the inequalities (7.5) are applicable to them, in view of which, arguing as in the derivation of (7.31), we have

$$\begin{aligned} J_2 &\le C_{16}\|Q_n^*(e^{i\theta})/[\pi(\Phi,R_n^3e^{i\theta})\pi(\Psi^2G^{-2};R_n^2e^{i\theta})]\|_r \\ &\le C_{16}\|Q_n^*(R_n^{-1}e^{i\theta})/[\pi(\Phi;R_n^2e^{i\theta})\pi(\Psi^2G^{-2};R_ne^{i\theta})\|_r \\ &\le C_{16}R_n^{-n}\|Q_n(R_ne^{i\theta})/[\pi(\Phi;R_n^2e^{i\theta})\pi(\Psi^2G^{-2};R_ne^{i\theta})]\|_r \\ &\le C_{16}R_n^{-n}\|Q_n(e^{i\theta})\Psi(\theta)/[\pi(\Phi;R_ne^{i\theta})G(\theta)]\|_r. \end{aligned} \tag{7.36}$$

Now (7.33) follows from (7.34)–(7.36).

In view of the representation $t_n(\theta)=e^{-in\theta}Q_{2n}(e^{i\theta})$, Theorems 7.4 and 7.5, after obvious adjustments in their formulations, remain valid also for the trigonometric polynomials $t_n(\theta)$. Theorems 7.1 and 7.5, under the additional assumption that the weights considered are even, are easily transformed into corresponding results for algebraic polynomials $p_n(t)$ on $[-1,1]$ with the aid of the relation

$$z^np_n((z+z^{-1})/2) = Q_{2n}(z). \tag{7.37}$$

The next theorem, an analog of Theorem 7.4 for $p_n(t)$, is proved by the same method, although it is not obtained as a simple corollary from the latter.

Theorem 7.6. *Let the conditions of Theorem 7.4 be satisfied, with the weight functions φ, ψ, g being even. Then there is a constant C_{17}, dependent on c, q, r, φ, ψ, g, such that if $n>c/2$, $R_n=1-cn^{-1}$, and $p_n(t)$ is a polynomial of degree at most n, then*

$$\begin{aligned} &\left\{\int_{-1}^1\left|\frac{p_n^{(j)}(t)\psi(a(t))(\sqrt{1-t^2}+n^{-1})^j}{\pi(\varphi;R_{2n}e^{ia(t)})g(a(t))}\right|^r\frac{dt}{\sqrt{1-t^2}}\right\}^{1/r} \\ &\qquad\le C_{17}n^{j+1/q-1/r}\left\{\int_{-1}^1\left|\frac{p_n(t)\psi(a(t))}{\pi(\varphi;R_{2n}e^{ia(t)})g(a(t))}\right|^q\frac{dt}{\sqrt{1-t^2}}\right\}^{1/q}, \end{aligned} \tag{7.38}$$

where $a(t)=\arccos t$.

Proof. It suffices to consider the case $j=1$. Indeed, if (7.38) is proved for some $j=j_0\in\mathbf{N}$, then, using the relations (2.22) and

$$(\sqrt{1-t^2}+n^{-1})^{j_0} \asymp |\pi(|\sin\tau|^{-2j_0};R_{2n}e^{ia(t)})|,$$

we may consider the weight $\varphi(\tau)|\sin\tau|^{2j_0}$ instead of $\varphi(\tau)$, thereby doing away with the inconvenient factor $(\sqrt{1-t^2}+n^{-1})^{j_0}$ on the left in (7.38), after which it becomes obvious that the validity of (7.38) for $j=j_0$ and $j=1$ implies its validity for

$j = j_0 + 1$. For $j = 1$ (7.38) is proved as follows. Differentiating (7.37), we get the equality

$$zQ'_{2n}(z) - nQ_{2n}(z) = q_{2n-2}(z)(z^2 - 1)/2, \tag{7.39}$$

where

$$q_{2n-2}(z) = z^{n-1}p'_n((z + z^{-1})/2). \tag{7.40}$$

Using (7.39), (7.40), and (7.27), and arguing as in the derivation of (7.31), we have

$$\begin{aligned}
&\|p'_n(\cos\theta)(|\sin\theta| + n^{-1})\psi(\theta)/[\pi(\varphi; R_{2n}e^{i\theta})g(\theta)]\|_r \\
&\le C_{18}\left\|\frac{q^*_{2n-2}(e^{i\theta})}{\pi(\varphi; R^3_{2n}e^{i\theta})\pi(\psi^2 g^{-2}\sin^2\tau; R^2_{2n}e^{i\theta})}\right\|_r \\
&\le C_{18}\left\|\frac{q^*_{2n-2}(R^{-1}_{2n}e^{i\theta})}{\pi(\varphi; R^2_{2n}e^{i\theta})\pi(\psi^2 g^{-2}\sin^2\tau; R_{2n}e^{i\theta})}\right\|_r \\
&\le C_{18}\left\|\frac{[e^{i\theta}Q'_{2n}(e^{i\theta}) - nQ_{2n}(e^{i\theta})]\psi(\theta)}{\pi(\varphi; R_{2n}e^{i\theta})g(\theta)}\right\|_r R^{2-2n}_{2n} \\
&\le C_{19}n^{1+1/q-1/r}\|p_n(\cos\theta)\psi(\theta)/[\pi(\varphi; R_{2n}e^{i\theta})g(\theta)]\|_q.
\end{aligned} \tag{7.41}$$

From (7.41), using the fact that the expressions under the "norms" are even, after the change $t = \cos\theta$ we obtain (7.38) for $j = 1$.

Remark 7.2. Directly from Theorem 7.4 (with the aid of (7.37)) one obtains an inequality weaker than (7.38) with the factor $(\sqrt{1-t^2})^j$ instead of $(\sqrt{1-t^2} + n^{-1})^j$ on the left-hand side.

Konyagin [46] obtained estimates uniform in n and j for the quantity

$$\sup_{q_n}\{\|q_n^{(j)}(t)(1-t^2)^\alpha\|_{L^r[-1,1]}/\|q_n(t)(1-t^2)^\beta\|_{L^q[-1,1]}\},$$

where q_n is a nonzero polynomial of degree at most n. This result cannot be obtained as a consequence of our theorems, since the nature of the dependence of C_{17} on j is unclear. The results of Lebed′ [47], [48], which have a slightly different form, do not follow from Theorems 7.4–7.6 either (for similar results on curves in $\mathbf{C}$ see [27]). At the same time, it is possible to derive as corollaries from Theorems 7.4–7.6 some results (obtained by different methods) of Potapov [49], Khalilova [50], and Nevai [52], where the weights φ, ψ, g, Φ, Ψ, G may have only power singularities and the exponents q and r satisfy the inequalities $1 \le q \le r \le \infty$. Along these lines, it is also possible to obtain some results of Daugavet and Rafal′son [51], where inequalities between the quantities

$$\|q_n^{(j)}(t)(1-t)^\alpha(1+t)^\beta\|_{L^r[-1,1]}, \qquad \|q_n(t)(1-t)^\gamma(1+t)^\delta\|_{L^q[-1,1]}$$

were established under the condition $q, r \in [1, \infty]$, and in some cases the correctness in order with respect to n of those results.

8. Proof and corollaries of Theorem 1.1

For $j = 0$ Theorem 1.1 was proved in [9]. Thus, if the conditions (1.7), (1.13), and (1.14) are satisfied, then the uniform asymptotic representation (1.11) holds on Γ_1. We introduce the function

$$\lambda(\rho) = \Lambda(1-\rho), \quad \text{where } \Lambda(\rho) = \sup_{0<u\le\rho}(1-u)\max_{|z|=u}\left|\frac{\pi'(\sigma'; z)}{\pi(\sigma'; z)}\right|. \tag{8.1}$$

It follows from (1.14) that

$$\lambda(\rho)\uparrow \quad (0<\rho<1), \qquad \lambda(\rho)\to 0 \quad (\rho\to +0). \tag{8.2}$$

We may assume that $\lambda(\rho)>0$ in $(0, 1)$. Otherwise the arguments are trivial. Put $a_k = \lambda(c/k)$ $(k\in\mathbf{N},\ k>l=[c/2])$. By (8.2) $a_k\downarrow 0$ $(k\to\infty)$. Therefore, $b_k = ka_k^{-1}$ $(k>l)$ increases and tends to infinity as $k\to\infty$. If $n\in\mathbf{N}$, $n>b_{l+1}$, then there exists a $\nu=\nu(n)\in\mathbf{N}$ such that $b_{\nu-1}\le n<b_\nu$. Consequently, $b_{\nu(n)}\to\infty$ $(n\to\infty)$. But then $\nu(n)\to\infty$, $\lambda(c/\nu(n))\to 0$ $(n\to\infty)$, and

$$0<\nu(n)\le(n+1)\lambda(c/\nu(n)) = o(n) \qquad (n\to\infty). \tag{8.3}$$

It follows from (8.1), (8.2) that if $\nu=\nu(n)$, then

$$\left|\ln\frac{\pi(\sigma';(1-cn^{-1})e^{i\theta})}{\pi(\sigma';(1-c\nu^{-1})e^{i\theta})}\right| \le \int_{c/n}^{c/\nu}\frac{\lambda(\rho)}{\rho}\,d\rho \le \lambda\left(\frac{c}{\nu}\right)\ln\frac{n}{\nu}$$
$$<\lambda(c/\nu)\ln(b_\nu/\nu) = \lambda(c/\nu)\ln\lambda^{-1}(c/\nu) = o(1)$$

as $n\to\infty$. From this we find that, as $n\to\infty$,

$$\pi(\sigma';(1-c\nu^{-1})e^{i\theta}) = \pi(\sigma';(1-cn^{-1})e^{i\theta})[1+o(1)] \tag{8.4}$$

uniformly in $\theta\in\mathbf{R}$. From (8.4) and (1.11) we conclude that

$$\|[\varphi^*_{\sigma,n}(e^{i\theta})-\varphi^*_{\sigma,\nu}(e^{i\theta})]/\pi(\sigma';(1-cn^{-1})e^{i\theta})\|_\infty = o(1)$$

as $n\to\infty$. This in view of Lemma 7.2 (for $r=q=\infty$) implies that, for fixed $j\in\mathbf{N}$ and $n\to\infty$,

$$\|[\varphi^{*(j)}_{\sigma,n}(e^{i\theta})-\varphi^{*(j)}_{\sigma,\nu}(e^{i\theta})]/\pi(\sigma';(1-cn^{-1})e^{i\theta})\|_\infty = o(n^j). \tag{8.5}$$

By (8.4), (8.3), (1.11), and Lemma 7.2

$$\begin{aligned}&\|\varphi^{*(j)}_{\sigma,\nu}(e^{i\theta})/\pi(\sigma';(1-cn^{-1})e^{i\theta})\|_\infty\\ &\quad = O(\nu^j)\|\varphi^*_{\sigma,\nu}(e^{i\theta})/\pi(\sigma';(1-c\nu^{-1})e^{i\theta})\|_\infty = o(n^j).\end{aligned} \tag{8.6}$$

It follows from (8.5), (8.6) that for $j\in\mathbf{N}$

$$\varphi^{*(j)}_{\sigma,n}(e^{i\theta}) = o(n^j)\pi(\sigma';(1-cn^{-1})e^{i\theta}) \qquad (n\to\infty) \tag{8.7}$$

uniformly in θ.

We now prove that for $j\in\mathbf{Z}_+$, $\rho\in(0, 1)$, and $\theta\in\mathbf{R}$ the expansion

$$\overline{\varphi^{(j)}_{\sigma,n}(\rho^{-1}e^{i\theta})} = \sum_{\nu=1}^{j} c_{j,n,\nu}(\rho e^{i\theta})^{j-n+\nu}\varphi^{*(\nu)}_{\sigma,n}(\rho e^{i\theta}) \tag{8.8}$$

holds, where the coefficients $c_{j,n,\nu}$ are real, independent of ρ, θ, and satisfy the inequality

$$|c_{j,n,\nu}|\le C_1(\sigma, j)n^{j-\nu}. \tag{8.9}$$

For $j=0$, (8.8) follows easily from (1.3) and has the form

$$\overline{\varphi_{\sigma,n}(\rho^{-1}e^{i\theta})} = (\rho e^{i\theta})^{-n}\varphi^*_{\sigma,n}(\rho e^{i\theta}).$$

In this case $\nu=j=0$, $c_{0,n,0}=1$, so that (8.9) also holds. Suppose that (8.8), (8.9) hold for some $j\in\mathbf{Z}_+$. Then, differentiating both sides of (8.8) with respect to ρ, after simple transformations we obtain

$$\begin{aligned}\overline{\varphi^{(j+1)}_{\sigma,n}(\rho^{-1}e^{i\theta})} &= \sum_{\nu=0}^{j}(n-j-\nu)c_{j,n,\nu}(\rho e^{i\theta})^{j+1+\nu-n}\varphi^{*(\nu)}_{\sigma,n}(\rho e^{i\theta})\\ &\quad-\sum_{\nu=1}^{j+1}c_{j,n,\nu-1}(\rho e^{i\theta})^{j+1+\nu-n}\varphi^{*(\nu)}_{\sigma,n}(\rho e^{i\theta}).\end{aligned}$$

Thus, we obtain a formula of type (8.8); (8.9) implies the estimate $|c_{j+1,n,\nu}| \le C_1(\sigma, j+1)n^{j+1-\nu}$. Thus, formulas (8.8), (8.9) are proved by induction.

It is not difficult to see that in (8.8)

$$c_{j,n,0} = n(n-1)\cdots(n-j+1) \qquad (j \in \mathbf{N}). \tag{8.10}$$

Indeed, multiplying both sides of (8.8) by ρ^{n-j}, for $\theta = 0$ we have

$$\rho^{n-j}\overline{\varphi_{\sigma,n}^{(j)}(\rho^{-1})} = \sum_{\nu=0}^{j} c_{j,n,\nu}\rho^{\nu}\varphi_{\sigma,n}^{*(\nu)}(\rho). \tag{8.11}$$

We pass to the limit as $\rho \to 0$ in (8.11); we find that

$$\kappa_{\sigma,n}n(n-1)\cdots(n-j+1) = c_{j,n,0}\varphi_{\sigma,n}^{*}(0).$$

Since $\varphi_{\sigma,n}^{*}(0) = \kappa_{\sigma,n} > 0$, this implies (8.10).

Passing in (8.8) to complex conjugates, for $\rho = 1$ we get

$$\varphi_{\sigma,n}^{(j)}(z) = \sum_{\nu=0}^{j} c_{j,n,\nu}z^{n-j-\nu}\overline{\varphi_{\sigma,n}^{*(\nu)}(z)} \qquad (z = e^{i\theta}). \tag{8.12}$$

Now (8.12), (1.11), (8.7), (8.9), and (8.10) imply (1.15); that is, Theorem 1.1 is proved also for $j \in \mathbf{N}$.

Corollary 8.1. *Suppose the weight* φ *satisfies the conditions of Theorem* 4.2 *or of Remark* 4.1 *and* $d\sigma(\tau) = \varphi(\tau)d\tau$. *Then for fixed* $j \in \mathbf{Z}_+$ *and* $c \in (0, \infty)$ *the uniform asymptotic representation* (1.15) *holds on* Γ_1.

To verify this, it suffices to put together the formulations of Theorems 1.1, 2.1, and 4.2.

Note that Lemmas 3.7–3.10 give examples of moduli of continuity; these can be used to derive more concrete results from Corollary 8.1.

With the aid of formulas (1.4), (1.5), (8.7), one can obtain from Theorem 1.1 the following fact.

Corollary 8.2. *Under the conditions of Theorem* 1.1,

$$T_{\sigma,2n-1}^{(j)}(\theta) = \sqrt{2}n^j|\pi(\sigma'', R_ne^{i\theta})|\{\cos[n\theta + j\pi/2 - \gamma(\theta)] + o(1)\},$$
$$T_{\sigma,2n}^{(j)}(\theta) = \sqrt{2}n^j|\pi(\sigma'; R_ne^{i\theta})|\{\sin[n\theta + j\pi/2 - \gamma(\theta)] + o(1)\},$$

as $n \to \infty$, *uniformly in* $\theta \in \mathbf{R}$; *here* $\gamma_n = \arg\pi(\sigma'; R_ne^{i\theta})$, $R_n = 1 - cn^{-1}$.

9. Two-sided estimates for Christoffel functions

For the polynomials $\{\varphi_{\sigma,n}(z)\}_{n=0}^{\infty}$ the Christoffel-Darboux formula ([2], formula (1.7)] is known:

$$\begin{aligned} K_{\sigma,n}(z,\zeta) &\overset{\text{def}}{=} \varphi_{\sigma,0}(z)\overline{\varphi_{\sigma,0}(\zeta)} + \cdots + \varphi_{\sigma,n}(z)\overline{\varphi_{\sigma,n}(\zeta)} \\ &= (1 - z\overline{\zeta})^{-1}[\varphi_{\sigma,n}^{*}(z)\overline{\varphi_{\sigma,n}^{*}(\zeta)} - z\overline{\zeta}\varphi_{\sigma,n}(z)\overline{\varphi_{\sigma,n}(\zeta)}] \\ &= (1 - z\overline{\zeta})^{-1}[\varphi_{\sigma,n+1}^{*}(z)\overline{\varphi_{\sigma,n+1}^{*}(\zeta)} - \varphi_{\sigma,n+1}(z)\overline{\varphi_{\sigma,n+1}(\zeta)}]. \end{aligned} \tag{9.1}$$

The functions

$$\omega_n(d\sigma; z) = K_{\sigma,n-1}^{-1}(z,z) \qquad (n \in \mathbf{N}) \tag{9.2}$$

are called Christoffel functions. It is known ([2], formula (3.20)) that

$$\omega_{n+1}(d\sigma\,;\,z) = \inf_{Q_n}(2\pi)^{-1}\int_0^{2\pi}|Q_n(e^{i\tau})/Q_n(e^{i\theta})|^2 d\sigma(\tau)\,, \tag{9.3}$$

where Q_n is an arbitrary polynomial of degree $\le n$. For a review of studies concerning the functions (9.2) see, for example, [8]. The following theorem gives an upper estimate for $\omega_n(d\sigma\,;\,e^{i\theta})$.

Theorem 9.1. *Suppose the weight* $\varphi \in L^1$ *is of the form*

$$\varphi(\tau) = h(\tau)\prod_{\nu=1}^{m} w_\nu(|\sin[(\tau-\theta_\nu)/2]|)\,, \tag{9.4}$$

where $-\pi < \theta_1 < \cdots < \theta_m \le \pi$, h *satisfies conditions* (1.18),

$$w_\nu(u) = \prod_{\mu=1}^{l_\nu}[g_{\mu,\nu}(u)]^{\alpha(\mu,\nu)}\,, \tag{9.5}$$

$m\,,\,l_\nu \in \mathbf{N}$, $\alpha(\mu\,,\,\nu) \in \mathbf{R}$, *and the* $g_{\mu,\nu}(u)$ *are concave moduli of continuity* ($\mu = 1\,,\,\ldots\,,\,l_\nu\,;\,\nu = 1\,,\,\ldots\,,\,m$). *Let* $d\sigma(\tau) = \varphi(\tau)d\tau$. *Then there is a constant* $C_1(\varphi) > 0$ *such that*

$$\omega_n(d\sigma\,;\,e^{i\theta}) \ge C_1(\varphi)n^{-1}\prod_{\nu=1}^{m} w_\nu(|\sin[(\theta-\theta_\nu)/2]| + n^{-1}) \tag{9.6}$$

for all $\theta \in \mathbf{R}$ *and* $n \in \mathbf{N}$.

Proof. For a polynomial Q_n of degree n the equality $|Q_n^*(z)| = |Q_n(z)|$ holds on Γ_1 by (1.3). Assume that $Q_n(z) = 0$ in $\overline{D}$. Then by the maximum modulus principle $|Q_n^*(z)| \le |Q_n(z)|$ everywhere in $\overline{D}$, so that, taking (1.3) into account, we find that

$$\rho^n|Q_n(\rho^{-1}e^{i\theta})| \le |Q_n(\rho e^{i\theta})| \qquad (0 < \rho < 1\,,\,\theta \in \mathbf{R})\,,$$

whence by the maximum modulus principle it follows that

$$|Q_n(z)| \le \rho^{-n}|Q_n(\rho^2 z)| \qquad (|z| \le \rho^{-1})\,. \tag{9.7}$$

It is easy to see that if $\rho^2 = 1 - n^{-1}$ and $n \ge 2$, then $\rho^{-n} \le 2$ and (9.7) implies that

$$|Q_n(z)| \le 2|Q_n((1-n^{-1})z)| \qquad (|z| \le 1 + (2n)^{-1}\,,\,n \ge 2)\,. \tag{9.8}$$

From (9.1) for $z = \zeta \in D$ we find that $|\varphi_{\sigma,n}(z)| < |\varphi^*_{\sigma,n}(z)|$ for all $z \in D$ and $n \in \mathbf{N}$ and, consequently, if the points $z\,,\,\zeta \in \overline{D}$ and one of them is not on Γ_1, then $K_{\sigma,n}(z\,,\,\zeta) \ne 0$. Therefore, using (9.8), we obtain that if $|z| \le 1 + (2n)^{-1}$, $|\zeta| \le 1$, $n \ge 2$, then

$$|K_{\sigma,n}(z\,,\,\zeta)| \le 2|K_{\sigma,n}((1-n^{-1})z\,,\,\zeta)|\,. \tag{9.9}$$

It follows from (9.9) and (9.1) that

$$K_{\sigma,n}(e^{i\theta}\,,\,e^{i\theta}) \le 4n|\varphi^*_{\sigma,n}((1-n^{-1})e^{\theta})\varphi^*_{\sigma,n}(e^{i\theta)})| \tag{9.10}$$

for all $\theta \in \mathbf{R}$ and $n \ge 2$. By Lemma 4.2 and Theorems 4.1 and 2.1, the right-hand side of (9.10) for the weight (9.4) with $h \equiv 1$ does not exceed the reciprocal of the right-hand side of (9.6), which together with (9.2) gives (9.6). As is known ([2], §3.4), under the conditions (1.18) the order of the left-hand side of (9.6) does not depend on h. Therefore, Theorem 9.1 may be considered proven.

Remark 9.1. The assertions (9.7) and (9.9) are modifications of the corresponding assertions in Rakhmanov [22]; the proofs differ only in the method of obtaining (9.7). Theorem 9.1 also follows from Theorem 2.1 and the inequality

$$\omega_n(d\sigma\,;e^{i\theta}) \geq (2/e)n^{-1}|\pi(\sigma'\,;(1-n^{-1})e^{i\theta})|^{-2} \qquad (\theta \in \mathbf{R},\, n \geq 2)$$

established in [8], p. 25.

The following theorem, complementary to Theorem 9.1, is proved under somewhat stiffer restrictions.

Theorem 9.2. *Let the weight* φ *satisfy the conditions of Theorem* 9.1 *and also the condition* (1.22). *Then there is a constant* $C_2(\varphi) > 0$ *such that*

$$\omega_n(d\sigma\,;e^{i\theta}) \leq C_2(\varphi)n^{-1}\prod_{\nu=1}^{m} w_\nu(|\sin[(\theta-\theta_\nu)/2]| + n^{-1}) \tag{9.11}$$

for all $\theta \in \mathbf{R}$ *and* $n \in \mathbf{N}$.

Proof. Let $[a\,,b] \subset [-\pi\,,\pi]$ and $\theta_\nu \notin [a\,,b]$, $\nu = 1\,,\dots\,,m$. Then, for a sufficiently small $\varepsilon > 0$, $\varphi \in \operatorname{Lip} 1$ on $[a-\varepsilon\,,a+\varepsilon]$ and is bounded away from zero. Therefore (see, for example, [6]), $\varphi^*_{\sigma\,,n}(e^{i\theta})$ tends to $\pi(\varphi\,;e^{i\theta})$ uniformly in $\theta \in [a\,,b]$. Whence, by (1.10) and the fact ([2], §1.1, 8) that all zeros of $\varphi_n(z)$ lie in D, we arrive at the conclusion that there exist constants C_3 and C_4 such that

$$0 < C_3 \leq |\varphi_{\sigma\,,k}(e^{i\theta})| \leq C_4 \qquad (k \in \mathbf{Z}_+\,,\ a \leq \theta \leq b)\,.$$

By (9.2), it follows from here that $\omega_n(d\sigma\,;e^{i\theta}) \asymp n^{-1}$ uniformly in $\theta \in [a\,,b]$. Therefore, (9.11) holds on $[a\,,b]$, and it remains to estimate $\omega_n(d\sigma\,;e^{i\theta})$ in sufficiently small neighborhoods of the points θ_ν $(\nu = 1\,,\dots\,,m)$. Let θ_k be one of them. By a shift of the argument if necessary, we may assume that $\theta_k = 0$. Suppose that $|\theta| \leq \varepsilon$ and $|\theta_\nu| > 2\varepsilon$ if $\nu \neq k$. Consider the polynomials $\{\chi_n(s\,;z)\}_{n=0}^{\infty}$, orthonormal on Γ_1 with weight $|\sin(\tau/2)|^s$ $(s > -1)$. It is known ([3], Proposition 2.4) that

$$0 < C_5(s) \leq |\chi_n(s\,;e^{i\theta})|(|\sin(\theta/2)| + n^{-1})^{s/2} \leq C_6(s)\,, \tag{9.12}$$

where C_5 and C_6 do not depend on $\theta \in \mathbf{R}$ and $n \in \mathbf{N}$. We shall use the representation

$$|\chi_n(s\,;e^{i(\tau-\theta)})|^2 = |Q_n(s\,,\theta\,,e^{i\tau})|^2\,, \tag{9.13}$$

where $Q_n(s\,,\theta\,,z)$ is some polynomial of z-degree n. The formula (9.13) follows from Fejér's theorem [1a], p. 3, since the left-hand side is a nonnegative trigonometric polynomial with real coefficients. It follows from (9.12), (9.13) that

$$C_5^2(s)k_n(s\,,\theta\,,\tau) \leq |Q_n(s\,,\theta\,,e^{i\tau})|^2 \leq C_6^2(s)k_n(s\,,\theta\,,\tau) \tag{9.14}$$

for all $\theta\,,\tau \in \mathbf{R}$ and $n \in \mathbf{N}$, where

$$k_n(s\,,\theta\,,\tau) = (|\sin[(\tau-\theta)/2]| + n^{-1})^{-s}\,. \tag{9.15}$$

It follows from (9.3) that

$$\omega_{n+1}(d\sigma\,;e^{i\theta}) \leq \frac{1}{2\pi}\int_{-\pi}^{\pi}\left|\frac{Q_n(s\,,\theta\,,e^{i\tau})}{Q_n(s\,,\theta\,,e^{i\theta})}\right|^2 \varphi(\tau)\,d\tau\,. \tag{9.16}$$

Since $k_n(s\,,\theta\,,\theta) = n^s$, (9.14)–(9.16) imply

$$\omega_{n+1}(d\sigma\,;e^{i\theta}) \leq C_7(s)n^{-s}\int_{-\pi}^{\pi} k_n(s\,,\theta\,,\tau)\varphi(\tau)\,d\tau\,. \tag{9.17}$$

Since $k_n(s, \theta, \tau)$ is bounded for $\tau \in [-\pi, -2\varepsilon]$ and $\tau \in [2\varepsilon, \pi]$, it follows from (9.17) and (9.15) that

$$\omega_{n+1}(d\sigma; e^{i\theta}) \le C_8(\varepsilon, s)[n^{-s} + I_1(0, \theta) + I_1(\theta, 2\varepsilon)], \tag{9.18}$$

where

$$I_1(c, d) = n^{-s} \int_c^d (|\tau - |\theta|| + n^{-1})^{-s} w_k(\tau)\, d\tau. \tag{9.19}$$

It is clear from (9.19) that θ may be considered nonnegative.

Let $0 \le \theta \le n^{-1}$. Then by (1.22) and (9.19)

$$I_1(0, \theta) = O(n^{-1} w(n^{-1})) \qquad (n \to \infty), \tag{9.20}$$

$$I_1(\theta, 2\varepsilon) \le C_9(\varepsilon, s)[n^{-1} w(2n^{-1}) + I_2], \tag{9.21}$$

where

$$I_2 = n^{-s} \int_{2/n}^{2\varepsilon} \tau^{-s} w_k(\tau)\, d\tau.$$

In order to estimate I_2 we introduce the sets

$$e_1 = \{\mu\colon \mu \in \mathbf{N} \cap [1, l_k], \alpha(\mu, k) \ge 0\}, \qquad e_2 = \{1, \dots, l_k\} \setminus e_1.$$

Applying the inequality (2.2) to $g_{\mu,k}(\tau)$ for $\mu \in e_1$, using the fact that $g_{\mu,k}(\tau)$ does not decrease for $\mu \in e_2$, choosing s sufficiently large, and setting $\alpha = \sum_{\mu\in e_1} \alpha(\mu, k)$, we obtain

$$I_2 \le n^{\alpha - s} w_k(2/n) \int_{2/n}^{2\varepsilon} \tau^{\alpha - s} d\tau \le C_{10}(\varepsilon, s) n^{-1} w_k(n^{-1}). \tag{9.22}$$

It follows from (9.18) and (9.20)–(9.22) that

$$\omega_{n+1}(d\sigma; e^{i\theta}) \le C_{11}(\varepsilon, s) n^{-1} w_k(n^{-1}) \qquad (0 \le \theta \le n^{-1}). \tag{9.23}$$

Now let $n^{-1} \le \theta \le \varepsilon$. Then, setting $x = \theta - 1/(2n)$ and $y = \theta/2$, and accounting for (1.22), we have

(9.24)

$$\begin{aligned} I_1(0, \theta) &\le C_{12}(s, \varepsilon) \left\{ n^{-s} \int_0^x \frac{w_k(\tau) d\tau}{(\theta - \tau)^s} + \frac{w_k(\theta)}{2n} \right\} \\ &\le C_{13}(s, \varepsilon) \left\{ \frac{w_k(\theta)}{n} + (n\theta)^{-s} \int_0^y w_k(\tau) d\tau + \frac{w_k(\theta)}{2n} \right\} \le C_{14}(s, \varepsilon) n^{-1} w_k(\theta), \\ I_1(\theta, 2\varepsilon) &\le C_{15}(s, \varepsilon) \left\{ w_k(\theta) \left[\frac{1}{n} + n^{-s} \int_{\theta + 1/n}^{2\theta} \frac{d\tau}{(\tau - \theta + n^{-1})^s} \right] + I_3 \right\}, \end{aligned}$$

where

$$I_3 = n^{-s} \int_{2\theta}^{2\varepsilon} \tau^{-s} w_k(\tau) d\tau.$$

Arguing as in the proof of (9.22), we can easily see that $I_3 = O(n^{-1} w_k(\theta))$ and, consequently, the same estimate is valid also for $I_1(\theta, 2\varepsilon)$, whence, by (9.24) and (9.18),

$$\omega_{n+1}(d\sigma; e^{i\theta}) \le C_{16}(\varepsilon, s) n^{-1} w_k(\theta) \qquad (n^{-1} \le \theta \le \varepsilon). \tag{9.25}$$

Theorem 9.2 follows from (9.23) and (9.25).

10. Proof of Theorems 1.2 and 1.3

First we prove Theorem 1.2. If its conditions are fulfilled, then from Lemma 4.2 and Theorem 4.1 we conclude that

$$\|\varphi_{\sigma,n}(e^{i\theta})/\pi(\varphi;(1-(2n)^{-1})e^{i\theta})\|_\infty \le C_1(\varphi) \qquad (n\in\mathbf{N}). \tag{10.1}$$

Clearly, φ satisfies condition (7.1). By Lemma 6.1 condition (7.2) is also fulfilled. Therefore, it follows from Theorem 7.2 and inequality (10.1) that

$$\|\varphi^{(j)}_{\sigma,n}(e^{i\theta})/\pi(\varphi;(1-(2n)^{-1})e^{i\theta})\|_\infty \le C_2(\varphi,j)n^j \tag{10.2}$$

for all $n\in\mathbf{N}$. Theorem 1.2 follows from (10.2) and Theorem 2.1.

Turning to the proof of Theorem 1.3, we first consider the case $j=0$, $m=1$, $\theta_1=0$. Let $\{\psi_n\}_0^\infty$, $\{\varphi_n\}_0^\infty$ be systems of polynomials orthonormal on Γ_1 with the weights

$$\psi(\tau)=w_1(|\sin(\tau/2)|), \qquad \varphi(\tau)=\psi(\tau)h(\tau).$$

Then, by (10.1),

$$\|\psi_n(e^{i\theta})/\pi(\psi;(1-(2n)^{-1})e^{i\theta})\|_\infty \le C_3(\psi) \qquad (n\in\mathbf{N}). \tag{10.3}$$

We prove the analogous inequality for φ_n:

$$\|\varphi_n(e^{i\theta})/\pi(\varphi;(1-(2n)^{-1})e^{i\theta})\|_\infty \le C_4(\varphi) \qquad (n\in\mathbf{N}). \tag{10.4}$$

Put

$$F_{\varphi,n}(\tau)=|\varphi_n(e^{i\tau})/\pi(\varphi;(1-(2n)^{-1})e^{i\tau})|, \qquad M_n=\|F_{\varphi,n}\|_\infty. \tag{10.5}$$

Denote by τ_n a point in $[-\pi,\pi]$ such that $F_{\varphi,n}(\tau_n)=M_n$. Let us prove the existence of a number δ_0 such that for every $\delta\in(0,\delta_0]$

$$M_n\le C_5(\psi,h,\delta)+C_6(\psi,h)\Omega(\delta)M_n, \qquad \Omega(\delta)=\int_0^\delta\frac{\omega(h;\tau)_\infty d\tau}{\tau}, \tag{10.6}$$

and $C_6(\psi,h)$ does not depend on δ. This immediately gives (10.4), since it is possible to choose $\delta\in(0,\delta_0]$ so as to ensure the inequality $C_6(\psi,h)\Omega(\delta)\le 1/2$.

To prove (10.6), we estimate $|\varphi_n(e^{i\tau_n})|$. To this end consider the expansion of $\varphi_n(z)$ with respect to the system $\{\psi_n(z)\}$ at the point $e^{i\tau_n}$. In view of the orthogonality of φ_n with weight φ to a polynomial of smaller degree, we have

$$\varphi_n(e^{i\tau_n})=j_1+\frac{j_2}{h(\tau_n)}, \qquad j_1=\frac{\psi_n(e^{i\tau_n})}{2\pi}\int_0^{2\pi}\varphi_n(e^{i\tau})\overline{\psi_n(e^{i\tau})}\psi(\tau)\,d\tau,$$

$$j_2=\int_{\tau_n+\pi}^{\tau_n-\pi}\frac{\varphi_n(e^{i\tau})}{2\pi}\sum_{k=0}^{n-1}\psi_k(e^{i\tau_n})\overline{\psi_k(e^{i\tau})}\{h(\tau_n)-h(\tau)\}\varphi(\tau)\,d\tau.$$

Applying the Cauchy inequality, we get

$$|j_1|\le\|1/h\|_\infty^{1/2}|\psi_n(e^{i\tau_n})|. \tag{10.7}$$

Let us now estimate the integral j_2. We represent j_2 as $j_2=j_3+j_4$, where j_3 and j_4 are integrals over the sets $\delta_n=[\tau_n-\delta,\tau_n+\delta]$ $(0<\delta\le\pi/2)$ and $\Delta_n=[\tau_n-\pi,\tau_n+\pi]\setminus\delta_n$, respectively. Since $|\sin[(\tau-\tau_n)/2]^{-1}|\le C_7(\delta)$ $(\tau\in\Delta_n)$, by (9.1) and the Cauchy inequality we have

$$|j_4|\le C_8(h,\delta)|\psi_n(e^{i\tau_n})|. \tag{10.8}$$

Observe, further, that (10.5) implies the inequality

$$|\varphi_n(e^{i\tau})|\le M_n|\pi(\varphi;(1-(2n)^{-1})e^{i\tau})| \qquad (\tau\in\mathbf{R}), \tag{10.9}$$

and that (9.1) and (9.10) imply the inequality

$$(10.10)\qquad \left|\sum_{k=0}^{n-1} \psi_k(e^{i\tau_n})\overline{\psi_k(e^{i\tau})}\right| \le 4|\psi_n(e^{i\tau_n})\psi_n^*((1-n^{-1})e^{i\tau})|.$$

On the basis of (1.3), (10.3), and the maximum modulus principle we conclude that

$$(10.11)\qquad |\psi_n^*((1-n^{-1})e^{i\tau})| \le C_3(\psi)|\pi(\psi;\,(1-n^{-1})(1-(2n)^{-1})e^{i\tau})|.$$

From (10.9)–(10.11) and Theorem 2.1 we derive the inequality

$$(10.12)\qquad |j_3| \le C_9(\psi\,,\,h)|\psi_n(e^{i\tau_n})|j_5\,,$$

where

$$j_5 = \int_{\delta_n} \frac{\omega(h;\,|\tau-\tau_n|)}{|\tau-\tau_n|+n^{-1}} \frac{w_1(|\sin(\tau/2)|)\,d\tau}{w_1(|\sin(\tau/2)|+n^{-1})}.$$

We prove that

$$(10.13)\qquad j_5 \le C_{10}(\psi)\Omega(\delta).$$

Setting $\delta_{n,1} = \delta_n \cap \{\tau\colon |\sin(\tau/2)| \ge n^{-1}\}$, $\delta_{n,2} = \delta_n \setminus \delta_{n,1}$, we have

$$j_5 \le C_{11}(\psi)\int_{\delta_{n,1}} \frac{\omega(h;\,|\tau-\tau_n|)_\infty}{|\tau-\tau_n|}\,d\tau + C_{12}(\psi)\frac{n\omega(h;\,\delta)_\infty}{w_1(n^{-1})}\int_{\delta_{n,2}} w_1\left(\left|\sin\frac{\tau}{2}\right|\right)d\tau.$$

But $\delta_{n,2} \subset [-\pi n^{-1}\,,\,\pi n^{-1}]$; therefore, by (1.22),

$$j_5 \le 2C_{11}(\psi)\Omega(\delta) + C_{13}(\psi)\omega(h;\,\delta)_\infty\,,$$

whence (10.13) follows in view of the inequality $\omega(h;\delta)_\infty \le 2\omega(h;\delta/2))_\infty \le 4\Omega(\delta)$. From (10.7), (10.8), and (10.10)–(10.13) we derive (10.6) and, thus, (10.4).

In the case $m > 1$, $j = 0$ Theorem 1.3 easily follows from the case $m = 1$, $j = 0$ on the basis of Lemma 2 in our paper [18]. The case $j \ge 1$ is reduced to the case $j = 0$ by applying Theorem 7.3.

11. Proof of Theorem 1.4

In the estimation from below of $|\varphi_{\sigma,n}^{(j)}(e^{i\theta})|$ we use the following fact.

Lemma 11.1. *For any $\theta \in \mathbf{R}$ and $n \in \mathbf{N}$*

$$(11.1)\qquad 2|\varphi_{\sigma,n}'(e^{i\theta})\varphi_{\sigma,n}(e^{i\theta})| \ge n|\varphi_{\sigma,n}(e^{i\theta})|^2 + K_{\sigma,n-1}(e^{i\theta}\,,\,e^{i\theta}).$$

Proof. Multiply by $1-z\bar\zeta$ and then differentiate with respect to z the second of the expressions (9.1) for $K_{\sigma,n-1}(z\,,\,\zeta)$. Subsequently putting $\zeta = z = e^{i\theta}$, we get

$$(11.2)\qquad K_{\sigma,n-1}(e^{i\theta}\,,\,e^{i\theta}) = e^{i\theta}\varphi_{\sigma,n}'(e^{i\theta})\overline{\varphi_{\sigma,n}(e^{i\theta})} - e^{i\theta}\varphi_{\sigma,n}^{*\prime}(e^{i\theta})\overline{\varphi_{\sigma,n}^*(e^{i\theta})}.$$

Differentiating the equality (see (1.3))

$$(11.3)\qquad \varphi_{\sigma,n}^*(e^{i\theta}) = e^{in\theta}\overline{\varphi_{\sigma,n}(e^{i\theta})},$$

we find that

$$(11.4)\qquad e^{i\theta}\varphi_{\sigma,n}^{*\prime}(e^{i\theta}) = ne^{i\theta}\overline{\varphi_{\sigma,n}(e^{i\theta})} - e^{i(n-1)\theta}\overline{\varphi_{\sigma,n-1}'(e^{i\theta})}.$$

Substituting for $\varphi_{\sigma,n}^*(e^{i\theta})$ and $e^{i\theta}\varphi_{\sigma,n}^{*\prime}(e^{i\theta})$ on the right-hand side of (11.2) the right-hand sides of (11.3) and (11.4), we obtain the formula

$$(11.5)\qquad 2\,\mathrm{Re}\{e^{i\theta}\varphi_{\sigma,n}'(e^{i\theta})\overline{\varphi_{\sigma,n}(e^{i\theta})}\} = n|\varphi_{\sigma,n}(e^{i\theta})|^2 + |K_{\sigma,n-1}(e^{i\theta}\,,\,e^{i\theta})|.$$

Now (11.1) follows from (11.5).

Corollary 11.1. *If the zeros of a polynomial $Q_n(z)$ of degree $n \in \mathbf{N}$ are contained in $\overline{D}$, then*

$$|Q_n^{(j)}(e^{i\theta})| \geq 2^{-j} n(n-1)\cdots(n-j+1)|Q_n(e^{i\theta})| \tag{11.6}$$

for all $\theta \in \mathbf{R}$ and $j = 1, \dots, n$. Equality in (11.6) is achieved for $Q_n(z) = (z+1)^n$ for $\theta = 0$.

Proof. Note that the zeros of $\varphi_{\sigma,n}(z)$ lie in D ([2], §1.1, 8)), and that every $Q_n(z)$ with this property differs only by a constant factor from $\varphi_{\sigma,n}(z)$ for the measure $d\sigma(\tau) = |Q_n(e^{i\tau})|^{-2} d\tau$ ([1a], p. 289). Thus, dropping the second term on the right-hand side of (11.1), we obtain (11.6) for the case when $j = 1$ and all the zeros of Q_n lie in D. The case when some or all zeros of Q_n appear on Γ_1 is obtained by passing to the limit. Since Q_n' cannot have zeros outside the smallest convex polygon containing all the zeros of Q_n, the inequality (11.6), already proved for $j = 1$, can be applied successively to $Q_n', \dots, Q_n^{(n-1)}$.

Corolary 11.2. *In the conditions of Corollary* 11.1,

$$\|Q_n^{(j)}(e^{i\theta})\|_\infty \geq 2^{-j} n(n-1)\cdots(n-j+1)\|Q_n(e^{i\theta})\|_\infty . \tag{11.7}$$

Equality is achieved on $Q_n(z) = (z+1)^n$.

Remark 11.1. For $j = 1$ the inequality (11.7) was first obtained by Turán [53]. In a remark he noted the possibility of proving also the inequality (11.6) for $j = 1$ by the same method. Our proof, based on establishing the inequality (11.1), is different from that in [53] and is simpler in our view.

Dropping the first term on the right in (11.1), we obtain

Corollary 11.3. *For any $\theta \in \mathbf{R}$ and $n \in \mathbf{N}$*

$$|\varphi_{\sigma,n}'(e^{i\theta})\varphi_{\sigma,n}(e^{i\theta})| \geq 2^{-1} K_{\sigma,n-1}(e^{i\theta}, e^{i\theta}). \tag{11.8}$$

Our proof of Theorem 1.4 is based on Corollaries 11.1 and 11.3. By Theorems 1.2 and 1.3 the inequality (1.20) holds for $j = 1$. Therefore, using (11.8) and Theorem 9.2, we obtain (1.23) for $j = 0$.

Now let $j \in \mathbf{N}$, $j \leq n$. Then, estimating from above every factor on the left-hand side of (11.8) by $|\varphi_{\sigma,n}^{(j)}(e^{i\theta})|$ with the aid of (11.6), we obtain the inequality

$$|\varphi_{\sigma,n}^{(j)}(e^{i\theta})|^2 \geq [2^{-j} n \cdots (n-j+1)]^2 n^{-1} |K_{\sigma,n-1}(e^{i\theta}, e^{i\theta})|,$$

which, in view of (9.2) and Theorem 9.2, implies Theorem 1.4.

12. Condition (1.22) is essential for Theorem 1.4

We give an example showing that without condition (1.22) Theorem 1.4, in general, is false.

Consider polynomials $\{\varphi_n(z)\}_0^\infty$ orthonormal on Γ_1 with the weight

$$\varphi(\tau) = |\sin \tau| \qquad (\tau \in \mathbf{R}), \tag{12.1}$$

and polynomials of the second kind ([2], §1.1, 9)) $\psi_0 = \varphi_0$,

$$\psi_n(z) = \frac{1}{4}\int_0^{2\pi} \frac{e^{i\tau} + z}{e^{i\tau} - z}[\varphi_n(e^{i\tau}) - \varphi_n(z)]\varphi(\tau)\, d\tau \qquad (n \in \mathbf{N}). \tag{12.2}$$

It is known ([2], §8.1) that the system $\{\psi_n(z)\}_0^\infty$ is orthonormal on Γ_1 with respect to some measure $ds(\tau)$, and

$$\lim_{n\to\infty} \frac{\psi_n^*(z)}{\varphi_n^*(z)} = F(z) \stackrel{\text{def}}{=} \frac{1}{4}\int_0^{2\pi} \frac{e^{i\tau} + z}{e^{i\tau} - z}\varphi(\tau)\, d\tau . \tag{12.3}$$

Since the condition $\ln s'(\tau) \in L^1[0, 2\pi]$ holds simultaneously with the condition $\ln \varphi \in L^1$ (see [2], §§8.1, 8.2), and the latter does hold, by Szegő's theorem (see (1.8)) we see that

$$\varphi_n^*(z) \to \pi(\varphi; z), \qquad \psi_n^*(z) \to \pi(s'; z) \qquad (n \to \infty)$$

uniformly inside D.

Therefore, by (1.10), (11.9), and (11.10)

$$s'(\tau) = |F(e^{i\tau})|^{-2}|\sin\tau| \tag{12.3}$$

a.e. on $[0, 2\pi]$. We show that $s \in C[0, 2\pi]$. Indeed, let $0 < \theta \leq 2\pi$. Then, for any polynomial $G_n(z)$ of degree n such that $G_n(e^{i\theta}) = 1$, for $0 < \varepsilon < \theta$ we have

$$\begin{aligned}\frac{1}{2\pi}\int_0^{2\pi} |G_n(e^{i\tau})|^2 ds(\tau) &\geq \frac{1}{2\pi}\int_{\theta-\varepsilon}^{\theta} |G_n(e^{i\tau})|^2\, ds(\tau) \\ &\geq \min_{\theta-\varepsilon\leq\tau\leq\theta} |G_n(e^{i\tau})|^2[s(\theta) - s(\theta-\varepsilon)]/2\pi.\end{aligned}$$

Passing to the limit as $\varepsilon \to 0$, with (9.1) and (9.3) taken into account, we obtain the inequality

$$\frac{s(\theta) - s(\theta - 0)}{2\pi} \leq \left\{\sum_{k=0}^{\infty} |\psi_k(e^{i\theta})|^2\right\}^{-1}. \tag{12.5}$$

We can prove similarly that if $0 \leq \theta < 2\pi$ then

$$\frac{s(\theta+0) - s(\theta)}{2\pi} \leq \left\{\sum_{k=0}^{\infty} |\psi_k(e^{i\theta})|^2\right\}^{-1}. \tag{12.6}$$

In the derivation of (12.5) and (12.6) we have used known arguments; about these consult [2], Notes, Chapter I, §1.1, where it is also shown that the right-hand side of (12.5) is equal to $[s(\theta+0) - s(\theta-0)]/(2\pi)$.

Using the formula ([2], §1.1, 10))

$$\psi_n(z)\varphi_n^*(z) + \psi_n^*(z)\varphi_n(z) = \pi z^n \qquad (n \in \mathbf{N}) \tag{12.7}$$

and Theorem 1.2, we conclude that

$$|\psi_n(e^{i\theta})| \geq \pi/|2\varphi_n(e^{i\theta})| \geq C_1(|\sin\theta| + n^{-1})^{1/2}, \tag{12.8}$$

where $C_1 > 0$ does not depend on $n \in \mathbf{N}$ and $\theta \in \mathbf{R}$. It follows from (12.5), (12.6), and (12.8) that $s \in C[0, 2\pi]$.

We prove that $s \in AC[0, 2\pi]$. Dividing both sides of (12.2) by z^n and passing to the limit as $z \to \infty$, we see that the coefficients $\kappa_n(\psi)$ and $\kappa_n(\varphi)$ at z^n of the polynomials ψ_n and φ_n coincide. Therefore, from (12.7) for $z = 0$ it follows that $\psi_n(0) = -\varphi_n(0)$. It is known (see, for example, formula (2.27) for $\alpha = \beta = 0$ in [3]) that $\varphi_n(0) = O(n^{-1})$. But then $\psi_n(0) = O(n^{-1})$ as well, so that by the Men'shov-Rademacher theorem the series

$$\overline{\psi_0(0)}\psi_0(e^{i\theta}) + \overline{\psi_1(0)}\psi_1(e^{i\theta}) + \cdots$$

converges to some function $A(\theta)$. By (9.1) this means that

$$\psi_n^*(e^{i\theta})\kappa_n(\psi) \to A(\theta) \quad \text{a.e.} \tag{12.9}$$

On the basis of (1.8)

$$\kappa_n(\psi) = \psi_n^*(0) \to \pi(s'; 0) \qquad (n \to \infty). \tag{12.10}$$

It is known ([2], Theorem 5.8) that $\|\psi_n^*(e^{i\theta})/\pi(s';e^{i\theta})-1\|_2\to 0$ $(n\to\infty)$ and, consequently, for some subsequence of indices n_k

$$\psi_{n_k}^*(e^{i\theta})\to\pi(s';e^{i\theta})\quad\text{a.e.}\quad (k\to\infty). \tag{12.11}$$

Comparing (12.9)–(12.11), we conclude that

$$\psi_n^*(e^{i\theta})\to\pi(s';e^{i\theta})\quad\text{a.e.}\quad (n\to\infty). \tag{12.12}$$

As is known ([2], proof of Theorem 3.10), for some subsequence n_k and any $\theta_1, \theta_2\in[0, 2\pi]$

$$\lim_{k\to\infty}\int_{\theta_1}^{\theta_2}\frac{d\theta}{|\psi_{n_k}(e^{i\theta})|^2}=s(\theta_2)-s(\theta_1). \tag{12.13}$$

It follows from (12.8), (12.12), (12.13), and (1.10) that if $0<\theta_1<\theta_2<2\pi$, then

$$\int_{\theta_1}^{\theta_2}s'(\theta)d\theta=s(\theta_2)-s(\theta_1). \tag{12.14}$$

Passing to the limit as $\theta_1\to 0$ or $\theta_2\to 2\pi$, we verify (12.14) for any $\theta_1, \theta_2\in[0, 2\pi]$, and, thus, $s\in AC[0, 2\pi]$.

To find the weight $\psi(\tau)=s'(\tau)$ we calculate the right-hand side of (12.3), setting $z=\rho\in(0, 1)$ for convenience:

$$F(\rho)=\frac{1}{2}(1-\rho^2)\int_0^\pi\frac{\sin\theta d\theta}{1+\rho^2-2\rho\cos\theta}=\frac{1-\rho^2}{2\rho}\ln\frac{1+\rho}{1-\rho}.$$

Hence, in view of the principle of analytic continuation,

$$F(z)=\frac{1-z^2}{2z}\ln\frac{1+z}{1-z}\qquad(z\in D). \tag{12.15}$$

From (12.4) and (12.5) we obtain

$$\psi(\tau)=s'(\tau)=\frac{4}{|\sin\tau|}\left[\ln^2\frac{1+\cos\tau}{1-\cos\tau}+\pi^2\right]^{-1}\qquad(\tau\in\mathbf{R}). \tag{12.16}$$

Consider on $[0, \pi]$ the function

$$a(\tau)=\psi(\tau)|\sin\tau|/4=\left[\ln^2\left(\sin^{-2}\frac{\tau}{2}-1\right)+\pi^2\right]^{-1}\overset{\text{def}}{=}b\left(\left|\sin\frac{\tau}{2}\right|\right). \tag{12.17}$$

It is not difficult to verify that on a sufficiently small interval $(0, \varepsilon)$ we have $b'(u)>0$ and $b''(u)<0$; thus, $b(u)$ is a concave modulus of continuity on $[0, \varepsilon]$. Consider the modulus of continuity $g(u)$ coinciding with $b(u)$ on $[0, \varepsilon]$ and with $b(\varepsilon)$ for $u\geq\varepsilon$. It is easy to see that if $\delta=2\arcsin\varepsilon$, then

$$g(|\sin(\tau/2)|)g(|\cos(\tau/2)|)/b(\varepsilon)=\begin{cases}a(\tau), & \tau\in[0, \delta]\cup[\pi-\delta, \pi],\\ b(\varepsilon), & \tau\in[\delta, \pi-\delta],\end{cases}$$

and so by (12.16)–(12.17) the weight ψ satisfies all conditions of Theorem 1.4, except (1.22).

We show that there is no estimate of type (1.23) (for $j=0$) for $\psi_n(1)$. In fact, (12.7) implies $\psi_n(1)=\pi/[2\varphi_n(1)]$. By Theorems 1.2 and 1.4 (see also the formulas (2.12) and (2.13) in [3]), $\varphi_n(1)\asymp n^{1/2}$. Therefore, $\psi_n(1)\asymp n^{-1/2}$. However, a formal application of Theorem 1.4 to $\psi_n(z)$ would have given us the estimate $\psi_n(1)\geq C_2 n^{-1/2}\ln n$.

13. Some applications of Theorems 1.2–1.4

We show that for a wide class of weights φ the inequality (7.20) is order-sharp with respect to the variable n.

Theorem 13.1. *Let* $j \in \mathbf{Z}_+$, $0 < q \le r \le \infty$, $0 < c < \infty$, $R_n = 1 - cn^{-1}$, *let the weight* φ *satisfy the conditions of Theorem* 1.2 *and the condition* (1.22), *let* $\theta_0 \in [-\pi, \pi]$ *not coincide with any of the points* $\theta_1, \dots, \theta_m$, *and let* $\{\chi_n(z)\}_{n=0}^{\infty}$ *be a system of polynomials orthonormal on* Γ_1 *with the weight*

$$\chi(\tau) = \varphi(\tau)|1 - e^{i(\tau-\theta_0)}|^{\gamma} \qquad (\tau \in \mathbf{R}),$$

where $\gamma > 2/q$. *Then*

$$\left\| \frac{\chi_n^{(j)}(e^{i\theta})}{\pi(\varphi\,;\, R_n e^{i\theta})} \right\|_r \asymp n^{j+1/q-1/r} \left\| \frac{\chi_n(e^{i\theta})}{\pi(\varphi\,;\, R_n e^{i\theta})} \right\|_q . \tag{13.1}$$

Proof. By Theorems 1.2, 1.4, and 2.1

$$|\chi_n^{(j)}(e^{i\theta})| \asymp n^j |\pi(\chi\,;\, R_n e^{i\theta})|.$$

From this and (2.22) we find that

$$\begin{aligned} &\|\chi_n^{(j)}(e^{i\theta})/\pi(\varphi\,;\, R_n e^{i\theta})\|_r \asymp n^j \|\pi(\chi\varphi^{-1}\,;\, R_n e^{i\theta})\|_r \\ &\quad = n^j \|(1 - R_n e^{i(\theta-\theta_0)})^{-\gamma/2}\|_r \asymp n^j \|(|\sin(\theta/2)| + n^{-1})^{-\gamma/2}\|_r \asymp n^{j+\gamma/2-1/r}. \end{aligned} \tag{13.2}$$

In particular, putting $j = 0$, $r = q$ here, we get

$$\|\chi_n(e^{i\theta})/\pi(\varphi\,;\, R_n e^{i\theta})\|_q \asymp n^{\gamma/2-1/q}. \tag{13.3}$$

Now (13.1) follows from (13.2) and (13.3).

Similar arguments can be carried out with respect to inequality (7.27).

It is easy to see that Theorems 1.2 and 1.3 hold also for the system $\{T_{\sigma,n}(\tau)\}_{n=0}^{\infty}$. For $j = 0$ this follows from the inequalities

$$2^{-1} \le \frac{|T_{\sigma,2n-1}(\theta)| + |T_{\sigma,2n}(\theta)|}{|\varphi_{\sigma,2n-1}(e^{i\theta})| + |\varphi_{\sigma,2n}(e^{i\theta})|} \le 2 \tag{13.4}$$

(see [3], Corollary 2.4). The case $j \in \mathbf{N}$ reduces to the case $j = 0$ with the aid of Corollary 7.1 (for $r = q = \infty$). It follows from (13.4) that in Theorem 1.4 for $j = 0$, one can replace $|\varphi_{\sigma,n}^{(j)}(e^{i\theta})|$ by $|T_{\sigma,2n-1}(\theta)| + |T_{\sigma,2n}(\theta)|$.

The results of §§1 and 7 may be applied to obtain estimates for polynomials $\{p_n(t)\}_{n=0}^{\infty}$ orthonormal on $[-1, 1]$ with weight $p(t)$.

Theorem 13.2. *Let the weight* $p(t)$ *be defined by the formula*

$$p(t) = H(t)(1-t^2)^{-1/2} w_1(\sqrt{1-t})\, w_m(\sqrt{1+t}) \prod_{\nu=2}^{m-1} w_\nu(|t - x_\nu|)$$

$$(-1 \le t \le 1, \quad -1 = x_1 < x_2 < \cdots < x_m = 1, \quad m \ge 3).$$

If the function $h(\tau) \overset{\text{def}}{=} H(\cos\tau)$ *and the* $w_\nu(u)$ $(\nu = 1, \dots, m)$ *satisfy the conditions of Theorem* 1.2 *or of Theorem* 1.3, *then there exists a constant* $C_1(p, j)$ *such that for all* $x \in [-1, 1]$ *and* $n \in \mathbf{N}$

$$\begin{aligned} &n^{-j}(\sqrt{1-x^2} + n^{-1})^j |p_n^{(j)}(x)| \\ &\quad \le C_1(p, j) \left\{ w_1\left(\sqrt{1-x} + \frac{1}{n}\right) w_m\left(\sqrt{1-x} + \frac{1}{n}\right) \prod_{\nu=2}^{m-1} w_\nu\left(|x - x_\nu| + \frac{1}{n}\right) \right\}^{-1/2}. \end{aligned}$$

For $j = 0$ this theorem is deduced from Theorems 1.2 and 1.3 (after an obvious modification of them with account taken of Corollary 2.3 and Remarks 4.1 and 6.1) with the aid of Szegő's formula (1.6). The case $j \in \mathbf{N}$ is reduced to the case $j = 0$ with the aid of Theorem 7.6, in which we should set $r = q = \infty$.

BIBLIOGRAPHY

1a. Gábor Szegő, *Orthogonal polynomials*, rev. ed., Amer. Math. Soc., Providence, RI, 1959.

1b. Ya. L. Geronimus, *Orthogonal polynomials*, Appendix to the Russian transl. of [1a], Fizmatgiz, Moscow, 1961, pp. 414–494; English transl. in Amer. Math. Soc. Transl. (2), vol. 108, 1977.

2. ——, *Polynomials, orthogonal on a circumference and on an interval*, Fizmatgiz, Moscow, 1958; English transl., *Orthogonal polynomials: estimates, asymptotic formulas, and series of polynomials orthogonal on the unit circle and on an interval*, Consultants Bureau, New York, 1961.

3. V. M. Badkov, *Approximations of functions in the uniform metric by Fourier sums of orthogonal polynomials*, Trudy Mat. Inst. Steklov. **145** (1980), 20–62; English transl. in Proc. Steklov Inst. Math. **1981**, no. 1 (145).

4. Gábor Szegő, *On bi-orthogonal systems of trigonometric polynomials*, Magyar Tud. Akad. Mat. Kutató Int. Közl. **8** (1963/64), 255–273.

5. P. K. Suetin, *A problem of V. A. Steklov in the theory of orthogonal polynomials*, Itogi Nauki i Tekhniki: Mat. Anal., vol. 15, VINITI, Moscow, 1977, pp. 5–82; English transl. in J. Soviet Math. **12** (1979), no. 6.

6. V. M. Badkov, *The asymptotic behavior of orthogonal polynomials*, Mat. Sb. **109 (151)** (1979), 46–59; English transl. in Math. USSR Sb. **37** (1980).

7. B. L. Golinskiĭ, *Asymptotic representation of orthogonal polynomials*, Uspekhi Mat. Nauk **35** (1980), no. 2 (212), 145–196; English transl. in Russian Math Surveys **35** (1980).

8. Paul G. Nevai, *Géza Freud, orthogonal polynomials and Christoffel functions. A case study*, J. Approx. Theory **48** (1986), no. 1.

9. V. M. Badkov, *Uniform asymptotic representations of orthogonal polynomials*, Approximation of Functions by Polynomials and Splines (Yu. N. Subbotin, ed.), Ural. Nauchn. Tsentr Akad. Nauk SSSR, Sverdlovsk, 1985, pp. 41–53. (Russian)

10. ——, *Uniform asymptotic representations of orthogonal polynomials and their derivatives*, Trudy Mat. Inst. Steklov. **180** (1987), 36–38; English transl. in Proc. Steklov Inst. Math. **1989**, no. 1 (180).

11. Christian Hörup, *An asymptotic formula for the derivatives of orthogonal polynomials on the unit circle*, Math. Scand. **20** (1967), 32–40.

12. B. L. Golinskiĭ, *Asymptotic formulas for the derviatives of orthogonal polynomials*, Izv. Akad. Nauk Armyan. SSR Ser. Mat. **11** (1976), 56–81. (Russian)

13. ——, *Asymptotic representation at a point of the derivatives of orthonormal polynomials*, Mat. Zametki **19** (1976), 659–672; English transl. in Math. Notes **19** (1976).

14. Paul G. Nevai, *An asymptotic formula for the derivatives of orthogonal polynomials*, SIAM J. Math. Anal. **10** (1979), 472–477.

15. I. A. Ibragimov and V. N. Solev, *Asymptotic behavior of the prediction error for a stationary sequence with a spectral density of special form*, Teor. Veroyatnost. i Primenen. **13** (1968), 746–751; English transl. in Theory Probab. Appl. **13** (1968).

16. V. M. Badkov, *The order of the best approximation of the Szegő function*, Approximation of Functions by Polynomials and Splines (Yu. N. Subbotin, ed.), Ural. Nauchn. Tsentr Akad. Nauk SSSR, Sverdlovsk, 1985, pp. 25–40. (Russian)

17. ——, *Generalization of a theorem of I. A. Ibragimov and V. N. Solev*, Approximation in Concrete and Abstract Banach Spaces (Yu. N. Subbotin, ed.), Ural. Nauchn. Tsentr Akad. Nauk SSSR, Sverdlovsk, 1987, pp. 15–30. (Russian)

18. ——, *Uniform asymptotic representations of orthogonal polynomials*, Trudy Mat. Inst. Steklov. **164** (1983), 3–36; English transl. in Proc. Steklov Inst. Math. **1985**, no. 2 (164).

19. Serge Bernstein, *Sur les polynomes orthogonaux relatifs à un segment fini*. I, II, J. Math. Pures Appl. (9) **9** (1930), 127–177; (9) **10** (1931), 219–286.

20. V. M. Badkov, *Systems of orthogonal polynomials expressed in explicit form in terms of Jacobi polynomials*, Mat. Zametki **42** (1987), 650–659; English transl. in Math. Notes **42** (1987).

21. E. A. Rakhmanov, *On Steklov's conjecture in the theory of orthogonal polynomials*, Mat. Sb. **108 (150)** (1979), 581–608; English transl. in Math. USSR Sb. **36** (1980).

22. ——, *Estimates of the growth of orthogonal polynomials whose weight is bounded away from zero*, Mat. Sb. **114 (156)** (1981), 269–298; English transl. in Math. USSR Sb. **42** (1982).

23. M. U. Ambroladze, *The possible growth of orthogonal polynomials with a continuous positive weight*, Mat. Zametki **45** (1989), no. 6, 99–101. (Russian)

24. S. M. Nikol'skiĭ, *Inequalities for entire functions of exponential type and their application to the theory of differentiable functions of several variables*, Trudy Mat. Inst. Steklov. **38** (1951), 244–278; English transl. in Amer. Math. Soc. Transl. (2) **80** (1969).

25. A. F. Timan, *Theory of approximation of functions of a real variable*, Fizmatgiz, Moscow, 1960; English transl., Pergamon Press, Oxford, and Macmillan, New York, 1964.

26. N. K. Bari and S. B. Stechkin, *Best approximations and differentiability properties of two conjugate functions*, Trudy Moskov. Mat. Obshch. **5** (1956), 483–522. (Russian)

27. V. K. Dzyadyk, *Introduction to the theory of uniform approximation of functions by polynomials*, "Nauka", Moscow, 1977. (Russian)

28. N. K. Bari, *Trigonometric series*, Fizmatgiz, Moscow, 1961; English transl., Vols. I, II, Pergamon Press, Oxford, and Macmillan, New York, 1964.

29. A. Zygmund, *Trigonmetric series*, 2nd rev. ed., Vol. I, Cambridge Univ. Press, Cambridge, 1959.

30. È. A. Storozhenko, *On a problem of Hardy-Littlewood*, Mat. Sb. **119 (161)** (1982), 564–583; English transl. in Math. USSR Sb. **47** (1984).

31. Dunham Jackson, *Certain problems of closest approximation*, Bull. Amer. Math. Soc. **39** (1933), 889–906.

32. N. K. Bari, *A generalization of inequalities of S. N. Bernstein and A. A. Markov*, Izv. Akad. Nauk SSSR Ser. Mat. **18** (1954), 159–176. (Russian)

33. V. V. Arestov, *Inequality of various metrics for trigonometric polynomials*, Mat. Zametki **27** (1980), 539–547; English transl. in Math. Notes **27** (1980).

34. S. N. Bernshteĭn [Bernstein], *On best approximation of continuous functions by polynomials of given degree*, Soobshch. Khar'kov. Mat. Obshch (2) **13** (1912), 47–194; reprinted in his *Collected works*. Vol. 1, Izdat. Akad. Nauk SSSR, Moscow, 1952, pp. 11–104; abridged French transl., Acad. Roy. Belg. Cl. Sci. Mém. (2) **14** (1912), fasc. 1.

35. A. Zygmund, *A remark on conjugate series*, Proc. London Math. Soc. (2) **34** (1932), 392–400.

36. S. B. Stechkin, *A generalization of some inequalities of S. N. Bernstein*, Dokl. Akad. Nauk SSSR **60** (1948), 1511–1514. (Russian)

37. V. I. Ivanov, *Some inequalities for trigonometric polynomials and their derivatives in various metrics*, Mat. Zametki **18** (1975), 489–498; English transl. in Math. Notes **18** (1975).

38. V. G. Krotov, P. Osval'd [Oswald], and È. A. Storozhenko, *Direct and inverse theorems of Jackson type in L^p spaces* $(0 < p < 1)$, Dokl. Akad. Nauk SSSR **226** (1976), 44–47; English transl. in Soviet Math. Dokl. **17** (1976).

39. È. A. Storozhenko, V. G. Krotov, and P. Osval'd [Oswald], *Direct and inverse theorems of Jackson type in L^p spaces,* $0 < p < 1$, Mat. Sb. **98 (140)** (1975), 395–415; English transl. in Math. USSR Sb. **27** (1975).

40. P. Osval'd [Oswald], *Some inequalities for trigonometric polynomials in the L_p-metric,* $0 < p < 1$, Izv. Vyssh. Uchebn. Zaved. Mat. **1976**, no. 7 (170), 65–75; English transl. in Soviet Math. (Iz. VUZ) **20** (1976).

41. V. V. Arestov, *On inequaliities of S. N. Bernstein for algebraic and trigonometric polynomials*, Dokl. Akad. Nauk SSSR **246** (1979), 1289–1292; English transl. in Soviet Math. Dokl. **20** (1979).

42. Attila Máté and Paul G. Nevai, *Bernstein's inequality in L^p for $0 < p < 1$ and $(C, 1)$ bounds for orthogonal polynomials*, Ann. of Math. (2) **111** (1980), 145–154.

43. V. I. Ivanov, *Some extremal properties of polynomials and inverse inequalities of approximation theory*, Trudy Mat. Inst. Steklov. **145** (1980), 79–110; English transl. in Proc. Steklov Inst. Math. **1981**, no. 1 (145).

44. V. V. Arestov, *On integral inequalities for trigonometric polynomials and their derivatives*, Izv. Akad. Nauk SSSR Ser. Mat. **45** (1981), 3–22; English transl. in Math. USSR Izv. **18** (1982).

45. A. Zygmund, *Trigonometric series*, 2nd rev. ed., Vol. II, Cambridge Univ. Press, Cambridge, 1959.

46. S. V. Konyagin, *Bounds on the derivatives of polynomials*, Dokl. Akad. Nauk SSSR **243** (1978), 1116–1118; English transl. in Soviet Math. Dokl. **19** (1978).

47. G. K. Lebed', *Inequalities for polynomials and their derivatives*, Dokl. Akad. Nauk SSSR **117** (1957), 570–572. (Russian)

48. ———, *Some inequalities for trigonometric and algebraic polynomials and their derivatives*, Trudy Mat. Inst. Steklov. **134** (1975), 142–160; English transl. in Proc. Steklov Inst. Math. **134** (1975).

49. M. K. Potapov, *Some inequalities for polynomials and their derivatives*, Vestnik Moskov. Univ. Ser. I Mat. Mekh. **1960**, no. 2, 10–20. (Russian)

50. B. A. Khalilova, *On certain estimates for polynomials*, Izv. Akad. Nauk Azerbaĭdzhan. SSR Ser. Fiz.-Tekhn Mat. Nauk 1974, no. 2, 46–55. (Russian)

51. I. K. Daugavet and S. Z. Rafal'son, *Some inequalities of Markov-Nikol'skiĭ type for algebraic polynomials*, Vestnik Leningrad. Univ. **1972**, no. 1 (Ser. Mat. Mekh. Astr. vyp. 1), 15–25; English transl. in Vestnik Leningrad Univ. Math. **5** (1978).

52. Paul G. Nevai, *Orthogonal polynomials*, Mem. Amer. Math. Soc. No. 213 (1979).

53. P. Turán, *Über die Ableitung von Polynomen*, Compositio Math. **7** (1939), 89–95.

Translated by S. IVANOV

Труды Матем. Инст. Стеклова
198 (1992)

Proceedings of the Steklov
Institute of Mathematics
1994, Issue 1

APPROXIMATION BY SPLINES ON AN INFINITE INTERVAL

UDC 517.5

YU. S. VASIL'EV

Let I be the half-line $[0, \infty)$ and let $0 < q \le \infty$. For a measurable set M denote the "norm" of a function $f(x)$ in the space $L_q(M)$ by

$$\|f\|_{q(M)} = \left(\int_M |f(x)|^q\,dx\right)^{1/q}, \qquad 0 < q < \infty,$$
$$\|f\|_{\infty(M)} = \operatorname*{ess\,sup}_{x\in M}|f(x)|, \qquad q = \infty.$$

Suppose that $k \ge 1$ is an integer, $p \ge 1$, $W_p^k(I)$ is the class of functions $f(x)$ on I having locally absolutely continuous derivative of order $k-1$ and kth derivative in the space $L_p(I)$, and $\widetilde{W}_p^k(I)$ is the set of functions in $W_p^k(I)$ with $\|f^{(k)}\|_{p(I)} \le 1$.

We consider the problem of approximating functions in $W_p^k(I)$ by splines of order l that do not have "too many nodes". This restriction can be made more precise, for example, as follows. With each spline $s_l(x)$ defined on I we associate the function $m_s(u)$ equal to the number of nodes of $s_l(x)$ on the interval $[0, u]$, and let $N(u)$ be a nondecreasing function on I with $N(u) \ge 1$. We define the following class of splines with nonfixed nodes:

$$S_{N,l} = \{s_l(x) : m_s(u) \le N(u),\ u \in I\}. \tag{0.1}$$

Let $T = \{t_0 = 0 < t_1 < t_2 < \cdots\}$ be a set of points on I, perhaps finite, having distribution function $\mathfrak{M}_T(u)$ equal to the number of points in T lying in the interval $[0, u]$. We define the following set of splines with fixed nodes:

$$\Phi_{\mathfrak{M},l} = \{s_l(x) : m_s(u) = \mathfrak{M}_T(u)\}. \tag{0.2}$$

In the study of approximation of the class $\widetilde{W}_p^k(I)$ by the classes of splines $S_{N,l}$ and $\Phi_{\mathfrak{M},l}$ in the $L_q(I)$-metric, i.e., in the study of the quantities

$$e_{N,l} = e_{N,l}(\widetilde{W}_p^k(I), q) = \sup_{f\in\widetilde{W}_p^k(I)} \inf_{s\in S_{N,l}} \|f - s\|_{q(I)}, \tag{0.3}$$

$$E_{\mathfrak{M},l} = E_{\mathfrak{M},l}(\widetilde{W}_p^k(I), q) = \sup_{I\in\widetilde{W}_p^k(I)} \inf_{s\in\Phi_{\mathfrak{M},l}} \|f - s\|_{q(I)} \tag{0.4}$$

the following question arises first: under what conditions on the functions $N(u)$ and $\mathfrak{M}(u)$ are the quantities $e_{N,l}$ and $E_{\mathfrak{M},l}$ finite? If they are finite, then estimates of these quantities should be given in terms of the parameters of the problem, and the possibility of simultaneously approximating a function and its derivatives should be investigated. These questions were considered by Chernykh in [3] and [4], where (noncoinciding) necessary conditions and sufficient conditions were given for the approximations to be finite.

1991 *Mathematics Subject Classification*. Primary 41A15; Secondary 41A28.

In this article we derive necessary and sufficient conditions for approximations by splines on the half-line to be finite.

We employ restrictions not in terms of a function $N(u)$ limiting the number of nodes on an interval $[0, u]$, but in terms of a set P restricting the location of the nodes. Let $\mathrm{P} = \{\rho_0 = 0 \leq \rho_1 \leq \rho_2 \leq \cdots\}$ be a countable set, in nondecreasing order, of points on I, and let $l_1 = \max\{1, l\}$. With each spline $s_l(x)$ we associate its set $\mathrm{T}_s = \{\tau_0 < \tau_1 < \tau_2 < \cdots\}$ of nodes and the set $\mathrm{T}_s^l = \{\tau_{il_1} : i = 0, 1, 2, \dots\}$ of part of the nodes. We define the following set of splines with nonfixed nodes:

$$S_{\mathrm{P},l} = \{s_l(x) : \tau_{il_1} \geq \rho_i, \ i = 0, 1, 2, \dots\}.$$

For $\mathrm{P} = \{\rho_0 = 0 < \rho_1 < \rho_2 < \cdots\}$ we define the following set of splines with partially fixed nodes:

$$\Phi_{\mathrm{P},l} = \{s_l(x) : \tau_{il_1} = \rho_i, \ i = 0, 1, 2, \dots\}.$$

We study the approximations $(r = 0, 1, \dots, k-1)$

$$e^r_{\mathrm{P},l} = e^r_{\mathrm{P},l}(\widetilde{W}^k_p(I), q) = \sup_{f\in\widetilde{W}^k_p(I)} \inf_{s\in S_{\mathrm{P},l}} \|f^{(r)} - s^{(r)}\|_{q(I)}, \tag{0.5}$$

$$E^r_{\mathrm{P},l} = E^r_{\mathrm{P},l}(\widetilde{W}^k_p(I), q) = \sup_{f\in\widetilde{W}^k_p(I)} \inf_{s\in \Phi_{\mathrm{P},l}} \|f^{(r)} - s^{(r)}\|_{q(I)}. \tag{0.6}$$

The imposition of restrictions only on part of the nodes has to do mostly with the approximation apparatus being used, and is not essential for finiteness of the approximations.

1. A procedure for approximating functions by Hermitian interpolation splines with extra nodes

Let $l \geq k-1 \geq 0$ and $l_1 = \max\{1, l\}$. In the construction of a spline approximating a function $f(x) \in W^k_p(I)$ we construct a set

$$\Xi = \{0 = \xi_0 < \xi_1 < \xi_2 < \cdots\} \tag{1.1}$$

of interpolation nodes, which is finite or infinite, depending on the situation. From the set Ξ we construct the set T of nodes of the spline by dividing each interval $[\Delta\xi_i] = [\xi_{i-1}, \xi_i]$ of length $\Delta\xi_i = \xi_i - \xi_{i-1}$ into l_1 parts:

$$T = \{t^i_j : t^i_j = \xi_{i-1} + \tau_j\Delta\xi_i, \ j = 0, 1, \dots, l_1, \ i = 0, 1, \dots\}, \tag{1.2}$$

where the numbers $0 = \tau_0 < \tau_1 < \cdots < \tau_{l_1} = 1$ form a given partition of the interval $[0, 1]$. For a given function $f(x) \in W^k_p$ we construct a spline $s(x) = s_l(x; \Xi, f)$ of order l and defect 1 with nodes in T that interpolates at the points in Ξ:

$$s^{(r)}(\xi_i) = \hat{f}^{(r)}(\xi_i), \qquad r = 0, 1, \dots, l_1 - 1,$$

where

$$\hat{f}^{(r)}(x) = \begin{cases} f^{(r)}(x) & \text{for } r = 0, 1, \dots, k-1, \\ 0 & \text{for } r = k, \ k+1, \dots. \end{cases} \tag{1.3}$$

It was shown in [1]–[3] that such a spline is uniquely determined on each interval $[\Delta\xi_i]$ by the formula

$$\begin{aligned} s(x) = {} & \sum_{j=0}^{k-1} f^{(j)}(\xi_i)\frac{(x-\xi_i)^j}{j!} \\ & - \frac{(-1)^k}{l!}\int_{[\Delta\xi_i]} \frac{d^{l+1-k}}{dt^{l+1-k}} \sum_{j=0}^{l_1-1} \frac{\omega_i(t)(x-t_j)^l_+}{(t-t_j)\omega_i'(t_j)} f^{(k)}(t)\,dt, \end{aligned} \tag{1.4}$$

where $t_j = t^i_j$, $\omega_i(t) = (t-t_0)(t-t_1)\cdots(t-t_{l_1})$, and

$$x^l_+ = \begin{cases} x^l & \text{for } x > 0, \\ 0 & \text{for } x \le 0. \end{cases}$$

If the set Ξ turns out to be bounded above, then the spline is determined on $[\sup_i \xi_i, \infty)$ by the formula

$$s_l(x; \Xi, f) = \sum_{j=0}^{k-1} f^{(j)}(\xi)(x-\xi)^j/j!, \qquad \text{where } \xi = \sup_i \xi_i. \tag{1.5}$$

The estimate

$$\|f^{(r)} - s^{(r)}\|_{q\,[\Delta\xi_i]} \le C_r \Delta\xi_i^{k-r+1/q-1/p} \|f^{(k)}\|_{p\,[\Delta\xi_i]} \qquad (r = 0, 1, \dots, k-1) \tag{1.6}$$

was derived for this spline in [3], and formulas for the constants $C_r = C_r(p, q, k, l)$ were presented ([3], Russian pp. 178–183, English pp. 191–197).

In [1] Subbotin and Chernykh considered approximations by such splines on a closed interval with $l = k-1$, $k \ge 2$, $1 \le q \le \infty$, and with the nodes τ_i taken as the roots of the corresponding Tchebycheff polynomial, carried over to $[0, 1]$ and moved apart in such a way that the extreme roots coincide with 0 and 1:

$$\tau_i = \frac{1}{2}\left\{1 - \cos\frac{(2i+1)\pi}{2(l+1)} \Big/ \cos\frac{\pi}{2(l+1)}\right\}, \qquad i = 0, 1, \dots, l. \tag{1.7}$$

It was proved that for such nodes

$$\|f^{(r)} - s^{(r)}\|_{q\,(0,1)} \le \frac{C(k-1)}{n^{k-r}} \|f^{(k)}\|_{p\,(0,1)}, \qquad r = 0, 1, \dots, k-1, \tag{1.8}$$

where n is the total number of nodes of the splines on $(0, 1]$, and

$$C(k-1) = \left(9 + \frac{4}{\pi}\ln k\right)\sqrt{\frac{2(k-1)}{\pi}}\left(\frac{e}{4}\right)^{k-1}.$$

Substituting $n = l = k-1$ here, making the change of argument to go over to the interval $[\Delta\xi_i]$, and using the fact that the proof of (1.8) given there is also valid for $q > 0$, we get that for $k \ge 2$, $l = k-1$, and the nodes (1.7) the constant C_r in (1.6) can be taken in the form

$$C_r = \left(9 + \frac{4}{\pi}\ln k\right)\sqrt{\frac{2(k-1)}{\pi}}\left(\frac{e}{4}\right)^{k-1}(k-1)^{-(k-r)}. \tag{1.9}$$

2. Auxiliary assertions

Below it will be assumed without mention that $1/p = 0$ for $p = \infty$, and $1/q = 0$ for $q = \infty$.

Let $f(x) \in W^k_p(I)$. For $p < \infty$ we define the function

$$F(x) = \int_0^x |f^{(k)}(t)|^p \, dt. \tag{2.1}$$

This is a nondecreasing continuous function, with $F(0) = 0$, $F(\infty) = \lim_{x\to\infty} F(x) = \|f^{(k)}\|^p_{p\,(I)}$. We let $\Delta F_{\xi_i} = F(\xi_i) - F(\xi_{i-1})$, and we estimate the deviation

$\|f^{(r)}(x) - s_l^{(r)}(x\,;\,\Xi,\, f)\|_{q\,(I)}$. Using (1.6), we get that for $q < \infty$

(2.2)

$$\|f^{(r)} - s^{(r)}\|_{q\,(I)} = \left(\sum_i \|f^{(r)} - s^{(r)}\|^q_{q\,[\Delta\xi_i]}\right)^{1/q}$$

$$\le \left(\sum_i C_r \Delta\xi_i^{q(k-r+1/q-1/p)} \|f^{(k)}\|^q_{p\,[\Delta\xi_i]}\right)^{1/q} = C_r \left(\sum_i \Delta\xi_i^{q(k-r+1/q-1/p)} \Delta F_{\xi_i}^{q/p}\right)^{1/q}.$$

If $\beta = q/p$ and

$$d_r = d_r(p\,,\,q\,,\,k) = \frac{q(k-r+1/q-1/p)}{1-q/p} = 1 + \frac{k-r}{1/q-1/p} \tag{2.3}$$
$$(r = 0\,,\,1\,,\,\dots\,,\,k-1)$$

for $p \ne q$, then this inequality can be rewritten as

$$\|f^{(r)} - s^{(r)}\|_{q\,(I)} \le C_r \left(\sum_i \Delta\xi_i^{d_r(1-\beta)} \Delta F_{\xi_i}^{\beta}\right)^{1/q}. \tag{2.4}$$

For $p = q \le \infty$ the numbers d_r are not defined. In this case

$$\|f^{(r)} - s^{(r)}\|_{q\,(I)} \le C_r \sup_i\{\Delta\xi_i^{k-r}\} \|f^{(k)}\|_{p\,(I)}. \tag{2.5}$$

If $q = \infty$ and $p < \infty$, then it follows from (1.6) that

$$\|f^{(r)} - s^{(r)}\|_{\infty\,(I)} = \sup_i \|f^{(r)} - s^{(r)}\|_{\infty\,[\Delta\xi_i]} \le C_r \sup_i\{\Delta\xi_i^{k-r-1/p} \Delta F_{\xi_i}^{1/p}\}. \tag{2.6}$$

In both the last cases ($q = p \le \infty$ or $q = \infty$, $p < \infty$)

$$\|f^{(r)} = s^{(r)}\|_{q\,(I)} \le C_r \|f^{(k)}\|_{p\,(I)} \sup_i\{\Delta\xi_i^{k-r+1/q-1/p}\}. \tag{2.7}$$

In the case of the estimate (2.4) it is clear that for nonfixed nodes the possibility of a finite deviation depends on whether it is possible to arrange the points ξ_i so that the series in (2.4) converges, with the restrictions on the nodes taken into account. If this is possible, then at the same time it is desirable to make this sum as small as possible. It will be clear from what follows that the finiteness conditions and the approximation estimates have different forms in the cases $p \le q$ and $p > q$.

For the lower estimates we need

Lemma 2.1. *Suppose that* $l \ge k-1 \ge 0$, $l_1 = \max\{1\,,\,l\}$, $1 \le p \le \infty$, $0 < q \le \infty$, $m \ge 1$, $a \ge 0$, $[\Delta) = [\alpha\,,\,\beta)$, $\Delta = \beta - \alpha$. *There exists a function* $\varphi(x) = \varphi(x\,;\,[\Delta)\,,\,m\,,\,a) \in W_p^k([\Delta))$ *with a given value*

$$\|\varphi^{(k)}(x\,;\,[\Delta)\,,\,m\,,\,a)\|_{p\,[\Delta)} = a$$

of the norm of the k*th derivative such that for any spline* $s_l(x)$ *with at most* m *nodes on* $[\alpha\,,\,\beta)$ *the inequality*

$$\|\varphi^{(r)} - s_l^{(r)}\|_{q\,[\Delta)} \ge \frac{C}{m^{k-r}} \Delta^{k-r+1/q-1/p} \|\varphi^{(k)}\|_{p\,[\Delta)}\,, \qquad r = 0\,,\,1\,,\,\dots\,,\,k-1\,, \tag{2.8}$$

holds, where the positive constant $C = C(p\,,\,q\,,\,k\,,\,l)$ *is independent of* $[\Delta)\,,\,a\,,\,r$, *and* m.

Proof. We take some function $\varphi_1(x) \in W_p^k([0\,,\,1))$ not identically equal to any polynomial of degree at most l and such that $\|\varphi_1^{(k)}\|_{p\,[0\,,\,1]} = 1$. Let

$$\min_{0\le r\le k-1} \inf_{p_l} \|\varphi_1^{(r)} - p_l^{(r)}\|_{q\,[0\,,\,1]} = b > 0. \tag{2.9}$$

Let us divide $[\alpha, \beta]$ into $2m$ equal intervals of length $h = \Delta/2m$ by the points $t_i = \alpha + ih$ $(i = 0, 1, \dots, 2m)$ and define on $[\alpha, \beta)$ the function

$$\psi(t) = a\Delta^{-1/p}\varphi_1^{(k)}\left(\frac{t - t_{i-1}}{h}\right), \qquad t \in [t_{i-1}, t_i), \quad i = 1, 2, \dots, 2m. \tag{2.10}$$

We define the desired function $\varphi(x)$ to be the continuous function with $\psi(t)$ as its kth derivative almost everywhere:

$$\varphi(x) = \frac{1}{(k-1)!}\int_a^x \psi(t)(x-t)^{k-1}\,dt. \tag{2.11}$$

Then for $p < \infty$

$$\begin{aligned}\|\varphi^{(k)}\|_{p[\Delta)} &= \left(\int_\alpha^\beta |\psi(t)|^p\,dt\right)^{1/p} = \left(\sum_{i=1}^{2m}\int_{t_{i-1}}^{t_i}\left|a\Delta^{-1/p}\varphi_1^{(k)}\left(\frac{t-t_{i-1}}{h}\right)\right|^p dt\right)^{1/p}\\ &= a\Delta^{-1/p}\left(\sum_{i=1}^{2m}\int_0^1 |\varphi_1^{(k)}(v)|^p h\,dv\right)^{1/p} = a\Delta^{-1/p}(2mh)^{1/p}\|\varphi_1^{(k)}\|_{p[0,1]} = a,\end{aligned}$$

and for $p = \infty$

$$\|\varphi^{(k)}\|_{\infty[\Delta)} = \max_{1\le i\le 2m}\|\psi\|_{\infty[\Delta t_i)} = a\|\varphi_1^{(k)}\|_{\infty[0,1]} = a.$$

Since $s_l(x)$ has at most nodes on $[\alpha, \beta)$ and since there are exactly $2m$ of the half-open intervals $[t_{i-1}, t_i)$, at least m of them are free of nodes of the spline. On each such half-open interval the spline coincides with some polynomial of degree at most l. Denoting any one of them by $[\delta) = [t_{j-1}, t_j)$, we have for $q < \infty$

$$\begin{aligned}\|\varphi^{(r)} - s_l^{(r)}\|_{q[\delta)} &\ge \inf_{p_l}\|\varphi^{(r)} - p_l^{(r)}\|_{q[\delta)} = \inf_{p_l}\left(\int_{[\delta)}|\varphi^{(r)}(x) - p_l^{(r)}(x)|^q\,dx\right)^{1/q}\\ &= \inf_{p_l}\left(\int_{[\delta)}\left|\frac{1}{(k-r-1)!}\int_{[\delta)}\psi(t)(x-t)_+^{k-r-1}\,dt - p_l^{(r)}(x)\right|^q dx\right)^{1/q}.\end{aligned}$$

We make the changes of variables $x = t_{j-1} + uh$ and $t = t_{j-1} + vh$ in the two integrals and use the fact that under this transformation and in the successive ones the degree of a polynomial does not change upon multiplication by a constant; the presence of the infimum allows us to leave the notation the same and to continue the formula as follows:

$$\begin{aligned}&= \inf_{p_l}\left(h\int_0^1\left|\frac{h^{k-r}}{(k-r-1)!}\int_0^1 a\Delta^{-1/p}\varphi_1^{(k)}(v)(u-v)_+^{k-r-1}\,dv - p_l^{(r)}(u)\right|^q du\right)^{1/q}\\ &= a\Delta^{-1/p}h^{k-r+1/q}\inf_{p_l}\left(\int_0^1|\varphi_1^{(r)}(u) - p_l^{(r)}(u)|^q\,du\right)^{1/q}\\ &= a\Delta^{-1/p}h^{k-r+1/q}\inf_{p_l}\|\varphi_1^{(r)} - p_l^{(r)}\|_{q[0,1]} \ge a\Delta^{-1/p}h^{k-r+1/q}b.\end{aligned} \tag{2.12}$$

Now

$$
\begin{aligned}
\|\varphi^{(r)} - s_l^{(r)}\|_{q\,[\Delta)} &= \left(\sum_{i=1}^{2m} \|\varphi^{(r)} - s_l^{(r)}\|^q_{q\,[t_{i-1},\,t_i)}\right)^{1/q} \\
&\ge \left(\sum_{[\delta)} \|\varphi^{(s)} - s_l^{(r)}\|^q_{q\,[\delta)}\right)^{1/q} \ge (m(a\Delta^{-1/p}h^{k-r+1/q}b)^q)^{1/q} \\
&= b2^{-(k-r+1/q)}m^{-(k-r)}\Delta^{k-r+1/q-1/p}a \\
&\ge C(p\,,q\,,k\,,l)m^{-(k-r)}\Delta^{k-r+1/q-1/p}\|\varphi^{(k)}\|_{p\,[\Delta)}.
\end{aligned} \tag{2.13}
$$

Here the sum $\sum_{[\delta)}$ is over the half-open intervals without nodes and contains at least m terms.

If $q = \infty$, then the same transformations of polynomials and of the integral for representing the derivative yield

$$
\begin{aligned}
\|\varphi^{(r)} - s_l^{(r)}\|_{\infty\,[\Delta)} &\ge \inf_{p_l} \|\varphi^{(r)} - p_l^{(r)}\|_{\infty\,[\delta)} \\
&= \inf_{p_l} \left\| \frac{1}{(k-r-1)!}\int_{[\delta)} \psi(t)(x-t)_+^{k-r-1}\,dt - p_l^{(r)}(x)\right\|_{\infty\,[\delta)} \\
&= \inf_{p_l} \left\| \frac{h^{k-r}}{(k-r-1)!}\int_0^1 a\Delta^{-1/p}\varphi_1^{(k)}(v)(u-v)^{k-r-1}\,dv - p_l^{(r)}(u)\right\|_{\infty\,[0\,,1]} \\
&= h^{k-r}a\Delta^{-1/p}\inf_{p_l}\|\varphi_1^{(r)}\|_{\infty\,[0\,,1]} \ge \left(\frac{\Delta}{2m}\right)^{k-r} a\Delta^{-1/p}b \\
&\ge C(p\,,q\,,k\,,l)m^{-(k-r)}\Delta^{k-r+1/q-1/p}\|\varphi^{(k)}\|_{p\,[\Delta)}.
\end{aligned} \tag{2.14}
$$

The lemma is proved.

The case $p = 1$, $q = \infty$, $k = 1$ will be special for the arguments. Then $f(x) \in W_1^1(I)$, and f is bounded on I, since it is a function with bounded variation, and $\sup_x f(x) - \inf_x f(x) \le \|f'\|_{1\,(I)}$. For any class of splines with fixed or nonfixed nodes we get that $\|f - s\|_{\infty\,(I)} \le \|f^{(1)}\|_{1\,(I)}/2$ by setting the approximating spline equal to the constant

$$s_l(x) = \tfrac{1}{2}\left[\inf_u f(u) + \sup_u f(u)\right],$$

and the approximation is finite. In particular, all the quantities (0.3)–(0.6) are finite and do not exceed $1/2$. In the case $p = 1$, $q = \infty$, $r = k-1$ we have by the estimate (2.7) that

$$\|f^{(k-1)}(x) - s_l^{(k-1)}(x\,;\Xi\,,f)\|_{\infty\,(I)} \le C_{k-1}(1\,,\infty\,,k\,,l)\|f^{(k)}\|_{1\,(I)}. \tag{2.15}$$

3. Approximation by splines with partially fixed nodes

3.1. Finiteness and estimates of the approximation for $p \le q$.

Theorem 3.1. *Suppose that* $0 \le k-1 \le l$, $1 \le p \le q \le \infty$, $l_1 = \max\{1\,,l\}$, *the equalities* $k = 1$, $p = 1$, *and* $q = \infty$ *do not hold simultaneously, and* $\mathrm{P} = \{\rho_0 = 0 < \rho_1 < \rho_2 < \cdots\}$ *is a given unbounded set of points on* I.

If the sequence $\{\Delta\rho_i\} = \{\rho_i - \rho_{i-1}\}$ *is bounded,* $\Delta\rho_i \le c$, *then for any function* $f(x) \in W_p^k(I)$ *there is a spline* $s(x) \in \Phi_{\mathrm{P}\,,l}$ *such that*

$$\|f^{(r)} - s^{(r)}\|_{q\,(I)} \le C_r c^{k-r+1/q-1/p}\|f^{(k)}\|_{p\,(I)}, \qquad r = 0\,,1\,,\ldots\,,k-1, \tag{3.1}$$

where the C_r *are the constants in the inequality* (1.6) *in* [3]. *But if the sequence* $\{\Delta\rho_i\}$ *is unbounded, then there exists a function* $f(x) \in W_p^k(I)$ *(and even* $f(x) \in \widetilde{W}_p^k(I)$*) such that for any spline* $s(x) \in \Phi_{\mathrm{P},l}$

$$f^{(r)} - s^{(r)} \notin L_q(I),$$

where $r = 0, \dots, k-2$ *for* $p = 1$ *and* $q = \infty$, *and* $r = 0, 1 \dots, k-1$ *for other values of* p *and* q.

Proof. Suppose first that all the $\Delta\rho_i$ are $\le c$. From a given function we construct the spline $s(x) = s_l(x; \mathrm{P}, f)$ by taking the interpolation nodes ξ_i to be the points $\xi_i = \rho_i$. If $q < \infty$, then the inequality (2.2) holds, and because $\Delta\xi_i = \Delta\rho_i$, $q/p \ge 1$, and the exponent for $\Delta\xi_i$ is positive, it can be extended as follows:

$$\begin{aligned}\|f^{(r)} - s^{(r)}\|_{q(I)} &\le C_r \left(\sum_i \Delta\rho_i^{q(k-r+1/q-1/p)} \Delta F_{\rho_i}^{q/p}\right)^{1/q} \\ &\le C_r c^{k-r+1/q-1/p} \left(\sum_i \Delta F_{\rho_i}^{q/p}\right)^{1/q}.\end{aligned}$$

Since

$$\begin{aligned}\left(\sum_i \Delta F_{\rho_i}^{q/p}\right)^{1/q} &= F^{1/p}(\infty) \left(\sum_i (\Delta F_{\rho_i}/F(\infty))^{q/p}\right)^{1/q} \\ &\le F^{1/p}(\infty) \left(\sum_i \Delta F_{\rho_i}/F(\infty)\right)^{1/q} \\ &= F^{1/p}(\infty) = \|f^{(k)}\|_{p(I)},\end{aligned}$$

it follows that

$$\|f^{(r)} - s^{(r)}\|_{q(I)} \le C_r c^{k-r+1/q-1/p} \|f^{(k)}\|_{p(I)}.$$

For $q = \infty$ we use the inequality (2.7), which gives (3.1) at once.

Suppose now that the sequence $\{\Delta\rho_i\}$ is unbounded. We take some convergent series $A = \sum_{i=1}^{\infty} a_i$ $(a_i > 0)$ and an unbounded positive sequence $\{\delta_n\}$. From the unbounded sequence $\{\Delta\rho_i\}$ choose a subsequence $\{\Delta\rho_{m_n}\}_{n=1}^{\infty}$ such that $\Delta\rho_{m_n} > 1$ and

$$\Delta\rho_{m_n}^{\mu+1/q-1/p} a_n^{1/p} \ge \delta_n, \tag{3.2}$$

where $\mu = 2$ for $p = 1$, $q = \infty$, while $\mu = 1$ for the remaining values of p and q. We define a function $\varphi_0(x)$ on I by setting

$$\varphi_0(x) = \begin{cases} \varphi(x; [\Delta\rho_{m_n}), l_1, a_n^{1/p}) & \text{for } x \in [\Delta\rho_{m_n}),\ n = 1, 2, \dots, \\ 0 & \text{otherwise.} \end{cases} \tag{3.3}$$

We define a continuous function $f(x)$ such that $f^{(k)}(x)$ and $\varphi_0^{(k)}(x)$ coincide almost everywhere:

$$f(x) = \frac{1}{(k-1)!} \int_0^x \varphi_0^{(k)}(t)(x-t)^{(k-1)}\, dt. \tag{3.4}$$

We have that $f(x) \in W_p^k(I)$, because $f^{(k-1)}(x)$ is absolutely continuous on each interval $[\Delta\rho_i]$, and for $p < \infty$

$$\|f^{(k)}\|_{p(I)} = \|\varphi_0^{(k)}\|_{p(I)} = \left(\sum_{n=1}^{\infty} \|\varphi^{(k)}(x; [\Delta\rho_n), l_1, a_n^{1/p})\|_{p[\Delta\rho_n)}^p\right)^{1/p}$$
$$= \left(\sum_{n=1}^{\infty} a_n\right)^{1/p} = A^{1/p},$$

while for $p = \infty$

$$\|f^{(k)}\|_{\infty(I)} = \sup_n \|\varphi_0^{(k)}\|_{\infty[\Delta\rho_n)} = \sup_n a_n^0 = 1.$$

If we take $A = 1$, then $f(x) \in \widetilde{W}_p^k(I)$.

Suppose now that $s(x)$ is any spline in $\Phi_{\mathrm{P},l}$. Since on each half-open interval $[\Delta\rho_i)$ the spline $s(x)$ has exactly l_1 nodes, and on each $[\Delta\rho_{m_n})$ the function $f(x)$ differs from $\varphi(x; [\Delta\rho_{m_n}), l_1, a_n^{1/p})$ by a polynomial $p_{k-1}(x)$ of degree at most $k-1$, it follows from Lemma 2.1 and the inequality (3.2) that

$$\begin{aligned} \|f^{(r)} - s^{(r)}\|_{q[\Delta\rho_{m_n})} &= \|\varphi^{(r)} - (s - p_{k-1})^{(r)}\|_{q[\Delta\rho_{m_n})} \\ &\geq C(p, q, k, l) l_1^{-(k-r)} \Delta\rho_{m_n}^{k-r+1/q-1/p} a_n^{1/p} \\ &\geq C_1(p, q, k, l) \Delta\rho_{m_n}^{\mu+1/q-1/p} a_n^{1/p} \geq C_1\delta_n, \\ & r = 0, 1, \dots, k-\mu. \end{aligned} \tag{3.5}$$

From this,

$$\|f^{(r)} - s^{(r)}\|_{q(I)} \geq \sup_s \|f^{(r)} - s^{(r)}\|_{q[\Delta\rho_{m_n})} \geq C_1 \sup_n \delta_n = \infty,$$
$$r = 0, 1, \dots, k-\mu.$$

The theorem just proved gives us

Theorem 3.2. *Suppose that* $0 \leq k-1 \leq l$, $1 \leq p \leq q \leq \infty$, *the equalities* $k = 1$, $p = 1$, *and* $q = \infty$ *do not hold simultaneously,* $\mathrm{P} = \{\rho_0 = 0 < \rho_1 < \rho_2 < \cdots\}$ *is an unbounded set of points on* I, $\mu = 2$ *if* $p = 1$ *and* $q = \infty$, *and* $\mu = 1$ *in the remaining cases.*

Then the quantities

$$E_{\mathrm{P},l}^r = E_{\mathrm{P},l}^r(\widetilde{W}_p^k(I), q) = \sup_{f \in \widetilde{W}_p^k(I)} \inf_{s \in \Phi_{\mathrm{P},l}} \|f^{(r)} - s^{(r)}\|_{q(I)}, \quad r = 0, 1, \dots, k-\mu, \tag{3.6}$$

are finite if and only if the sequence $\{\Delta\rho_i\}$ *is bounded. If* $\Delta\rho_i \leq c$, *then*

$$E_{\mathrm{P},l}^r \leq C_r c^{k-r+1/q-1/p}, \qquad r = 0, 1, \dots, k-1. \tag{3.7}$$

Moreover,

$$\sup_{f \in \widetilde{W}_p^k(I)} \inf_{s \in \Phi_{\mathrm{P},l}} \left\{ \max_{0 \leq r \leq k-1} \frac{1}{C_r c^{k-r+1/q-1/p}} \|f^{(r)} - s^{(r)}\|_{q(I)} \right\} \leq 1. \tag{3.8}$$

3.2. Finiteness and estimates for $p > q$.

Theorem 3.3. *Suppose that* $0 \le k-1 \le l$, $p \ge 1$, $0 < q < p \le \infty$, $\mathrm{P} = \{0 = \rho_0 < \rho_1 < \rho_2 < \cdots\}$ *is a given unbounded sequence of points on* $I = [0, \infty)$, $d_r = 1 + (k-r)/(1/q - 1/p)$, *and* $0 \le \nu \le k-1$. *If the series*

$$R_\nu = \sum_{i=I}^{\infty} \Delta\rho_i^{d_\nu} \tag{3.9}$$

converges, then for any $f(x) \in W_p^k(I)$ *there is a spline* $s(x) \in \Phi_{\mathrm{P},l}$ *such that*

$$\|f^{(r)} - s^{(r)}\|_{p(I)} \le C_r R_r^{1/q - q/p} \|f^{(k)}\|_{p(I)}, \qquad r = 0, 1, \dots, \nu. \tag{3.10}$$

But if the series R_ν *diverges, then there is a function* $f(x) \in \widetilde{W}_p^k(I)$ *such that for any spline* $s(x) \in \Phi_{\mathrm{P},l}$

$$f^{(\nu)} - s^{(\nu)} \notin L_q(I).$$

Proof. Suppose that the series $R_\nu = \sum_{i=1}^{\infty} \Delta\rho_i^{d_\nu}$ converges. We construct a spline $s(x) = s_l(x; \mathrm{P}, f)$ by taking the interpolation nodes ξ_i to be the points ρ_i and defining it by the formulas (1.4). Since all the d_r are > 1 and $0 \le \beta = q/p < 1$, we extend the inequality (2.4) by the using Hölder's inequality:

$$\|f^{(r)} - s^{(r)}\|_{q(I)} \le C_r \left(\sum_{i=1}^{\infty} \Delta\rho_i^{d_r(1-\beta)} \Delta F_{\rho_i}^{\beta} \right)^{1/q}$$

$$\le C_r \left(\sum_{i=1}^{\infty} \Delta\rho_i^{d_r} \right)^{(1-\beta)/q} \left(\sum_{i=1}^{\infty} \Delta F_{\rho_i} \right)^{\beta/q} = C_r \left(\sum_{i=1}^{\infty} \Delta\rho_i^{d_r} \right)^{1/q - 1/p} \|f^{(k)}\|_{p(I)}.$$

The exponents d_r decrease with increasing r; therefore, it follows from the convergence of the series R_ν that the series R_r converge for $r = 0, \dots, \nu$. This implies (3.10).

Suppose now that the series R_ν diverges. We construct the function $f(x)$ we need as follows. Let $\Delta\rho_i^{d_\nu} = \eta_i$ and $S_n = \sum_{i=1}^{n} \eta_i$, and define the numbers

$$a_n = \begin{cases} K\eta_n / S_n^{p/q}, & p < \infty, \\ 1, & p = \infty, \end{cases} \qquad n = 1, 2, \dots. \tag{3.11}$$

Here and below we use the known fact that if the series $\sum_{i=1}^{\infty} b_i$ of positive terms diverges and $B_n = \sum_{i=1}^{n} b_i$, then the series $\sum_{n=1}^{\infty} b_n / B_n^{\sigma}$ converges for $\sigma > 1$ and diverges for $\sigma \le 1$. If $p < \infty$, then for the series $\sum_{n=1}^{\infty} a_n$ the exponent $\sigma = p/q$ is greater than 1, and it converges. By suitably choosing the factor K, we ensure that $\sum_{n=1}^{\infty} a_n = 1$.

We define the functions $\varphi_0(x)$ and $f(x)$ in terms of the functions in Lemma 2.1:

$$\begin{gathered} \varphi_0(x) - \varphi(x; [\Delta\rho_n), l_1, a_n^{1/p}), \qquad x \in [\rho_{n-1}, \rho_n), \quad n = 1, 2, \dots; \\ f(x) = \frac{1}{(k-1)!} \int_0^x \varphi_0^{(k)}(t)(x-t)^{k-1}\, dt. \end{gathered} \tag{3.12}$$

For $p < \infty$,

$$\|f^{(k)}\|_{p(I)} = \|\varphi_0^{(k)}\|_{p(I)} = \left(\sum_{n=1}^{\infty} \|\varphi^{(k)}(x; [\Delta\rho_n), l_1, a_n^{1/p})\|_{p[\Delta\rho_n)}^p \right)^{1/p} = \left(\sum_{n=1}^{\infty} a_n \right)^{1/p} = 1.$$

For $p = \infty$,

$$\|f^{(k)}\|_{\infty(I)} = \sup_n \|\varphi_0^{(k)}\|_{\infty[\Delta\rho_n)} = 1.$$

Thus, $f(x) \in \widetilde{W}_p^k(I)$.

Suppose now that $s(x)$ is any spline in $\Phi_{\mathrm{P},l}$. Then we get from Lemma 2.1 that

$$\|f^{(\nu)} - s^{(\nu)}\|_{q(I)} = \left(\sum_{n=1}^{\infty} \|\varphi^{(\nu)} - (s - p_{k-1})^{(\nu)}\|_{q[\Delta\rho_n)}^q\right)^{1/q}$$

$$\geq C l_1^{-k+\nu} \left(\sum_{n=1}^{\infty} \Delta\rho_n^{q(k-\nu+1/q-1/p)} a_n^{q/p}\right)^{1/q} = C l_1^{-k+\nu} \left(\sum_{n=1}^{\infty} \Delta\rho_n^{d_\nu(1-q/p)} a_n^{q/p}\right)^{1/q}.$$

For $p < \infty$ the last series is equal to $K^{q/p} \sum_{n=1}^{\infty} \eta_n / S_n$, while for $p = \infty$ it is equal to $\sum_{n=1}^{\infty} \eta_n$. In both cases it diverges, and

$$\|f^{(\nu)} - s^{(\nu)}\|_{q(I)} = \infty.$$

From the point of view of approximation of the class $\widetilde{W}_p^k(I)$ this theorem can be formulated as follows.

Theorem 3.4. *Suppose that* $p \geq 1$, $0 < q < p \leq \infty$, $0 \leq k - 1 \leq l$,

$$\mathrm{P} = \{0 = \rho_0 < \rho_1 < \cdots\}, \quad \Delta\rho_i = \rho_i - \rho_{i-1}, \quad d_r = 1 + (k - r)/(1/q - 1/p).$$

Then for finiteness of any of the quantities

$$E_{\mathrm{P},l}^r = E_{\mathrm{P},l}^r(\widetilde{W}_p^k(I), q) = \sup_{f \in \widetilde{W}_p^k(I)} \inf_{s \in \Phi_{\mathrm{P},l}} \|f^{(r)} - s^{(r)}\|_{q(I)}$$

$$(r = 0, 1, \ldots, k-1)$$

it is necessary and sufficient that the corresponding series $R_r = \sum_{i=1}^{\infty} \Delta\rho_i^{d_r}$ *converge. If for some* ν $(0 \leq \nu \leq k-1)$ *the series* R_ν *converges, then*

$$E_{\mathrm{P},l}^r \leq C_r R_r^{1/q-1/p}, \qquad r = 0, 1, \ldots, \nu.$$

Moreover,

$$\sup_{f \in \widetilde{W}_p^k(I)} \inf_{s \in \Phi_{\mathrm{P},l}} \max_{0 \leq r \leq \nu} \left\{ \frac{1}{C_r R_r^{1/q-1/p}} \|f^{(r)} - s^{(r)}\|_{q(I)} \right\} \leq 1.$$

4. Approximation by splines with nonfixed nodes

4.1. Finiteness and estimates for $p \leq q$.

Theorem 4.1. *Suppose that* $1 \leq p \leq q \leq \infty$, *the equalities* $k = 1$, $p = 1$, *and* $q = \infty$ *do not hold simultaneously,* $0 \leq k - 1 \leq l$, $\mathrm{P} = \{0 = \rho_0 \leq \rho_1 \leq \rho_2 \leq \cdots\}$ *is a given infinite sequence on* I, *and*

$$\gamma = \gamma_0 = 1 + \frac{1/p - 1/q}{k + 1/q - 1/p}. \tag{4.1}$$

If the restricting set P *is such that* $\rho_m = O(m^\gamma)$ *and* $c = \sup_m\{\rho_m/m^\gamma\}$, *then for any function* $f(x) \in W_p^k(I)$ *there is a spline* $s(x) \in S_{\mathrm{P},l}$ *such that*

$$\|f^{(r)} - s^{(r)}\|_{q(I)} \leq C_r c^{k-r+1/q-1/p} \|f^{(k)}\|_{p(I)}, \qquad r = 0, 1, \ldots, k-1. \tag{4.2}$$

But if $\rho_m \neq O(m^\gamma)$, *then the class* $\widetilde{W}_p^k(I)$ *contains a function* $f(x)$ *with infinite deviation in* $L_q(I)$ *from any spline* $s(x) \in S_{\mathrm{P},l}$.

Proof. Let $d_r = 1 + (k-r)/(1/q - 1/p)$ $(r = 0, 1 \dots, k-1)$, and $d = d_0$. Assume that $\rho_m \leq cm^\gamma$ $(m = 1, 2, \dots)$. If $\|f^{(k)}\|_{p(I)} = 0$, then $f^{(k)}(x)$ is equal to zero almost everywhere, and $f(x)$ coincides with a polynomial of degree at most $k-1$. Taking $s(x) = f(x)$ as an approximating spline, we get that the estimates (4.2) hold. Thus, it will be assumed that $\|f^{(k)}\|_{p(I)} > 0$.

Suppose first that $p < q < \infty$. For a given function $f(x)$ we choose the interpolation points ξ_i, depending on the behavior of $f^{(k)}(x)$, in such a way that $\xi_i \geq \rho_i$, and we estimate the deviation $\|f^{(r)} - s^{(r)}\|_{q(I)}$. In this case the inequality (2.4) holds:

$$\|f - s\|_{q(I)} \leq C_0 \left(\sum_i \Delta\xi_i^{d(1-\beta)} \Delta F_{\xi_i}^\beta \right)^{1/q}. \tag{4.3}$$

We remark that the exponent β here is greater than 1. The desire to apply Hölder's inequality to this sum with the goal of making it as small as possible (with it taken into account that equality holds when the $\Delta\xi_i^d$ are proportional to the ΔF_{ξ_i}) leads to the following spline construction.

Suppose that $p < q \leq \infty$. Let $\xi_0 = 0$. If the point ξ_{i-1} has already been defined, then the point ξ_i is defined as a root of the equation

$$F(x) - F(\xi_{i-1}) = K(x - \xi_{i-1})^d \qquad (x \in (\xi_{i-1}, \infty))$$

with respect to x. The value of the constant K will be determined later. Since the left-hand side of the equation is continuous and nondecreasing from zero to $F(\infty) - F(\xi_{i-1})$ for $x > \xi_{i-1}$, while the right-hand side decreases from ∞ to 0 (the exponent d is negative), there is a unique root in the interval (ξ_{i-1}, ∞). If the next point ξ_i is such that $F(\xi_i) = F(\infty)$, then the construction process is over. In any case the point ξ_1 is constructed, because $F(\infty) - F(0) > 0$. Thus, we get a sequence $\Xi = \{\xi_i\}$ (possibly finite) such that

$$\Delta F_{\xi_i} = K \Delta\xi_i^d, \qquad \Delta\xi_i = K^{-1/d} \Delta F_{\xi_i}^{1/d}. \tag{4.4}$$

Employing the procedure described above, we use the set Ξ to construct a set T of intermediate nodes and a spline $s(x) = s_l(x; \Xi, f)$ satisfying the interpolation conditions (1.3) at the points ξ_i and defined by the formulas (1.4) and perhaps (1.5). If it turns out that the sequence $\{\xi_i\}$ is finite and ξ_N is its rightmost point, then this means that $f^{(k)}(x) = 0$ almost everywhere on the interval (ξ_N, ∞), and $f(x)$ coincides with a polynomial of degree at most $k-1$. Then the spline $s(x) = s_l(x; \Xi, f)$ constructed on this part according to the formula (1.5) coincides with $f(x)$, and $\|f^{(r)} - s^{(r)}\|_{q(I)} = \|f^{(r)} - s^{(r)}\|_{q[0, \xi_N)}$. Therefore, the subsequent sums, without indication of the limits of summation, will be assumed in this case to be over the indices from 1 to N.

Using Hölder's inequality and the fact that $d < 0$ and $\gamma = 1 - 1/d$, we get the estimate

$$\begin{aligned} \xi_m &= \sum_{i=1}^m \Delta\xi_i = K^{-1/d} \sum_{i=1}^m \Delta F_{\xi_i}^{1/d} = K^{-1/d} \sum_{i=1}^m \Delta F_{\xi_i}^{1/d} \cdot 1^{1-1/d} \\ &\geq K^{-1/d} \left(\sum_{i=1}^m \Delta F_{\xi_i} \right)^{1/d} m^{1-1/d} \geq K^{-1/d} F^{1/d}(\infty) m^\gamma, \end{aligned}$$

$m = 1, 2, \dots$; $m = 1, \dots, N$ for a finite set $\{\xi_i\}$.

If in the construction we let

$$K = c^{-d} F(\infty), \tag{4.5}$$

then $\xi_m \geq cm^\gamma \geq \rho_m$, and no point ξ_i is to the left of ρ_i; hence the constructed spline is in the class $S_{P,l}$. From (4.4) and (4.5),

$$\Delta\xi_i = c(\Delta F_{\xi_i}/F(\infty))^{1/d}. \tag{4.6}$$

Since $d_r/d \leq 1$ and $\beta > 1$, it follows that

$$(1-\beta)\,d_r/d + \beta \geq 1. \tag{4.7}$$

Using these relations, we extend the inequality (2.4) for $q < \infty$:
(4.8)

$$\begin{aligned}
\|f^{(r)} - s^{(r)}\|_{q(I)} &\leq C_r \left(\sum_i \Delta\xi_i^{d_r(1-\beta)} \Delta F_{\xi_i}^{\beta}\right)^{1/q} \\
&= C_r c^{d_r(1-\beta)/q} F^{1/p}(\infty) \left(\sum_i (\Delta F_{\xi_i}/F(\infty))^{(1-\beta)\,d_r/d+\beta}\right)^{1/q} \\
&= C_r c^{k-r+1/q-1/p} \|f^{(k)}\|_{p(I)} \left(\sum_i (\Delta F_{\xi_i}/F(\infty))^{(1-\beta)\,d_r/d+\beta}\right)^{1/q} \\
&\leq C_r c^{k-r+1/q-1/p} \|f^{(k)}\|_{p(I)} \left(\sum_i \Delta F_{\xi_i}/F(\infty)\right)^{1/q} = C_r c^{k-r+1/q-1/p} \|f^{(k)}\|_{p(I)}.
\end{aligned}$$

For $q = \infty$ the inequality (2.6) can be extended similarly:

$$\begin{aligned}
\|f^{(r)} - s^{(r)}\|_{\infty(I)} &\leq C_r \sup_i \{\Delta\xi_i^{k-r-1/p} \Delta F_{\xi_i}^{1/p}\} \\
&= C_r c^{k-r-1/p} \sup_i \{\Delta F_{\xi_i}^{(k-r-1/p)/d+1/p} F_{(\infty)}^{-(k-r-1/p)/d}\} \\
&= C_r c^{k-r-1/p} F^{1/p} \sup_i \{(\Delta F_{\xi_i}/F(\infty))^{q/(kp-1)}\} \leq C_r c^{k-r-1/p} \|f^{(k)}\|_{p(I)}.
\end{aligned} \tag{4.9}$$

Suppose now that $q = p = \infty$. Then $\gamma = 1$ and $\rho_m \leq cm$. We construct the spline $s(x) = s_l(x; \Xi, f)$ by taking the points $\xi_i = ic$ as the set Ξ of interpolation nodes, $i = 0, 1, \ldots$. Then $\Delta\xi_i = c$, and (2.5) implies that

$$\|f^{(r)} - s^{(r)}\|_{q(I)} \leq C_r \sup_i \{\Delta\xi_i^{k-r}\} \|f^{(k)}\|_{p(I)} = C_r c^{k-r} \|f^{(k)}\|_{p(I)}.$$

Thus, for any function $f(x) \in W_p^k(l)$ we have constructed a spline $s(x) \in S_{P,l}$ such that

$$\|f^{(r)} - s^{(r)}\|_{q(I)} \leq C_r c^{k-r+1/q-1/p} \|f^{(k)}\|_{p(I)}, \qquad r = 0, 1, \ldots, k-1.$$

We prove the second part of the theorem. Suppose that the sequence ρ_m/m^γ is unbounded. Then the sequence $(\rho_m - \rho_n)/m^\gamma$ is also unbounded as $m \to \infty$. Taking the positive power $k + 1/q - 1/p$, we get that the expression $(\rho_m - \rho_n)^{k+1/q-1/p} m^k$ is also unbounded with respect to m. Let $\sum_{i=1}^{\infty} a_i$ be a convergent series with positive terms and sum 1, and let $\{\delta_n\}$ be a positive unbounded sequence. We choose a subsequence $\{\rho_{m_n}\}$ of the sequence $\{\rho_i\}$ according to the following rule. Let $m_0 = 0$. If m_{n-1} has already been defined, then m_n is chosen so that

$$\frac{(\rho_{m_n} - \rho_{m_{n-1}})^{k+1/q-1/p}}{m_n^k} a_n^{1/p} \geq \delta_n. \tag{4.10}$$

Using Lemma 2.1, we now construct the necessary function $f(x)$ by defining $\varphi_0(x) = \varphi(x; [\rho_{m_{n-1}}, \rho_{m_n}), l_1 m_n, a_n^{1/p})$, $x \in [\rho_{m_{n-1}}, \rho_{m_n})$, $n = 1, 2, \dots$, and

$$f(x) = \frac{1}{(k-1)!} \int_0^x \varphi_0^{(k)}(t)(x-t)^{k-1}\, dt.$$

This function has locally absolutely continuous derivative $f^{(k-1)}(x)$, and for $p < \infty$

$$\|f^{(k)}\|_{p\,(I)} = \|\varphi_0^{(k)}\|_{p\,(I)} = \left(\sum_{n=1}^{\infty} \|\varphi_0^{(K)}\|_{p\,[\rho_{m_{n-1}}, \rho_{m_n})}\right)^{1/p} = \left(\sum_{n=1}^{\infty} a_n\right)^{1/p} = 1,$$

while for $p = \infty$

$$\|f^{(k)}\|_{\infty\,(I)} = \sup_n \{\alpha_n^0\} = 1.$$

Hence, $f(x) \in \widetilde{W}_p^k(I)$.

Suppose now that there is a spline $s(x) \in S_{\mathrm{P},l}$ with nodes $\{t_i\}$. Since $t_j l_1 \geq \rho_j$, $j = 0, 1, 2, \dots$, there are at most $l_1 m_n$ nodes of the spline on each half-open interval $[0, \rho_{m_n})$ and, all the more so, on $[\rho_{m_{n-1}}, \rho_{m_n})$. Again using Lemma 2.1 and the condition (4.10) of the construction, we get a lower estimate for the deviation. On each interval $[\rho_{m_{n-1}}, \rho_{m_n}) = [\Delta_n)$ the function $f(x)$ differs from $\varphi(x; [\Delta_n), l_1 m_n, a_n^{1/p})$ by some polynomial p_{k-1} of degree at most $k-1$, and

$$\begin{aligned} \|f - s\|_{q\,[\Delta_n)} &= \|\varphi(\cdot\,; [\Delta_n), l_1 m_n, a_n^{1/p}) - (s - p_{k-1})\|_{q\,(\Delta_n)} \\ &\geq \frac{C}{(l_1 m_n)^k} (\rho_{m_n} - \rho_{m_{n-1}})^{k+1/q-1/p} a_n^{1/p} \geq C l_1^{-k} \delta_n. \end{aligned}$$

Then

$$\|f - s\|_{q\,(I)} \geq \sup_n \|f - s\|_{q\,[\Delta n)} \geq C l_1^{-k} \sup_n \delta_n = \infty.$$

Let

$$\gamma_r = 1 - \frac{1/p - 1/q}{k - r + 1/p - 1/q}, \qquad r = 0, 1, \dots, k-1. \tag{4.11}$$

Theorem 4.2. *Suppose that* $1 \leq p \leq q \leq \infty$, $0 \leq k-1 \leq l$, $\mathrm{P} = \{0 = \rho_0 \leq \rho_1 \leq \rho_2 \leq \cdots\}$ *is the restricting set for the class* $S_{\mathrm{P},l}$, $0 \leq \nu \leq k-1$, *and* $k - \nu + 1/q - 1/p \neq 0$. *The quantities*

$$e_{\mathrm{P},l}^{\nu} = e_{\mathrm{P},l}^{\nu}(\widetilde{W}_p^k(I), q) = \sup_{f \in \widetilde{W}_p^k(I)} \inf_{s \in S_{\mathrm{P},l}} \|f^{(\nu)} - s^{(\nu)}\|_{q\,(I)}$$

are finite if and only if $\rho_m = O(m^{\gamma_\nu})$. *If* $\rho_m = O(m^{\gamma_\nu})$ *and* $c = \sup_m\{\rho_m m^{-\gamma_\nu}\}$, *then*

$$e_{\mathrm{P},l}^{r} \leq C_{r-\nu}(p, q, k-\nu, l) c^{k-r+1/q-1/p}, \qquad r = \nu, \nu+1, \dots, k-1. \tag{4.12}$$

Moreover,

$$\sup_{f \in \widetilde{W}_p^k(I)} \inf_{s \in S_{\mathrm{P},l}} \max_{\nu \leq r \leq k-1} \left\{ \frac{1}{C_{r-\nu}(p, q, k-\nu, l) c^{k-r+1/q-1/p}} \|f^{(r)} - s^{(r)}\|_{q\,(I)} \right\} \leq 1.$$

Proof. If $f(x) \in \widetilde{W}_p^k(I)$, then $f^{(\nu)} \in \widetilde{W}_p^{k-\nu}(I)$. Applying Theorem 4.1 to the function $f^{(\nu)}(x)$, we get the assertion of the theorem.

Remark. If $k-\nu+1/q-1/p=0$, then this means that $\nu=k-1$, $p=1$, and $q=\infty$, and since $f(x)\in\widetilde{W}_1^k(I)$, it follows that $f^{(\nu)}(x)=f^{(k-1)}(x)\in\widetilde{W}_1^1(I)$. Then

$$e_{\mathrm{P},l}^{k-1}(\widetilde{W}_1^k(I),\infty)\le 1/2.$$

Thus, it can happen that the set P restricting the location of the nodes of the spline is such that for some function $f(x)\in W_p^k(I)$ a finite deviation from splines in $S_{\mathrm{P},l}$ in the $L_q(I)$-metric is possible only for some of the higher derivatives, whereas we cannot ensure finiteness of the deviation of the function itself in this metric for any spline in $S_{\mathrm{P},l}$. No matter how we switch the coefficients of the spline in moving to infinity, the restrictions on the switching times can make it impossible for us to follow the function, while at the same time we can ensure a finite deviation for the higher deviatives by switching only the highest derivative of the spline. We remark that this situation does not arise for $p=q$. In that case all the γ_r are equal to 1, and there are no differences in the conditions for finiteness of the deviation for the function and its derivatives.

4.2. **Finiteness and estimates of the deviation for** $p>q$. Suppose that $0=\rho_0\le\rho_1\le\rho_2\le\cdots$, $\rho_i\to\infty$, $\Delta\rho_i=\rho_i-\rho_{i-1}$, and $d>1$. The following assertions are needed.

The inequality of Hardy and Landau ([5], Theorem 326):

$$\sum_{i=1}^{\infty}\left(\frac{\rho_i}{i}\right)^d\le\frac{d}{d-1}\sum_{i=1}^{\infty}\left(\frac{\rho_i}{i}\right)^{d-1}\Delta\rho_i\le\left(\frac{d}{d-1}\right)^d\sum_{i=1}^{\infty}\Delta\rho_i^d. \tag{4.13}$$

Lemma 4.1. *The following representation is valid* $(0\le n\le m)$:

$$\sum_{i=n+1}^{m}\left(\frac{\rho_i}{i}\right)^{d-1}\Delta\rho_i=\sum_{i=n+1}^{m-1}\rho_i\left[\left(\frac{\rho_i}{i}\right)^{d-1}-\left(\frac{\rho_{i+1}}{i+1}\right)^{d-1}\right]+\frac{\rho_m^d}{m^{d-1}}-\frac{\rho_{n+1}^{d-1}\rho_n}{(n+1)^{d-1}}. \tag{4.14}$$

The proof is obtained by changing the limits of summation:

$$\begin{aligned}\sum_{i=n+1}^{m}\left(\frac{\rho_i}{i}\right)^{d-1}\Delta\rho_i&=\sum_{i=n+1}^{m}\left(\frac{\rho_i}{i}\right)^{d-1}(\rho_i-\rho_{i-1})=\sum_{i=n+1}^{m}\left(\frac{\rho_i}{i}\right)^{d-1}\rho_i-\sum_{i=n+1}^{m}\left(\frac{\rho_i}{i}\right)^{d-1}\rho_{i-1}\\&=\sum_{i=n+1}^{m}\left(\frac{\rho_i}{i}\right)^{d-1}\rho_i-\sum_{i=n}^{m-1}\left(\frac{\rho_{i+1}}{i+1}\right)^{d-1}\rho_i\\&=\sum_{i=n+1}^{m-1}\rho_i\left[\left(\frac{\rho_i}{i}\right)^{d-1}-\left(\frac{\rho_{i+1}}{i+1}\right)^{d-1}\right]+\frac{\rho_m^d}{m^{d-1}}-\frac{\rho_{n+1}^{d-1}\rho_n}{(n+1)^{d-1}}.\end{aligned}$$

Lemma 4.2. *If the series* $\sum_{i=1}^{\infty}(\rho_i/i)^d$ *converges, then* $\rho_i^d i^{1-d}\to 0$.

Proof. Since the sequence ρ_i is nondecreasing, the remainder of this series can be estimated as follows:

$$R_m=\sum_{i=m}^{\infty}\left(\frac{\rho_i}{i}\right)^d>\rho_m^d\sum_{i=m}^{\infty}\left(\frac{1}{i}\right)^d>\rho_m^d\int_m^{\infty}x^{-d}\,dx=\frac{1}{d-1}\rho_m^d m^{1-d}.$$

From this, $\rho_m^d m^{1-d}<(d-1)R_m$, and since $R_m\to 0$, it follows that $\rho_m^d m^{1-d}\to 0$.

Lemma 4.3. *The series* $\sum_{i=1}^{\infty}(\rho_i/i)^d$ *and* $\sum_{i=1}^{\infty}(\rho_i/i)^{d-1}\Delta\rho_i$ *either both converge or both diverge, and*

$$\sum_{i=1}^{\infty}(\rho_i/i)^{d-1}\Delta\rho_i\le(d-1)\sum_{i=1}^{\infty}(\rho_i/i)^d. \tag{4.15}$$

Proof. Suppose that the series $\sum_{i=1}^{\infty}(\rho_i/i)^d$ converges. Taking the representation (4.14) with $n=0$ and replacing ρ_{i+1} by ρ_i in it, we get

$$\sum_{i=1}^{m}\left(\frac{\rho_i}{i}\right)^{d-1}\Delta\rho_i=\sum_{i=1}^{m-1}\rho_i\left[\left(\frac{\rho_i}{i}\right)^{d-1}-\left(\frac{\rho_{i+1}}{i+1}\right)^{d-1}\right]+\rho_m^d m^{1-d}$$
$$\le\sum_{i=1}^{m-1}\rho_i^d\left[\left(\frac{1}{i}\right)^{d-1}-\left(\frac{1}{i+1}\right)^{d-1}\right]+\rho_m^d m^{1-d}\le(d-1)\sum_{i=1}^{m-1}\left(\frac{\rho_i}{i}\right)^d+\rho_m^d m^{1-d}.$$

Letting m go to infinity in this inequality and using Lemma 4.2, we obtain the inequality (4.15), which implies convergence of the series $\sum_{i=1}^{\infty}(\rho_i/i)^{d-1}\Delta\rho_i$. If the series $\sum_{i=1}^{\infty}(\rho_i/i)^d$ diverges, then the divergence of the series $\sum_{i=1}^{\infty}(\rho_i/i)^{d-1}\Delta\rho_i$ follows from the inequality (4.13).

Theorem 4.3. *Suppose that* $0<q<p\le\infty$, $p\ge 1$, $0\le k-1\le l$, $\mathrm{P}=\{\rho_0=0\le\rho_1\le\rho_2\le\cdots\}$ *is an unbounded sequence, and*

$$d_r=1+\frac{k-r}{1/q-1/p},\qquad U_r=\sum_{i=1}^{\infty}\left(\frac{\rho_i}{i}\right)^{d_r},\qquad r=0,1,\ldots,k-1. \tag{4.16}$$

If for some ν $(0\le\nu\le k-1)$ *the series* U_ν *converges, then for any function* $f(x)\in W_p^k(I)$ *the class* $S_{\mathrm{P},l}$ *contains a spline such that*

$$\|f^{(r)}-s^{(r)}\|_{q(I)}\le C_r[(d_r-1)]^{1/q-1/p}\|f^{(k)}\|_{p(I)},\qquad r=0,1,\ldots,\nu. \tag{4.17}$$

But if the series U_ν *diverges, then there is a function* $f(x)\in\widetilde{W}_p^k(I)$ *such that* $\|f^{(r)}-s^{(r)}\|_{q(I)}=\infty$ $(r=\nu,\nu+1,\ldots,k-1)$ *for any spline in* $S_{\mathrm{P},l}$.

Proof. Suppose that the restricting set $\mathrm{P}=\{\rho_i\}$ is such that $\rho_i/i\to 0$. We construct the set $\mathrm{H}=\{\eta_i\}$ as follows. Let $\theta_0=0$ and $m_0=0$. If θ_{n-1} and m_{n-1} have already been determined, then the next index m_n is taken equal to an i that maximizes the expression $(\rho_i-\theta_{n-1})/(i-m_{n-1})$ for $i>m_{n-1}$. This can be done, because the given expression also tends to zero as $i\to\infty$. If there are several such i, then take the largest of them. Let $\theta_n=\rho m_n$. We get a sequence of points $0=\theta_0<\theta_1<\cdots$ such that

$$\frac{\theta_n-\theta_{n-1}}{m_n-m_{n-1}}=\max_{i>m_{n-1}}\frac{\rho_i-\theta_{n-1}}{i-m_{n-1}},\qquad n=1,2,\ldots. \tag{4.18}$$

Geometrically, this means that in the set of points in the plane with coordinates (i,ρ_i) we have chosen the vertices (m_n,θ_n) of the smallest convex polygon containing the points (i,ρ_i). Let $\Delta\theta_n=\theta_n-\theta_{n-1}$ and $\Delta m_n=m_n-m_{n-1}$. We remark that in this construction

$$\frac{\Delta\theta_n}{\Delta m_n}\le\frac{\theta_n}{m_n},\qquad n=1,2,\ldots, \tag{4.19}$$

$$\frac{\Delta\theta_1}{\Delta m_1}>\frac{\Delta\theta_2}{\Delta m_2}>\frac{\Delta\theta_3}{\Delta m_3}>\cdots. \tag{4.20}$$

The points η_i are located uniformly between the points θ_n:

$$\eta_i=\theta_{n-1}+(i-m_{n-1})\Delta\theta_n/\Delta m_n,\qquad m_{n-1}\le i<m_n,\qquad n=1,2,\ldots. \tag{4.21}$$

By the condition (4.18) of the construction, the inequalities $\eta_i\ge\rho_i$ $(i=1,2,\ldots)$ hold, and $\eta_{m_n}=\theta_n=\rho_{m_n}$ $(n=1,2,\ldots)$. The sequence $\{\Delta\eta_i\}=\{\eta_i-\eta_{i-1}\}$ is nonincreasing in view of (4.20).

Let $d > 1$. We prove that the series $U = \sum_{i=1}^{\infty}(\rho_i/i)^d$ converges if and only if the series $R = \sum_{i=1}^{\infty} \Delta\eta_i^d$ converges.

Suppose that U converges. The sequence ρ_i is nondecreasing; therefore, a segment of U can be estimated from below as follows:

$$\sum_{i=m_{n-1}}^{m_n-1} \left(\frac{\rho_i}{i}\right)^d \geq \theta_{n-1}^d \sum_{i=m_{n-1}}^{m_n-1} \left(\frac{1}{i}\right)^d > \theta_{n-1}^d \int_{m_{n-1}}^{m_n} x^{-d}\,dx = \frac{1}{d-1}(m_{n-1}^{1-d} - m_n^{1-d}), \qquad n = 2, 3, \dots. \tag{4.22}$$

Using this estimate, changing the limits of summation, and taking into account that $\theta_{\mu-1} < \theta_\mu$, we then get for $\mu \geq 2$

(4.23)

$$\begin{aligned}\sum_{i=1}^{m_\mu-1} \left(\frac{\rho_i}{i}\right)^d &\geq \sum_{n=2}^{\mu} \sum_{i=m_{n-1}}^{m_n-1} \left(\frac{\rho_i}{i}\right)^d > \frac{1}{d-1} \sum_{n=2}^{\mu} \theta_{n-1}^d (m_{n-1}^{1-d} - m_n^{1-d}) \\ &= \frac{1}{d-1} \sum_{n=2}^{\mu} (\theta_{n-1}^d m_{n-1}^{1-d} - \theta_{n-1}^d m_n^{1-d}) = \frac{1}{d-1}\left(\sum_{n=1}^{\mu-1} \theta_n^d m_n^{1-d} - \sum_{n=2}^{\mu} \theta_{n-1}^d m_n^{1-d}\right) \\ &= \frac{1}{d-1}\left[\sum_{n=1}^{\mu-1} (\theta_n^d - \theta_{n-1}^d) m_n^{1-d} - \theta_{\mu-1}^d m_\mu^{1-d}\right] \\ &> \frac{1}{d-1}\left[\sum_{n=1}^{\mu-1} (\theta_n^d - \theta_{n-1}^d) m_n^{1-d} - \rho_{m_\mu}^d m_\mu^{1-d}\right].\end{aligned}$$

By the inequalities $\theta_{n-1} < \theta_n$ and (4.19), the terms in the last sum can be estimated as follows:

$$\begin{aligned}(\theta_n^d - \theta_{n-1}^d) m_n^{1-d} &= \left(\frac{\theta_n}{m_n}\right)^{d-1} \theta_n - \left(\frac{\theta_{n-1}}{m_n}\right)^{d-1} \theta_{n-1} \\ &> \left(\frac{\theta_n}{m_n}\right)^{d-1} \theta_n - \left(\frac{\theta_n}{m_n}\right)^{d-1} \theta_{n-1} \\ &= \left(\frac{\theta_n}{m_n}\right)^{d-1} (\theta_n - \theta_{n-1}) \geq \left(\frac{\Delta\theta_n}{\Delta m_n}\right)^{d-1} (\theta_n - \theta_{n-1}) = \Delta\theta_n^d \Delta m_n^{1-d}.\end{aligned} \tag{4.24}$$

We substitute this estimate in (4.23) and let μ go to infinity. By Lemma 4.2, $\rho_{m_\mu}^d m_\mu^{1-d} \to 0$. We obtain

$$\sum_{n=1}^{\infty} \Delta\theta_n^d \Delta m_n^{1-d} \leq (d-1) \sum_{i=1}^{\infty} (\rho_i/i)^d. \tag{4.25}$$

From the definition of the points η_i in (4.21),

$$\Delta\theta_n^d \Delta m_n^{1-d} = \sum_{i=m_{n-1}+1}^{m_n} \Delta\eta_i^d.$$

Substituting in (4.25), we conclude that the series $R = \sum_{i=1}^{\infty} \Delta\eta_i^d$ converges if U converges, and

$$R \leq (d-1)U. \tag{4.26}$$

Assume that U diverges. Since $\eta_i \geq \rho_i$, the series $\sum_{i=1}^{\infty}(\eta_i/i)^d$ diverges. It follows from the inequality (4.13) of Hardy and Landau that the series $\sum_{i=1}^{\infty}\Delta\eta_i^d$ also diverges.

Assume that the series U_ν $(0 \leq \nu \leq k-1)$ converges. The exponent d_ν is positive; therefore, the series U_r $(r = 0, 1, \dots, \nu)$ and $R_r = \sum_{i=1}^{\infty}\Delta\eta_i^{d_r} \leq (d_r - 1)U_r$ converge. Since $\eta_i \geq \rho_i$ $(i = 1, 2, \dots)$, it follows that $\Phi_{\mathrm{H},i} \subset S_{\mathrm{P},l}$. By Theorem 3.3, there is a spline $s(x) \in \Phi_{\mathrm{H},l} \subset S_{\mathrm{P},l}$ such that

$$\|f^{(r)} - s^{(r)}\|_{q(I)} \leq C_r[(d_r - 1)U_r]^{1/q-1/p}\|f^{(k)}\|_{p(I)}, \qquad r = 0, 1, \dots, \nu.$$

Suppose now that the series U_ν diverges. Assume first that $\rho_i/i \not\to 0$. This means that there exist an $\varepsilon > 0$ and a sequence $0 < m_1 < m_2 < \cdots$, $m_i \to \infty$, such that $\rho_{m_i}/m_i \geq \varepsilon$ $(i = 1, 2, \dots)$. Since $(\rho_{m_i} - \rho_{m_j})/m_i \sim \rho_{m_i}/m_i$ as $i \to \infty$, for any ε_1 with $0 < \varepsilon_1 < \varepsilon$ there exist a subsequence $\{n_j\}$ of the sequence $\{m_i\}$ and corresponding points $\theta_j = \rho_{n_j}$ such that

$$\frac{\Delta\theta_j}{n_j} = \frac{\theta_j - \theta_{j-1}}{n_j} > \varepsilon_1, \qquad j = 1, 2, \dots.$$

Take some series $\sum_{j=1}^{\infty} a_j$ with positive terms and sum 1 such that $\sum_{j=1}^{\infty} a_j^{q/p}$ diverges. With the help of the functions in Lemma 2.1 we construct a function $f(x)$:

$$\varphi_0(x) = \varphi(x; [\Delta\theta_j), l_1 n_j, a_j^{1/p}), \qquad x \in [\Delta\theta_j), \quad j = 1, 2, \dots,$$
$$f(x) = \frac{1}{(k-1)!}\int_0^x \varphi_0^{(k)}(x)(x-t)^{k-1}\,dt.$$

This function $f(x)$ is in the class $\widetilde{W}_p^k(I)$, because for $p < \infty$

$$\|f^{(k)}\|_{p(I)} = \left(\sum_{j=1}^{\infty}\|f^{(k)}\|_{p[\Delta\theta_j)}^p\right)^{1/p} = \left(\sum_{j=1}^{\infty}\|\varphi_0^{(k)}\|_{p[\Delta\theta_j)}^p\right)^{1/p} = \left(\sum_{j=1}^{\infty} a_j\right)^{1/p} = 1,$$

while for $p = \infty$

$$\|f^{(k)}\|_{\infty(I)} = \sup_j \|\varphi_0^{(k)}\|_{\infty[\Delta\theta_j)} = 1.$$

For any spline $s(x) \in S_{\mathrm{P},l}$ there are at most $l_1 n_j$ nodes on the half-open interval $[\Delta\theta_j)$; therefore, by the same Lemma 2.1,

(4.27)
$$\|f^{(r)} - s^{(r)}\|_{q[\Delta\theta_j)} \geq \frac{C}{(l_1 n_j)^{k-r}}\Delta\theta_j^{k-r+1/q-1/p}a_j^{1/p} \geq C\varepsilon_1^{k-r+1/q-1/p}l_1^{r-k}n_j^{1/q-1/p}a_j^{1/p}$$
$$\geq C\varepsilon_1^{k-r+1/q-1/p}l_1^{r-k}a_j^{1/p} \geq C_1 a_j^{1/p},$$
$$\|f^{(r)} - s^{(r)}\|_{q(I)} = \left(\sum_{j=0}^{\infty}\|f^{(r)} - s^{(r)}\|_{q[\Delta\theta_j)}^q\right)^{1/q} > C_1\left(\sum_{j=1}^{\infty} a_j^{q/p}\right)^{1/q} = \infty,$$
$$r = 0, 1, \dots, k-1.$$

We infer that if $\rho_i/i \not\to 0$, then there is an $f(x) \in \widetilde{W}_p^k(I)$ such that the deviations of the function itself and all its derivatives from the corresponding derivatives of any spline in $S_{\mathrm{P},l}$ are infinite.

Assume now that $\rho_i/i \to 0$. Then the procedure for constructing the sequence $\{m_n\}$ and the points $\{\theta_n\}$ is applicable. Let $d = d_\nu$ and define the numbers ($n = 1, 2, \dots$)

$$a_n = \begin{cases} K\Delta\theta_n^d \Delta m_n^{1-d}/(\sum_{i=1}^n \Delta\theta_i^d \Delta m_i^{1-d})^{p/q} & \text{for } p < \infty, \\ K\Delta\theta_n^d \Delta m_n^{1-d} & \text{for } p = \infty. \end{cases} \tag{4.28}$$

Since the series $\sum_n \Delta\theta_n^d \Delta m_n^{1-d}$ diverges and $p/q > 1$, the series $\sum a_n$ converges for $p < \infty$. The factor K is chosen so that $\sum_{n=1}^\infty a_n = 1$ for $p < \infty$, while $K = 1$ for $p = \infty$. We divide each half-open interval $[\Delta\theta_n)$ into $2l_1\Delta m_n$ half-open intervals $[\Delta\tau_j)$ of the same length $\Delta\tau_j = \Delta\theta_n/(2l_1\Delta m_n)$ by the points

$$\begin{gathered} \tau_j = \theta_{n-1} + (j - 2l_1 m_{n-1})\Delta\theta_n/(2l_1\Delta m_n), \\ n = 1, 2, \dots, \qquad j = 2l_1 m_{n-1}, \dots, 2l_1 m_n - 1. \end{gathered} \tag{4.29}$$

Just as in Lemma 2.1, we take some function $\varphi(x)$ on $[0, 1]$ that is not a polynomial $p_l(x)$ of degree at most l, and we let

$$\min_{0 \le r \le k-1} \inf_{p_l} \|\varphi^{(r)} - p_l^{(r)}\|_{q\,[0,1]} = \kappa. \tag{4.30}$$

On I we define the functions

$$\begin{gathered} \psi(x) = \left(\frac{a_n}{\Delta\theta_n}\right)^{1/p} \varphi^{(k)}\left(\frac{x - \tau_{j-1}}{\Delta\tau_j}\right), \qquad x \in [\Delta\tau_j) \subset [\Delta\theta_n), \\ n = 1, 2, \dots, \qquad j = 2l_1 m_{n-1}, \dots, 2l_1 m_n - 1, \\ f(x) = \frac{1}{(k-1)!}\int_0^x \psi(t)(x - t)^{k-1}\, dt. \end{gathered} \tag{4.31}$$

For $p < \infty$

$$\begin{aligned} \|f^{(k)}\|^p_{p\,(\Delta\tau_j)} = \|\psi\|^p_{p\,[\Delta\tau_j)} &= \frac{a_n}{\Delta\theta_n}\int_{[\Delta\tau_j)} \left|\varphi^{(k)}\left(\frac{x - \tau_{j-1}}{\Delta\tau_j}\right)\right|^p dx \\ &= \frac{a_n}{2l_1\Delta m_n \Delta\tau_j}\Delta\tau_j \int_0^1 |\varphi^{(k)}(u)|^p\, du = \frac{a_n}{2l_1\Delta m_n}, \end{aligned}$$

from which

$$\|f^{(k)}\|_{p\,(I)} = \left(\sum_{n=1}^\infty \sum_{j=2l_1 m_{n-1}+1}^{2l_1 m_n} \|f^{(k)}\|^p_{p\,[\Delta\tau_j)}\right)^{1/p} = \left(\sum_{n=1}^\infty a_n\right)^{1/p} = 1.$$

For $p = \infty$

$$\|f^{(k)}\|_{\infty\,(I)} = \sup_n \max_{2l_1 m_{n-1} < j \le 2l_1 m_n} \|\psi\|_{\infty\,[\Delta\tau_j)} = \sup_n \|\varphi^{(k)}\|_{\infty\,[0,1]} = 1.$$

Thus, $f(x)$ is in the class $\widetilde{W}_p^k(I)$.

For $\nu \le r \le k-1$ we estimate the deviation of the rth derivative from the corresponding derivative of any $p_l(x)$ of degree at most l on the interval $[\Delta\tau_j) \subset [\Delta\theta_n)$:

(4.32)

$$\inf_{p_l} \|f^{(r)} - p_l^{(r)}\|^q_{q\,[\Delta\tau_j)} = \inf_{p_l} \int_{[\Delta\tau_j)} \left|\frac{1}{(k-r-1)!}\int_{[\Delta\tau_j)} \psi(t)(x-t)_+^{k-r-1}\, dt - p_l^{(r)}(x)\right|^q dx$$

(just as in Lemma 2.1, we make the change of variables $x = \tau_{j-1} + u\Delta\tau_j$, $t = \tau_{j-1} + v\Delta\tau_j$)

$$= \inf_{p_l} \Delta\tau_j \int_0^1 \left| \left(\frac{a_n}{\Delta\theta_n}\right)^{1/p} \frac{\Delta\tau_j^{k-r}}{(k-r-1)!} \int_0^1 \varphi^{(k)}(v)(u-v)_+^{k-r-1}\,dv - p_l^{(r)}(u) \right|^q du$$

$$= \left(\frac{a_n}{\Delta\theta_n}\right)^{q/p} \Delta\tau_j^{q(k-r)+1} \inf_{p_l} \|\varphi^{(r)} - p_l^{(r)}\|_{q\,[0,1]}^q \ge \kappa^q \left(\frac{a_n}{\Delta\theta_n}\right)^{q/p} \left(\frac{\Delta\theta_n}{2l_1\Delta m_n}\right)^{q(k-r)+1}$$

$$= \kappa^q (2l_1)^{q(r-k)+1} \Delta\theta_n^{q(k-r)+1-q/p} \Delta m_n^{-q(k-r)+1} a_n^{q/p}$$

$$= \kappa^q (2l_1)^{q(r-k)+1} \left(\frac{\Delta\theta_n}{\Delta m_n}\right)^{q(\nu-r)} \Delta\theta_n^{q(k-\nu+I/q-1/p)} \Delta m_n^{-q(k-\nu+1/q)} a_n^{q/p}$$

(use the facts that $r \ge 0$, the sequence $\Delta\theta_n/\Delta m_n$ is decreasing, and $\nu \le r$)

$$\ge \kappa^q (2l_1)^{-qk-1} \left(\frac{\Delta\theta_1}{\Delta m_1}\right)^{q(\nu-r)} \Delta\theta_n^{d(1-q/p)} \Delta m_n^{-d(1-q/p)-q/p} a_n^{q/p}$$

$$\ge C\Delta\theta_n^{d(1-q/p)} \Delta m_n^{-d(1-q/p)-q/p} a_n^{q/p},$$

where the constant C is independent of j and n.

For $p = \infty$

$$\inf_{p_l} \|f^{(r)} - p_l^{(r)}\|_{q\,[\Delta\tau_j)}^q \ge C \left(\frac{\Delta\theta_n}{\Delta m_n}\right)^d, \qquad [\Delta\tau_j) \subset [\Delta\theta_n),$$

and for $p < \infty$

$$\inf_{p_l} \|f^{(r)} - p_l^{(r)}\|_{q\,[\Delta\tau_j)}^q \ge CK^{q/p} \left(\frac{\Delta\theta_n}{\Delta m_n}\right)^d \frac{1}{\sum_{i=1}^n \Delta\theta_i^d \Delta m_i^{1-d}}. \tag{4.33}$$

Let

$$b_j = \begin{cases} (\Delta\theta_n/\Delta m_n)^d / \sum_{i=1}^n \Delta\theta_i^d \Delta m_i^{1-d}, & p < \infty, \\ (\Delta\theta_n/\Delta m_n)^d, & p = \infty; \end{cases} \tag{4.34}$$
$$2l_1 m_{n-1} < j \le 2l_1 m_n, \qquad n = 1, 2, \dots.$$

The series $B = \sum_{j=1}^\infty b_j$ consists of groups of $2l_1\Delta m_n$ identical terms. The terms of this series are nonincreasing, and it diverges. Indeed, since $\sum_{n=1}^\infty \Delta\theta_n^d \Delta m_n^{1-d}$ diverges, the series

$$\sum_{j=1}^\infty b_j = \sum_{n=1}^\infty \frac{\Delta\theta_n^d \Delta m_n^{1-d}}{\sum_{i=1}^n \Delta\theta_i^d \Delta m_i^{1-d}}$$

diverges for $p < \infty$, and for $p = \infty$ we have that $\sum_{j=1}^\infty b_j = \sum_{n=1}^\infty \Delta\theta_n^d \Delta m_n^{1-d}$.

Let $s(x)$ be any spline in $S_{\mathrm{P},l}$. On each half-open interval $[0, \theta_n)$ there are exactly $2l_1 m_n$ half-open intervals $[\Delta\tau_j)$, and at most $l_1 m_n$ nodes of the spline $s(x)$. Therefore, at least $l_1 m_n$ of the half-open intervals $[\Delta\tau_j)$ are free of nodes of the spline. On such half-open intervals the spline coincides with a polynomial of degree at most l. We estimate the deviation on $[0, \infty)$ by using (4.33):

$$\|f^{(r)} - s^{(r)}\|_{q\,(I)} = \left(\sum_{j=1}^\infty \|f^{(r)} - s^{(r)}\|_{q\,[\Delta\tau_j)}^q\right)^{1/q} \tag{4.35}$$
$$\ge \left(\sum{}' \|f^{(r)} - s^{(r)}\|_{q\,[\Delta\tau_j)}^q\right)^{1/q} \le \left(CK^{q/p} \sum{}' b_j\right)^{1/q}.$$

Here the sum $\sum'$ is over the indices j for which $[\Delta\tau_j)$ does not contain nodes of the spline. The series $\sum' b_j$ contains at least half the terms of the series B. More precisely, each partial sum $\sum_{j=1}^{2l_1 m_n} b_j$ contains at least $l_1 m_n$ terms of the series $\sum' b_j$. If the series $\sum' b_j$ contains all but finitely many terms of the series B, then the former series diverges. Suppose now that infinitely many terms of the series B are not present in $\sum' b_j$. Let $\lambda_1, \lambda_2, \dots$ be the indices of the terms of B that are contained in $\sum' b_j$. With each term b_{λ_i} we associate a term b_{μ_i} according to the following rule. With b_{λ_i} we associate the first term b_{μ_1} in $\{b_j\}_{j=1}^\infty$ that is not contained in $\{b_{\lambda_i}\}_{i=1}^\infty$. Next, with each b_{λ_i} we associate the first b_{μ_i} that is contained neither in the set $\{b_{\mu_j}\}_{j=1}^{i-1}$ of terms already associated, nor in $\{b_{\lambda_i}\}_{i=1}^\infty$. Under this correspondence the term b_{μ_i} associated with each term b_{λ_i} is always no smaller than the latter. Indeed, if $1 \le \lambda_i \le 2l_1 m_1$, then $b_{\lambda_i} = b_1 \ge b_{\mu_i}$. When we make this association for a term b_{λ_i} with $2l_1 m_{n-1} < \lambda_i \le 2l_1 m_n$, we have already made associations for all terms in the series $\sum' b_j$ that are in $\sum_{j=1}^{2l_1 m_{n-1}} b_j$, i.e., at least $l_1 m_{n-1}$ associations, and so at least $2l_1 m_{n-1}$ initial terms of B are involved in the pairing. Therefore, with the term b_{λ_i} we must associate a b_{μ_i} that is not smaller and has index μ_i larger than $2l_1 m_{n-1}$. From this,

$$\sum{}' b_j = \sum_{i=1}^\infty b_{\lambda_i} \ge \frac{1}{2}\sum_{i=1}^\infty (b_{\lambda_i} + b_{\mu_k}) = \infty,$$

since the last series contains all the terms of the divergent series $\sum_{j=1}^\infty b_j$. Thus, the series in (4.35) diverges, and

$$\|f^{(r)} - s^{(r)}\|_{q(I)} = \infty, \qquad r = \nu, \nu+1, \dots, k-1.$$

The theorem is proved.

We remark that the possibility of simultaneous approximation of a function and its derivatives when $p > q$ is opposite to the situation when $p < q$. If the restricting set P allows us to ensure finiteness of the deviation of the νth derivative of the function from the corresponding derivative of a spline in $S_{\mathrm{P},l}$, then it is also possible to simultaneously approximate, with finite deviations, all the lower-order derivatives.

Theorem 4.4. *Suppose that* $0 < q < p \le \infty$, $0 \le k-1 \le l$, *and* $\mathrm{P} = \{0 = \rho_0 \le \rho_1 \le \rho_2 \le \cdots\}$ *is an unbounded sequence on* I. *Then for finiteness of any of the quantities*

$$e^r_{\mathrm{P},l} = e^r_{\mathrm{P},l}(\widetilde{W}^k_p(I), q) = \sup_{f\in\widetilde{W}^k_p(I)} \inf_{s\in S_{\mathrm{P},l}} \|f^{(r)} - s^{(r)}\|_{q(I)}$$

$$(r = 0, 1, \dots, k-1)$$

it is necessary and sufficient that the corresponding series $U_r = \sum_{i=1}^\infty (p_i/i)^{d_r}$ *and/or* $V_r = \sum_{i=1}^\infty (p_i/i)^{d_r-1}\Delta\rho_i$ *converge. If for some* ν $(0 \le \nu \le k-1)$ *the series* U_ν *and/or* V_ν *converge, then*

$$e^r_{\mathrm{P},l} \le C_r[(d_r-1)U_r]^{1/q-1/d} \le C_r(d_r V_r)^{1/q-1/p}, \qquad r = 0, 1, \dots, \nu. \tag{4.36}$$

Moreover,

$$\sup_{f\in\widetilde{W}^k_p(I)} \inf_{s\in S_{\mathrm{P},l}} \max_{0\le r\le\nu} \left\{\frac{1}{C_r[(d_r-1)U_r]^{1/q-1/p}}\|f^{(r)} - s^{(r)}\|_{q(I)}\right\} \le 1.$$

Only the assertion and the estimate with the series V_r require proof, but they follow from Lemma 4.3 on simultaneous convergence or divergence of the series U_r and V_r and from the inequality (4.13) of Hardy and Landau.

BIBLIOGRAPHY

1. Yu. N. Subbotin and N. I. Chernykh, *The order of best spline approximation of certain classes of functions*, Mat. Zametki **7** (1970), 31–42; English transl. in Math. Notes **7** (1970).

2. Yu. N. Subbotin, *A connection between spline approximation and the problem of approximating a class by a class*, Mat. Zametki **9** (1971), 501–510; English transl. in Math. Notes **9** (1971).

3. N. I. Chernykh, *Approximation by splines with a given distribution density of the nodes*, Trudy Mat. Inst. Steklov. **138** (1975), 174–197; English transl. in Proc. Steklov Inst. Math. **138** (1975).

4. ———, *Approximation of classes of differentiable functions by splines in weighted spaces*, Trudy Mat. Inst. Steklov. **145** (1980), 160–247; English transl. in Proc. Steklov Inst. Math. **1981**, no. 1 (145).

5. G. G. Hardy, J. E. Littlewood, and G. Pólya, *Inequalities*, Cambridge Univ. Press, Cambridge, 1934.

Translated by H. H. McFADEN

Труды Матем. Инст. Стеклова
198 (1992)

Proceedings of the Steklov
Institute of Mathematics
1994, Issue 1

ON ESTIMATES OF GAUSSIAN SUMS AND WARING'S PROBLEM FOR A PRIME MODULUS

UDC 511.336

S. V. KONYAGIN

1. Introduction

Let p be a prime and k a natural number. We denote by $\gamma(k, p)$ the smallest γ such that for any integer y the congruence

$$y \equiv x_1^k + \cdots + x_\gamma^k \pmod p \tag{1}$$

is solvable in integers $x_1, \ldots, x_\gamma$, and by $\theta(k, p)$ the smallest θ such that the congruence

$$x_1^k + \cdots + x_\theta^k \equiv 0 \pmod p \tag{2}$$

has a nontrivial solution. The problem of finding $\gamma(k, p)$ is called *Waring's problem modulo* p. When $\gamma = \gamma(k, p)$, for certain integers $x_1, \ldots, x_\gamma$ we have

$$x_1^k + \cdots + x_\gamma^k \equiv -1 \pmod p, \quad \text{i.e. } 1 + x_1^k + \cdots + x_\gamma^k \equiv 0 \pmod p;$$

hence

$$\theta(k, p) \le \gamma(k, p) + 1. \tag{3}$$

Let $d = \mathrm{GCD}(k, p-1)$, $t = (p-1)/d$. It is well known that the sets of powers with exponents k and d are the same modulo p, hence

$$\gamma(k, p) = \gamma(d, p), \qquad \theta(k, p) = \theta(d, p). \tag{4}$$

Therefore, in considering $\gamma(k, p)$ and $\theta(k, p)$ we can limit ourselves to the case $p \equiv 1 \pmod k$.

Hardy and Littlewood [1], p. 533, showed that

$$\gamma(k, p) \le k. \tag{5}$$

By Fermat's Little Theorem, for $p = k + 1$ and for any $x \not\equiv 0 \pmod p$ we have $x^k \equiv 1 \pmod p$, hence congruence (1) has no solution with $\gamma < y = k$. In this case, therefore, the inequality (5) becomes an equality. Analogously, $\gamma(k, p) = k$ also in the case $p = 2k + 1$. However, for p that are large compared to k the estimate (5) can be significantly improved. The following inequalities are well known:

$$\gamma(k, p) \le \max(3, [32 \ln k] + 1) \quad \text{if } p > k^2 \quad [2];$$
$$\gamma(k, p) \le 68(\ln k)^2 k^{1/2} \quad \text{if } k \ge 2, \ p > 2k + 1 \quad [3];$$
$$\gamma(k, p) \le (1 + 2k^2/(p-1))(1 + [2\log_2 p]) \quad \text{if } p > k^{3/2} \quad [4];$$
$$\gamma(k, p) \le 170 \frac{k^{7/3}}{(p-1)^{4/3}} \ln p \quad \text{if } p \le k^{7/4} + 1 \quad [5].$$

1991 *Mathematics Subject Classification.* Primary 11L05, 11P05.

Note that $\theta(k, p) = k + 1$ if $p = k + 1$. For $p > k + 1$ we have the following results:

$$\theta(k, p) \le [\tfrac{1}{2}(k+4)] \quad [6],$$
$$\theta(k, p) \overset{\varepsilon}{\ll} k^{1-c+\varepsilon}, \quad c = (103 - 3\sqrt{641})/220 \quad [7],$$
$$\theta(k, p) \overset{\varepsilon}{\ll} k^{2/3+\varepsilon} \quad [2],$$
$$\theta(k, p) \overset{\varepsilon}{\ll} k^{1/2+\varepsilon} \quad [8].$$

Heilbronn [9] conjectured that for any $\varepsilon > 0$ and $t \ge t(\varepsilon)$ we have $\gamma(k, p) \overset{\varepsilon}{\ll} k^{\varepsilon}$. Heilbronn's conjecture implies that $\theta(k, p) \overset{\varepsilon}{\ll} k^{\varepsilon}$ for $p > k + 1$.

The quantities $\theta(k, p)$ can be estimated by means of the Gaussian sums

$$S_k(a, p) = \sum_{x=1}^{p} e(ax^k/p),$$

where $e(u) = \exp(2\pi i u)$, $(a, p) = 1$. It is well known that

$$|S_k(a, p)| < k\sqrt{p} \tag{6}$$

(see [10], Chapter XI, §1, proof of Lemma 1); this inequality is nontrivial for $p > k^2$. Little is known about estimates of Gaussian sums when $p < k^2$. In this direction we prove

Theorem 1. *For any $\varepsilon > 0$ there exists $c_\varepsilon > 0$ such that for any $k \ge 2$, $p \ge k \ln k/(\ln(\ln k + 1))^{1-\varepsilon}$, and $(a, p) = 1$ we have*

$$|S_k(a, p)| \le p(1 - c_\varepsilon)/(\ln k)^{1+\varepsilon}).$$

From this theorem we can deduce estimates for $\gamma(k, p)$.

Theorem 2. *For any $\varepsilon > 0$ there exists $c_\varepsilon > 0$ such that for any $k \ge 2$ and $p \ge k \ln k/(\ln(\ln k + 1))^{1-\varepsilon}$ we have $\gamma(k, p) \le c_\varepsilon(\ln k)^{2+\varepsilon}$.*

Theorem 2 implies an affirmative answer to Heilbronn's question.

Theorem 3. *For any $\varepsilon > 0$ there exists $c_\varepsilon > 0$ such that for any $k \ge 2$ and $p > k+1$ we have $\theta(k, p) \le c_\varepsilon(\ln k)^{2+\varepsilon}$.*

The constants c_ε in Theorems 1–3 are different.

In Theorems 1 and 2, the condition on p is essential and the exponent $1 - \varepsilon$ cannot be replaced by 1.

Theorem 4. *For any $A > 0$ there exist infinitely many natural numbers k and primes p such that $p \ge Ak \ln k/\ln\ln k$ and for some a, $(a, p) = 1$, we have*

$$|S_k(a, p)| \ge p(1 - 1/(\ln k)^A)$$

and $\gamma(k, p) \ge (\ln k)^A$.

The next theorem enables us to find lower bounds on $\gamma(k, p)$ and $\theta(k, p)$ for an infinite set of values of k and p satisfying the conditions of Theorem 2.

Theorem 5. *For any $A \ge 1$ and $\varepsilon > 0$ there exist infinitely many natural numbers k and primes p such that $p \ge k(\ln k)^A$ and $\theta(k, p) \ge (\ln k)^{1-\varepsilon}$, and if $A = 1$, then $\theta(k, p) \ge c \ln k$, where c is a positive absolute constant.*

In view of (3), for the k and p mentioned in Theorem 5 we have $\gamma(k, p) \ge (\ln k)^{1-\varepsilon} - 1$. The estimates in Theorems 2 and 3 can probably be improved by a factor of $\ln k$. We conjecture that under the conditions of Theorem 2 we have $\gamma(k, p) \le c_\varepsilon(\ln k)^{1+\varepsilon}$, and if $p > k \ln k$, $k \ge 2$, then $\gamma(k, p) \le c \ln k$, which easily implies that $\theta(k, p) \le c' \ln k$ if $k \ge 2$ and $p > k+1$, where c, c' are positive constants.

At the same time, it is likely that under the conditions of Theorem 1 the estimate of $|S_k(a, p)|$ cannot be significantly improved, i.e., the inequality $|S_k(a, p)| \le p(1 - c/\ln k)$ does not hold in general. We can expect a significant improvement in the estimate of $|S_k(a, p)|$ when p is large compared with $k \ln k$.

If u is a real number, we denote $\inf\{|u - kp| : k \in \mathbf{Z}\}$ by $\|u\|$.

2. Auxiliary algebraic results

Lemma 1. *A primitive mth root of unity is an algebraic number of degree* $\varphi(m)$.

This is well known (see [11], §42).

Lemma 2. *If* n *is a sufficiently large natural number,* $s \ge 2(\ln n)(\ln \ln n)$, *and* z_1 *is an algebraic integer of degree* n *with conjugates* $z_2, \dots, z_n$, *then there exists a prime* q *such that* $s < q \le 2s$ *and the numbers* $z_1^q, \dots, z_n^q$ *are distinct.*

(Here and below, we are speaking of conjugacy of algebraic numbers and irreducibility of polynomials over the field $\mathbf{Q}$.)

Proof. If a number z_i^q is algebraic of degree n, then the numbers z_j^q, where $1 \le j \le n$ and $j \ne i$, are its conjugates, hence the numbers z_j^q are distinct. By Lemma 3 of [12], the number of primes q for which z_j^q has degree less than n for some j is at most $\log_2 n$. At the same time, the number of primes $q \in (s, 2s]$ is equal to $(s/\ln s)(l + o(1))$ ([13], Chapter III, §1). For sufficiently large n and $s \ge 2(\ln n)(\ln\ln n)$ this number is greater than $\log_2 n$. Thus there exists a prime $q \in (s, 2s]$ such that all of the numbers $z_1^q, \dots, z_n^q$ have degree n and are therefore distinct. The lemma is proved.

For each natural number $n \ge 2$ we denote by λ_n the largest number such that for any algebraic integer z_1 of degree n with conjugates $z_2, \dots, z_n$ either $|z_1| = |z_2| = \cdots = |z_n| = 1$ or

$$\prod_{i=1}^{m} \max(1, |z_i|) \ge \lambda_n . \tag{7}$$

A conjecture of Lehmer [14] says that $\inf_{n \ge 2} \lambda_n > 1$. It is known that

$$\lambda_n \ge 1 + (1 - \varepsilon_n)(\ln\ln n/\ln n)^3 \qquad (n \ge 2,\ \varepsilon_n \to 0 \text{ as } n \to \infty) \quad [12]. \tag{8}$$

Lemma 3. *Suppose* n *and* N *are natural numbers,* $P(z)$ *is an irreducible polynomial of degree* n, $n \ge 2$, *with integral coefficients,*

$$Q(z) = \sum_{j=0}^{N} b_j z^j \not\equiv 0, \qquad b_j \in \{0, \pm 1\} \quad (j = 0, \dots, N),$$

$q > \ln(N+1)/\ln \lambda_n$, $(z')^q \ne (z'')^q$ *for distinct roots* z', z'' *of* P, *and* $P(z)$ *divides* $Q(z^q)$. *Then the modulus of each root of* P *is equal to* 1.

Proof. Let $R(z) = Q(z^q)$. It follows from the condition on the coefficients of R that its roots are algebraic integers; hence the same is true of the roots of P. Let $\{z_1, \dots, z_n\}$ be the roots of P, and $\{w_1, \dots, w_m\}$ the roots of Q, among which are $z_1^q, \dots, z_n^q$. If our assertion is not true, then (7) holds, and hence

$$\prod_{i=1}^m \max(1, |w_i|)) \ge \lambda_n^q > N+1.$$

On the other hand, it is easy to see that

$$\prod_{i=1}^m \max(1, |w_i|) = \exp\left(\frac{1}{2\pi}\int_{|z|=1} \ln|Q(z)|\,|dz|\right),$$

hence

$$\prod_{i=1}^m \max(1, |w_i|) \le \exp\left(\frac{1}{2\pi}\int_{|z|=1} \ln(N+1)\,|dz|\right) = N+1.$$

The lemma is proved.

Lemma 4. *Suppose* $n \ge 3$, z_1 *is an algebraic integer of degree* n *with conjugates* $z_2, \dots, z_n$, *and* $|z_1| = \dots = |z_n| = 1$. *Then* $z_1^m = 1$ *for some* $m \le cn\ln\ln n$, *where* c *is an absolute constant.*

That z_1 is a root of unity is the content of a theorem of Kronecker [15]. If z_1 is a primitive mth root of 1, then, by Lemma 1, $n = \varphi(m)$. But $\varphi(m) \ge c'm/\ln\ln m$, where c' is a positive absolute constant ([13], Chapter I, Theorem 5.1). This implies the desired inequality $m \le cn\ln\ln n$.

For a set of numbers $X = (x_0, \dots, x_m)$ and $n \le m$ we denote by $\Lambda_n(X)$ the set of nonzero polynomials $P(z) = \sum_{i=0}^n a_i z^i$ such that $\sum_{i=0}^n a_i x_{i+j} = 0$ for $j = 0, \dots, m-n$. Note that if $\psi(z) = \sum_{i=0}^m x_i z^{-i}$ and Q is a polynomial of degree at most $m-n$, then the coefficient of z^0 in the Laurent expansion of $\psi(z)P(z)Q(z)$ is equal to zero. Conversely, if the latter condition holds for all polynomials of the form $Q(z) = z^j$, $j = 0, \dots, m-n$, then $\sum_{i=0}^n a_i x_{i+j} = 0$ $(j = 0, \dots, m-n)$, i.e., $P \in \Lambda_n(X)$. Thus $P \in \Lambda_n(X)$ if and only if

$$\oint_{|z|=1} \frac{\psi(z)P(z)Q(z)}{z}\,dz = 0$$

for any polynomial Q of degree at most $m-n$.

Let $\Lambda(X)$ denote the set of all polynomials $P \in \Lambda_{\deg P}(X)$, $\deg P \le m$, where $\deg P$ is the degree of P.

Lemma 5. *Suppose* $X = (x_0, \dots, x_m)$, $P_1 \in \Lambda(X)$, $P_2 \in \Lambda(X)$, $\deg P_1 + \deg P_2 \le m$, *and* P *is the greatest common divisor of* P_1, P_2. *Then* $P \in \Lambda(X)$.

Proof. There exist polynomials R_1, R_2 such that $P = P_1R_1 + P_2R_2$. Suppose $n = \deg P$, $\deg Q \le m-n$. Then $PQ = P_1R_1Q + P_2R_2Q$. Let Q_1 be the remainder of the division of R_1Q by P_2; then $R_1Q = Q_1 + P_2T$, where T is some polynomial, and $PQ = P_1Q_1 + P_2Q_2$, $Q_2 = R_2Q + P_1T$. We have $\deg P_1Q_1 < \deg P_1 + \deg P_2 \le m$, $\deg P_2Q_2 \le \max(\deg PQ, \deg P_1Q_1) \le m$.

Since $P_1 \in \Lambda(X)$, $P_2 \in \Lambda(X)$, it follows that

$$\oint_{|z|=1} \frac{\psi(z)P_j(z)Q_j(z)}{z}\,dz = 0 \qquad (j = 1, 2),$$

hence

$$\oint_{|z|=1} \frac{\psi(z)P(z)Q(z)}{z}\,dz = 0.$$

Since the polynomial Q of degree at most $m-n$ is arbitrary, $P \in \Lambda(X)$. The lemma is proved.

Corollary. *If* $X = (x_0, \dots, x_m)$ *and there exists a polynomial* $P_0 \in \Lambda(X)$ *with* $\deg P_0 \le m/2$, *then a polynomial of minimal degree in* $\Lambda(X)$ *is unique to within a constant factor.*

3. Proof of the main lemma

Lemma 6. *Suppose* $0 < \varepsilon < 1$, $\delta = \varepsilon/4$, $k \ge k(\varepsilon)$ *is a natural number,* p *is a prime,*

$$p \ge k \ln k/(\ln(\ln k + 1))^{1-\varepsilon}, \tag{9}$$

α *is a primitive root modulo* p, $\beta = \alpha^k$, $m = [6 \ln p (\ln\ln p)^4]$, $x_0 \not\equiv 0 \pmod p$, $x_i \equiv x_0 \beta^i \pmod p$ *for* $i = 1, \dots, m$, *and* $A = p^2/(\ln p)^{3\delta}$. *Then*

$$\sum_{i=0}^{m} x_i^2 > A. \tag{10}$$

Proof. We assume that k is a sufficiently large natural number. Suppose (10) does not hold, i.e.

$$\sum_{i=0}^{m} x_i^2 \le A. \tag{11}$$

Let $m_1 = [3 \ln p (\ln\ln p)^4] \le m/2$, $X = (x_0, \dots, x_m)$, $n = [\ln p/(\ln\ln p)^{1-\delta}]$, $B = [(\ln p)^\delta]$. Also, let $\{\psi_i(x)\}_{i=0}^{\infty}$ be the Rademacher system on $(0, 1)$ ([16], Chapter 2, §1, Definition 3), $\mathcal{B} = \{(b_0, \dots, b_n) \in \mathbf{Z}^{n+1} : 1 \le b_i \le B \ (i = 0, \dots, n)\}$, $\mathcal{B}' = \{(b_0, \dots, b_n) \in \mathbf{Z}^{n+1} : |b_i| \le B \, (i = 0, \dots, n)\}$. For each $j = 0, \dots, m-n$ we have

$$\operatorname{meas}\left\{x \in (0,1) : \left|\sum_{i=0}^{n} b_i \psi_i(x) x_{i+j}\right| > \frac{p}{2}\right\} \le 2 \exp\left(-\left(\frac{p}{2}\right)^2 \Big/ \left(4 \sum_{i=0}^{n} b_i^2 x_{i+j}^2\right)\right) \tag{12}$$

([16], Chapter 2, §2, Theorem 5). (The right-hand side of (12) is taken to be zero if it contains a zero denominator.) Since

$$\sum_{i=0}^{n} b_i^2 x_{i+j}^2 \le B^2 \sum_{i=0}^{n} x_i^2 \le AB^2,$$

the right-hand side of (12) is bounded by

$$2\exp(-p^2/(16AB^2)) \le 2\exp(-(\ln p)^{3\delta}/(16(\ln p)^{2\delta})) = 2\exp(-(\ln p)^{\delta}/16) < 1/m,$$

hence

$$\begin{aligned}\operatorname{meas}&\left\{x \in (0,1) : \exists j \in \{0, \dots, m-n\} \left|\sum_{i=0}^{n} b_i \psi_i(x) x_{i+j}\right| > p/2\right\} \\ &\le \sum_{j=0}^{m-n} \operatorname{meas}\left\{x \in (0,1) : \left|\sum_{i=0}^{n} b_i \psi_i(x) x_{i+j}\right| > p/2\right\} < \frac{m-n+1}{m} \le 1.\end{aligned}$$

Thus for each set of numbers $(b_0, \dots, b_n) \in \mathcal{B}$ there exists $x \in (0,1)$ such that $\psi_i(x) \ne 0$ $(i = 0, \dots, n)$ and

$$\left|\sum_{i=0}^{n} b_i \psi_i(x) x_{i+j}\right| < p/2 \qquad (j = 0, \dots, m-n). \tag{13}$$

(We have used the fact that $\operatorname{meas}\{x \in (0, 1) : \exists i\ \psi_i(x) = 0\} = 0$, and also that p is odd, which implies that equality in (13) is impossible.)

Putting $b_i' = b_i\psi_i(x)$, we have $(b_0', \dots, b_n') \in \mathscr{B}'$, $|b_i'| = |b_i|$ for $i = 0, \dots, n$, and

$$\left|\sum_{i=0}^{n} b_i' x_{i+j}\right| < p/2 \qquad (j = 0, \dots, m-n).$$

In all there are B^{n+1} elements of $\mathscr{B}$; they correspond to the B^{n+1} different sets of numbers $(b_0', \dots, b_n') \in \mathscr{B}'$ satisfying (14). Since

$$\begin{aligned} B^{n+1} \geq \exp(\ln([\ln p]^\delta) \ln p/(\ln\ln p)^{1-\delta}) &\geq \exp(\delta \ln\ln p/2 \times \ln p/(\ln\ln p)^{1-\delta}) \\ &= \exp(\delta \ln p(\ln\ln p)^\delta/2) > p, \end{aligned}$$

it follows from Dirichlet's principle that there exist different sets $(b_0', \dots, b_n')$, $(b_0'', \dots, b_n'')$ in $\mathscr{B}'$ satisfying (14) such that

$$\sum_{i=0}^{n} b_i' x_i = \sum_{i=0}^{n} b_i'' x_i.$$

Putting $b_i = b_i' - b_i''$, we have $(b_0, \dots, b_n) \neq 0$, $\sum_{i=0}^{n} b_i x_i = 0$, and, in view of (14),

$$\left|\sum_{i=0}^{n} b_i x_{i+j}\right| < p \qquad (j = 0, \dots, m-n). \tag{15}$$

On the other hand, by definition of the x_i,

$$\sum_{i=0}^{n} b_i x_{i+j} \equiv \sum_{i=0}^{n} b_i x_i \beta^j = 0 \pmod{p}.$$

It now follows from (15) that

$$\sum_{i=0}^{n} b_i x_{i+j} = 0 \qquad (j = 0, \dots, m-n).$$

This means that the polynomial $P_1(z) = \sum_{i=0}^{n} b_i z^i$ belongs to $\Lambda_n(x_0, \dots, x_m)$. We cannot say, however, that $P_1 \in \Lambda(x_0, \dots, x_m)$, since $\deg P_1 = n_1$ can be less than n. In any event, since $n - n_1 < m/2 \leq m - m_1$, we obtain $P_1 \in \Lambda_n(X)$.

Since $\deg P_1 < m_1/2$, it follows from the corollary of Lemma 5 that there exists a unique (to within multiplication by a constant factor) polynomial $P \in \Lambda(X)$ of minimal degree, and $\deg P \leq n_1 \leq n$. If there exists a root z_1 of P such that z_1 is an algebraic integer and $|z_1| > 1$, and if $z_2, \dots, z_{n'}$ are the conjugates of z_1, then, by Lemma 2, there exists a prime $q \in (s, 2s]$, $s = (\ln p)^4$, such that the numbers $z_1^q, \dots, z_{n'}^q$ are distinct. In the contrary case we take as q any prime in the interval $(s, 2s]$.

Let $N = [\log_2 p] + 1$. By analogy with (12), for any $j = 0, \dots, m - qN$ we can write

$$\operatorname{meas}\left\{x \in (0, 1) : \left|\sum_{i=0}^{N} \psi_i(x) x_{j+iq}\right| > \frac{p}{2}\right\} \leq 2\exp\left(-\left(\frac{p}{2}\right)^2 \Big/ \left(4\sum_{i=0}^{N} x_{j+iq}^2\right)\right),$$

hence

$$\operatorname{meas}\left\{x \in (0,1) : \exists j \in \{0, \dots, m - Nq\} \left|\sum_{i=0}^{N} \psi_i(x) x_{j+iq}\right| > \frac{p}{2}\right\}$$
$$\leq \sum_{j=0}^{m-Nq} \exp\left(-p^2 \Big/ \left(16 \sum_{i=0}^{N} x_{j+iq}^2\right)\right) \leq 2(m - Nq + 1)\exp(-p^2/(16A))$$
$$\leq 2m \exp(-(\ln p)^{3\delta}/16) < 1/2.$$

This means there exist at least 2^N sets $(b_0, \dots, b_N) \in \{-1, +1\}^{N+1}$ such that

$$\left|\sum_{i=0}^{N} b_i x_{j+iq}\right| < \frac{p}{2} \qquad (j = 0, \dots, m - Nq).$$

By Dirichlet's principle, we can choose two sets $(b'_0, \dots, b'_N)$, $(b''_0, \dots, b''_N)$ such that

$$\sum_{i=0}^{N} b'_i x_{iq} = \sum_{i=0}^{N} b''_i x_{iq}.$$

Putting $b_i = (b'_i - b''_i)/2$, we have

$$\sum_{i=0}^{N} b_i x_{iq} = 0,$$

$$\left|\sum_{i=0}^{N} b_i x_{j+iq}\right| = \frac{1}{2}\left|\sum_{i=0}^{N} b'_i x_{j+iq} - \sum_{i=0}^{N} b''_i x_{j+iq}\right| < \frac{p}{2} \qquad (j = 0, \dots, m - Nq),$$

$$\sum_{i=0}^{N} b_i x_{j+iq} \equiv \sum_{i=0}^{N} b_i x_{iq} \beta^j = 0 \pmod p \qquad (j = 0, \dots, m - Nq),$$

and hence

$$\sum_{i=0}^{N} b_i x_{j+iq} = 0 \qquad (j = 0, \dots, m - Nq).$$

If we now let $Q(z) = \sum_{i=1}^{N} b_i z^i$ and $P_2(z) = Q(z^q)$, we have $P_2 \in \Lambda_{Nq}(x_0, \dots, x_m)$. Let $N_1 = \deg Q$. Since $Nq - N_1 q < m/2 \leq m - m_1$, it follows that $P_2 \in \Lambda(X)$.

Recall that we defined P to be a polynomial of minimal degree in $\Lambda(X)$. Since X is integral, the coefficients of P are commensurable. We may assume they are relatively prime integers. Since $\deg P + \deg P_2 \leq \deg P_1 + \deg P_2 = n_1 + N_1 q \leq n + 2Ns \leq m_1$, it follows from Lemma 5 that P divides P_2. Thus the leading coefficient of P is 1 or -1, and its roots are algebraic integers. Assume there exists a root z_1 of P such that $|z_1| > 1$ and $z_2, \dots, z_{n'}$ are its conjugates. We choose z_1 as in the definition of q. Then the numbers $z_1^q, \dots, z_{n'}^q$ are distinct. In view of (8), we have

$$\lambda_{n'} \geq 1 + \frac{1}{2}\left(\frac{\ln\ln n}{\ln n}\right)^3 \geq 1 + \frac{1}{2}\left(\frac{\ln\ln\ln p}{\ln\ln p}\right)^3,$$

$$\ln \lambda'_n \geq \frac{1}{3}\left(\frac{\ln\ln\ln p}{\ln\ln p}\right)^3 \geq \frac{2}{(\ln\ln p)^3},$$

$$\ln(N+1) \leq \ln(\log_2 p + 2) < 2\ln\ln p,$$

hence $\ln(N+1)/\ln\lambda_{n'} < (\ln\ln p)^4 = s < q$. Since $P_3(z) = \prod_{i=1}^{n'}(z - z_i)$ divides $P_2(z)$, it follows from Lemma 3 that the modulus of each root of P_3 is equal to 1, which contradicts the assumption $|z_1| > 1$.

Thus the modulus of each root of P is at most 1. Then $P(z) = z^l \prod_{i=1}^{I} P_i(z)$, where the P_i are (not necessarily distinct) irreducible polynomials with integral coefficients. Since, by Viète's theorem, the product of the roots of each P_i is an integer, it follows that the modulus of each root is exactly 1. By Lemma 4, each polynomial $P_i(z)$ divides a polynomial $z^{k_i} - 1$ with $k_i \le n_3 = [cn\ln\ln n]$, which implies that $P(z)$ divides

$$P_s(z) = z^n \prod_{j=1}^{n_3}(z^j - 1)^n .$$

If $P(z) = \sum_{i=0}^{n_2} b_i z^i$, then, since $P \in \Lambda(X)$, we have $\sum_{i=0}^{n_2} b_i x_i = 0$, hence

$$\sum_{i=0}^{n_2} b_i x_0 \beta^i \equiv 0 \pmod p, \qquad \sum_{i=0}^{n_2} b_i \beta^i \equiv 0 \pmod p,$$

i.e. $P(\beta) \equiv 0 \pmod p$ and therefore $P_3(\beta) \equiv 0 \pmod p$. Since $\beta \not\equiv 0 \pmod p$, it follows that $\beta^j - 1 \equiv 0 \pmod p$ for some $j \le n_3$, i.e., $\alpha^{kj} - 1 \equiv 0 \pmod p$. But α is a primitive root modulo p, hence $kj \ge p-1$. On the other hand, $kj \le ckn\ln\ln n < ck\ln p/(\ln\ln p)^{1-2\delta}$. In view of (9), it follows that $kj < k\ln k/(\ln\ln k)^{1-3\delta} < p-1$. This contradiction completes the proof of Lemma 6.

4. Upper bounds for $|S_k(a, p)|$, $\gamma(k, p)$, and $\theta(k, p)$

Proof of Theorem 1. We use the notation of Lemma 6. Since for any $i = 0, 1, \dots, m$ each number x, $1 \le x \le p-1$, can be uniquely represented modulo p in the form $y\alpha^i$, $1 \le y \le p-1$, we have

$$\sum_{x=1}^{p-1} \|(\beta - 1)ax^k\|^2 = \frac{1}{m+1} \sum_{i=0}^{m}\sum_{y=1}^{p-1} \|(\beta - 1)ay^k\alpha^{ik}\|^2$$

$$= \frac{1}{m+1} \sum_{y=1}^{p-1}\sum_{i=0}^{m} \|(\beta - 1)ay^k\beta^i\|^2,$$

hence, by Lemma 6, for k, p satisfying (9) we obtain

$$\sum_{x=1}^{p-1} \|(\beta - 1)ax^k\|^2 > \frac{1}{m+1}\sum_{y=1}^{p-1} A = \frac{(p-1)A}{m+1}.$$

If $x_0 \in \mathbf{R}$ is arbitrary, the last inequality can be rewritten in the form

$$\sum_{x=1}^{p-1} \|(a(\alpha x)^k - x_0) - (ax^k - x_0)\|^2 > \frac{(p-1)A}{m+1},$$

and since $\|u - v\|^2 \le 2\|u\|^2 + 2\|v\|^2$, we obtain

$$2\sum_{y=1}^{p-1} \|a(\alpha y)^k - x_0\|^2 + 2\sum_{y=1}^{p-1} \|ax^k - x_0\|^2 > \frac{(p-1)A}{m+1}. \tag{16}$$

The sums in the left-hand side of (16) differ only in the order of their terms, hence

$$\sum_{x=1}^{p-1} \|ax^k - x_0\|^2 > \frac{(p-1)A}{4(m+1)}. \tag{17}$$

To pass from (17) to an estimate of the exponential sums we need

Lemma 7. *For any* $u \in [0, 1/2]$ *we have* $\cos(2\pi u) \le 1 - 8u^2$.

Proof. The function $\psi(u) = \cos(2\pi u) - 1 + 8u^2$ has a double zero $u = 0$ and a simple zero $u = 1/2$. If ψ assumed the value 0 on $(0, 1/2)$, then, by Rolle's theorem, ψ'' would have at least two zeros on $(0, 1/2)$. But this is not so. Thus ψ has constant sign on $(0, 1/2)$. This sign is negative, inasmuch as $\psi(1/4) = -1/2$. The lemma is proved.

It follows from the lemma that for any $x \in \mathbf{R}$ we have $\operatorname{Re}(e(x/p)) \le 1 - 8\|x\|^2/p^2$. Then from (17) we obtain

$$\begin{aligned}
\operatorname{Re}\left(e(-x_0/p)\sum_{x=1}^{p} e(ax^k/p)\right) &= \operatorname{Re}\left(\sum_{x=1}^{p} e((ax^k - x_0)/p)\right) \\
&\le 1 + \sum_{x=1}^{p-1}(1 - 8\|ax^k - x_0\|^2/p^2) \\
&= p - \frac{8}{p^2}\sum_{x=1}^{p-1}\|ax^k - x_0\|^2 < p - \frac{2(p-1)A}{(m+1)p^2}.
\end{aligned}$$

Since $x_0 \in \mathbf{R}$ is arbitrary, it follows that

$$|S_k(a, p)| \le p - \frac{2(p-1)A}{(m+1)p^2};$$

hence, substituting the values of A and m, we obtain

$$|S_k(a, p)| \le p(1 - 1/(\ln p)^{1+\varepsilon}).$$

This implies the assertion of the theorem for $k \ge k(\varepsilon)$ and $p \le k^3$. If $2 \le k < k(\varepsilon)$ and $p \le k^3$, we can obtain the desired inequality by choosing c_ε sufficiently small. Finally, if $p > k^3$, the inequality follows from (6). Theorem 1 is proved.

Theorem 2 can be deduced from Theorem 1 in a standard way. Since for $x \in \mathbf{Z}$ we have

$$\frac{1}{p}\sum_{j=1}^{p} e(jx/p) = \begin{cases} 1, & x \equiv 0 \pmod p, \\ 0, & x \not\equiv 0 \pmod p, \end{cases}$$

the number of solutions of (1) is equal to

$$\begin{aligned}
N &= \frac{1}{p}\sum_{j=1}^{p}\sum_{x_1=1}^{p}\cdots\sum_{x_\gamma=1}^{p} e\left(\frac{j(x_1^k + \cdots + x_\gamma^k - y)}{p}\right) \\
&= p^{\gamma-1} + \frac{1}{p}\sum_{j=1}^{p-1}\sum_{x_1=1}^{p}\cdots\sum_{x_\gamma=1}^{p} e\left(\frac{j(x_1^k + \cdots + x_\gamma^k - y)}{p}\right) \\
&= p^{\gamma-1} + \frac{1}{p}\sum_{j=1}^{p-1} e(-jy/p)\left(\sum_{x=1}^{p} e(jx^k/p)\right)^{\gamma}.
\end{aligned}$$

By Theorem 1, under the conditions of Theorem 2 we have

$$\begin{aligned}
N &\ge p^{\gamma-1}\frac{p-1}{p}p^{\gamma}(1 - c_\varepsilon/(\ln k)^{1+\varepsilon})^{\gamma} = p^{\gamma-1}(1 - (p-1)(1 - c_\varepsilon/(\ln k)^{1+\varepsilon})^{\gamma}) \\
&\ge p^{\gamma-1}(1 - (p-1)\exp(-c_\varepsilon\gamma/(\ln k)^{1+\varepsilon})).
\end{aligned}$$

If $\gamma \geq (\ln k)^{1+\varepsilon} \ln p/c_\varepsilon$, then $(p-1)\exp(-c_\varepsilon \gamma/(\ln k)^{1+\varepsilon}) \leq (p-1)/p < 1$; hence $N > 0$, i.e., the congruence (1) has a solution. Thus $\gamma(k, p) \leq (\ln k)^{1+\varepsilon} \ln p/c_\varepsilon$. If $p \leq k^2$, we obtain the desired inequality. If $p > k^2$, it follows from the estimate of $\gamma(k, p)$ of Dodson [2] mentioned in the introduction. Theorem 2 is proved.

Corollary. *For any* $\varepsilon > 0$ *there exists a number* $c_\varepsilon > 0$ *such that for any* p, k *for which* $t \geq t(\varepsilon)$ *we have* $\gamma(k, p) \leq c_\varepsilon k^\varepsilon$.

Proof. We make use of (4). If $d = 1$, then $\gamma(d, p) = 1$ and everything is clear. Suppose $d > 1$. Let $l = [1/\varepsilon] + 1$. By a theorem of Bovey [4], there exist numbers $c(l) > 0$, $t_0(l) > 0$ such that for $t \geq t_0(l)$ we have $\gamma(d, p) \leq c(l)\varphi(t)d^{1/l}$. For $t \leq \ln d$ we have

$$\gamma(k, p) = \gamma(d, p) < c(l)d^{1/l} \ln d \leq c'_\varepsilon d^\varepsilon \leq c'_\varepsilon k^\varepsilon .$$

If $t > \ln d$, then, by Theorem 1, $\gamma(d, p) \leq c(\ln d)^3$, where c is an absolute constant; hence $\gamma(k, p) = \gamma(d, p) \leq c(\ln k)^3 \leq C''_\varepsilon k^\varepsilon$. The corollary is proved.

Proof of Theorem 3. If $d = 1$, then, in view of (4), $\theta(k, p) = \theta(d, p) = 2$ and there is nothing to prove. Suppose $d > 1$. When $t \leq \ln d$ we use the inequality $\theta(d, p) \leq t$ [2], and when $t > \ln d$ it follows from Theorem 2 and (3) that $\theta(d, p) \leq c_\varepsilon(\ln d)^{2+\varepsilon} + 1$. In either case, $\theta(d, p) \leq c'_\varepsilon(\ln k)^{2+\varepsilon}$. This together with (4) implies the assertion of the theorem.

5. Lower bounds for $|S_k(a, p)|$, $\gamma(k, p)$, and $\theta(k, p)$

The next lemma is a consequence of the asymptotic distribution law for primes in arithmetic progressions ([13], Chapter IV, Theorem 8.3).

Lemma 8. *For any* $B > 0$, $s \geq s(B)$, *and natural number* $t \leq (\ln s)^B$ *there exist at least* $s/(2\ln s)$ *primes* p *of the form* $kt + 1$, *where* k *is a natural number and* $s < k \leq 2s$.

Proof of Theorem 4. Let p_j be the jth prime $(j = 1, 2, \ldots)$. Choose n so that

$$\prod_{j=1}^{n}(1 - 1/p_j) \leq 1/8A^2 . \tag{18}$$

This is possible by Theorem 4.1 in [13], Chapter I. Suppose $Q = p_1 \cdots p_n$, T is a sufficiently large natural number, $t = QT$, $S = t^{t/(2A)}$. By Lemma 8, there exists a prime $p = kt + 1$, $s < k \leq 2s$. We show that p, k satisfy the demands of the theorem.

Note first that for $t \to \infty$, $s(t) = t^{t/(2A)}$, $s < k \leq 2s$ we have the asymptotic relations

$$\frac{\ln s}{\ln\ln s} \sim \frac{t \ln t/(2A)}{\ln t} = \frac{t}{2A}, \qquad \ln k \sim \ln s, \qquad \ln\ln k \sim \ln\ln s; \tag{19}$$

hence

$$t > A\ln k/\ln\ln k > \sqrt{\ln k}, \qquad p \geq Ak\ln k/\ln\ln k . \tag{20}$$

Suppose α is a primitive root modulo p and $\beta = \alpha^k$. Let $\mathscr{L}$ denote the set of numbers $l = \tau + Tm$, where $0 \leq \tau < T$, $0 \leq m < Q$, and $m \not\equiv 0 \pmod{p_j}$ for $j = 1, \ldots, n$. Using (18), we can estimate the number of elements in $\mathscr{L}$ as follows:

$$\operatorname{card}\mathscr{L} = T\prod_{j=1}^{n}(p_j - 1) = QT\prod_{j=1}^{n}\left(1 - \frac{1}{p_j}\right) \leq \frac{t}{8A^2} .$$

We partition the half-open interval $[0, p)$ into $[(\ln k)^{2A}]$ similar intervals of length $\Delta = p/[(\ln k)^{2A}]$. In view of (20),

$$[(\ln k)^{2A}]^{\operatorname{card}\mathscr{L}} \le (\ln k)^{2At/(8A^2)} = \ln k^{t(4A)} < (t^2)^{t/(4A)} = s < p .$$

Consequently, by Dirichlet's principle, there exist integers a', a'' such that $0 \le a' < a'' < p$ and, for any $l \in \mathscr{L}$, the remainders of the divisions of $a'\beta^l$ and $a''\beta^l$ by p lie in the same half-open interval. Putting $a = a' - a''$, we have $\|a\beta^l\| < \Delta$ for $l \in \mathscr{L}$.

We now show that if $0 \le r < n$, $l = \tau + Tm$, where $0 \le \tau < T$, $m \not\equiv 0 \pmod{p_j}$ for $j = r+1, \dots, n$, then $\|a\beta^l\| < \Delta p_1 \cdots p_r$. We proceed by induction on r. If $r = 0$, let m' be the remainder of the division of m by Q, and let $l' = \tau + Tm'$. Then $0 \le m' < Q$, $m' \not\equiv 0 \pmod{p_j}$ for $j = 1, \dots, n$, hence $l' \in \mathscr{L}$. Also,

$$l - l' = T(m - m') \equiv 0 \pmod{t}, \qquad kl \equiv kl' \pmod{p-1};$$

hence, by Fermat's Little Theorem, $\alpha^{kl} \equiv \alpha^{kl'} \pmod{p}$, i.e., $\|a\beta^l\| = \|a\beta^{l'}\| < \Delta$.

Now suppose our assertion is true for all $r < r_0$. Let us prove it for $r = r_0$. Suppose $l = \tau + Tm$, $0 \le \tau < T$, $m \not\equiv 0 \pmod{p_j}$ for $j = r+1, \dots, n$. If $m \not\equiv 0 \pmod{p_r}$, then, by the induction hypothesis, $\|\beta^l\| < \Delta p_1 \cdots p_{r-1}$; hence we assume $m \equiv 0 \pmod{p_r}$. Let $l_i = l + it/p_r$ $(i = 0, \dots, p_r - 1)$. We have

$$\sum_{i=0}^{p_r-1} \beta^{l_i} = \beta^l \sum_{i=0}^{p_r-1} \alpha^{i(p-1)/p_r} = \beta^l(\alpha^{p-1} - 1)/(\alpha^{(p-1)/p_r} - 1) \equiv 0 \pmod{p}. \tag{21}$$

We have used the fact that $\alpha^{(p-1)/p_r} \not\equiv 1 \pmod{p}$, inasmuch as α is a primitive root modulo p and $(p-1)/p_r < p-1$. For $i = 1, \dots, p_r - 1$ we have $l \not\equiv 0 \pmod{p_j}$ $(j = r, \dots, n)$; hence, by the induction assumption, $\|a\beta^{l_i}\| < \Delta p_1 \cdots p_{r-1}$. Therefore, in view of (21),

$$\|\alpha\beta^l\| = \left\| - \sum_{i=1}^{p_r-1} a\beta^{l_i} \right\| \le \sum_{i=1}^{p_r-1} \|a\beta^{l_i}\| < (p_r - 1)\Delta p_1 \cdots p_{r-1} < \Delta p_1 \cdots p_r .$$

Our assertion is proved.

It implies, in particular, that $\|a\beta^l\| < \Delta Q$ for any $l \in \mathbf{Z}$. Note that any number $x \not\equiv 0 \pmod{p}$ is congruent to α^l modulo p for some $l \in \mathbf{Z}$, hence $\|ax^k\| = \|a\beta^l\| < \Delta Q$. Since for $|u| \le u_0 = \Delta Q/p$ we have $\operatorname{Re}(e(u)) = \cos 2\pi u \ge 1 - 2\pi^2 u^2 \ge 1 - 2\pi^2 u_0^2$, it follows that

$$\begin{aligned}\operatorname{Re}(S_k(a, p)) &\ge \sum_{x=1}^{p} \operatorname{Re}(e(ax^k/p)) \ge p(1 - 2\pi^2 u_0^2) \\ &\ge p\left(1 - 2\pi^2 \left(\frac{Q}{(\ln k)^{2A} - 1}\right)^2\right) \ge p(1 - 1/(\ln k)^A).\end{aligned}$$

The first assertion of the theorem is proved.

To estimate $\gamma(k, p)$ we choose an integer b such that $ab \equiv 1 \pmod{p}$ and put $y = \frac{p-1}{2} b$. If (1) holds, then

$$\frac{p-1}{2} \equiv ax_1^k + \cdots + ax_\gamma^k \pmod{p},$$

hence

$$\frac{p-1}{2} = \left\| \frac{p-1}{2} \right\| \le \sum_{j=1}^{\gamma} \|ax_j^\gamma\| < \gamma\Delta Q, \qquad \gamma > \frac{p-1}{2\Delta Q} > \frac{(\ln k)^{2A} - 2}{3Q} > (\ln k)^A .$$

The theorem is proved.

Proof of Theorem 5. Choose a sufficiently large prime t, let $s = \frac{1}{2}\exp(t^{1/A})$, and let $\theta_0 = t/2$ if $A = 1$, $\theta_0 = t^{1/A}/\ln t$ if $A > 1$. We will show there exists a prime $p = kt + 1$, $s < k \le 2s$, such that $\theta(k, p) > \theta_0$. If $p = kt + 1$ is a prime, α is a primitive root modulo p, and $\beta = \alpha^k$, then each number of the form x^k is congruent to some β^j ($j \in \{0, \dots, t-1\}$) or to zero modulo p. If $\theta(k, p) \le \theta_0$, there exists a set of nonnegative integers $(l_0, \dots, l_{t-1})$ such that

$$0 < l_0 + \cdots + l_{t-1} = \theta \le \theta_0, \tag{22}$$

$$\sum_{j=0}^{t-1} l_j \beta_j \equiv 0 \pmod p. \tag{23}$$

To each such set there correspond $t-1$ different numbers $(l_0+1, l_0+l_1+2, \dots, l_0+\cdots+l_{t-2}+t-1)$ from the set $\{1, \dots, \theta+t-1\}$. Therefore, the total number of sets satisfying (22) is equal to

$$N_i = \sum_{0<\theta\le\theta_0} C_{\theta+t-1}^{t-1}.$$

Using Stirling's formula we easily obtain the estimate

$$N_1 \le \exp(c_A t^{1/A}), \qquad c_A < 1. \tag{24}$$

Let us fix one of the sets $(l_0, \dots, l_{t-1})$ satisfying (22). The roots of the polynomial $P_1(z) = 1+z+\cdots+z^{t-1}$ are the primitive tth roots of 1. It follows from Lemma 1 that P_1 is irreducible, hence it is relatively prime to the polynomial $P_2(z) = \sum_{j=0}^{t-1} l_j z^j$. Thus the resultant of these polynomials ([11], §34)

$$\Delta = \det \begin{pmatrix} l_{t-1} & \cdots & l_0 & & & \\ & l_{t-1} & \cdots & l_0 & & \\ & & \cdots\cdots & & & \\ & & & l_{t-1} & \cdots & l_0 \\ 1 & \cdots & 1 & & & \\ & 1 & \cdots & 1 & & \\ & & \cdots\cdots & & & \\ & & & 1 & \cdots & 1 \end{pmatrix}$$

(the entries that are not indicated are zeros) is different from 0. If (23) holds, then $P_1(\beta) = (\beta^k - 1)/(\beta - 1) \equiv 0 \pmod p$, $P_2(\beta) \equiv 0 \pmod p$, hence P_1, P_2 have a common root in $\mathbf{Z}_p$. Therefore, the resultant Δ is divisible by p. But

$$|\Delta| \le \left(\sum_{j=0}^{t-1} l_j^2\right)^{t/2} \left(\sum_{j=0}^{t-1}\right)^{t/2} \le \theta_0^t t^{1/2} \le t^{3t/2};$$

hence the number of distinct prime divisors of Δ is at most

$$N_2 = \log_2 |\Delta| < 3t \ln t, \tag{25}$$

and the number of primes p of the form $kt+1$ for which $\theta(k, p) \le \theta_0$ is at most $N_1 N_2$. It follows from (24) and (25) that $N_1 N_2 \le \exp(c_A t^{1/A}) 3t \ln t < s/(2\ln s)$. By Lemma 8, there exists a prime $p = kt+1$, $k \in (s, 2s]$, such that $\theta(k, p) > \theta_0$. It is clear from the choice of t, s, and θ_0 that $p > k(\ln k)^A$ and $\theta_0 \ge \ln k/2$ if $A = 1$, $\theta_0 \ge (\ln k)^{1-\varepsilon}$ if $A \ge 1$. Theorem 5 is proved.

Bibliography

1. G. H. Hardy and J. E. Littlewood, *Some problems of "Partitio Numerorum". VIII: The number* $\Gamma(k)$ *in Waring's problem*, Proc. London Math. Soc. (2) **28** (1927), 518–542.
2. M. M. Dodson, *On Waring's problem in* GF[p], Acta Arith. **19** (1971), 147–173.
3. M. M. Dodson and A. Tietäväinen, *A note on Waring's problem in* GF[p], Acta Arith. **30** (1976), 159–167.
4. J. D. Bovey, *A new upper bound for Waring's problem* $(\bmod p)$, Acta Arith. **32** (1977), 157–162.
5. A. Garsia and J. F. Voloch, *Fermat curves over finite fields*, J. Number Theory **30** (1988), 345–356.
6. S. Chowla, H. B. Mann, and E. G. Straus, *Some applications of the Cauchy-Davenport theorem*, Norske Vid. Selsk. Forh. Trondheim **32** (1959), 74–80 (1960).
7. Inder Chowla, *On Waring's problem* $(\bmod p)$, Proc. Ind. Nat. Acad. Sci. India Sect. A **13** (1943), 195–200.
8. A. Tietäväinen, *Proof of a conjecture of S. Chowla*, J. Number Theory **7** (1975), 353–356.
9. H. Heilbronn, *Lecture notes on additive number theory* $\bmod p$, Calif. Inst. Technology, Pasadena, CA, 1964.
10. A. A. Karatsuba, *Principles of analytic number theory*, 2nd ed., "Nauka", Moscow, 1983; Spanish transl. of 1st ed.,, "Mir", Moscow, 1979.
11. B. L. van der Waerden, *Algebra*, 8th ed., Springer-Verlag, Berlin, 1971.
12. E. Dobrowolski, *On a question of Lehmer and the number of reducible factors of a polynomial*, Acta Arith. **34** (1979), 391–401.
13. Karl Pracher, *Primzahlverteilung*, Springer-Verlag, Berlin, 1957.
14. D. H. Lehmer, *Factorization of certain cyclotomic functions*, Ann. of Math. (2) **34** (1933), 461–479.
15. R. Kronecker, *Zwei Sätze über Gleichungen mit ganzzahligen Koeffizienten*, J. Reine Angew. Math. **53** (1857), 133–175.
16. B. S. Kashin and A. A. Saakyan, *Orthogonal series*, "Nauka", Moscow, 1984; English transl., Amer. Math. Soc., Providence, RI, 1989.

Translated by G. A. KANDALL

Труды Матем. Инст. Стеклова
198 (1992)

Proceedings of the Steklov
Institute of Mathematics
1994, Issue 1

SPLINE INTERPOLATION OF FUNCTIONS OF SEVERAL VARIABLES, AND BASES IN SOBOLEV SPACES

UDC 517.518.85+517.518.34

O. V. MATVEEV

INTRODUCTION

This article is devoted to questions in the approximation of functions by D^m-splines. A *D^m-spline* is defined to be a function $u \in W_2^m(\Omega)$ minimizing the functional

$$\int_\Omega \sum_{|\beta|=m} \frac{m!}{\beta!}(D^\beta u)^2\,dx \tag{1}$$

under the conditions $u(x) = f(x)$, where $x \in \Delta$ (Ω is a domain in $\mathbf{R}^n$, $\Delta \subset \Omega$, $n < 2m$). Here $\beta = (\beta_1, \dots, \beta_n)$ is a multi-index, $|\beta| = \beta_1 + \cdots + \beta_n$, $\beta! = \beta_1! \cdots \beta_n!$. A spline defined in this way depends linearly on the function f being interpolated; moreover, the operator under the integral sign is invariant with respect to motions of the space. The definition of a D^m-spline was first given by Atteia [1].

For $n = 1$ a spline turns out to be a piecewise polynomial function if the set Δ is finite. For $n \geq 2$ this is not so, and one can only claim that a spline is a polyharmonic function of order m on the set $\Omega \backslash \overline{\Delta}$. The corresponding extremal property of piecewise polynomial functions of a single variable served as a basis for the indicated extension of the concept of a "spline" to functions of several variables. Splines of a single variable, which have been studied in considerably greater detail than splines of several variables, have a number of advantages over other tools for approximating functions and occupy an important place in approximation theory (Ahlberg, Nilson, and Walsh [2], Korneĭchuk [3]).

We remark that the definition given above for a D^m-spline is a special case of the definition of a spline in a Hilbert space (Atteia [4], Laurent [5], Vasilenko [6]).

Many authors have investigated convergence of D^m-splines to the function being interpolated in various cases as $\overline{h}(\Delta) = \sup_{x\in\Omega} \inf_{y\in\Delta} |x-y| \to 0$, and inequalities of the form

$$\max_{|\beta|=l} \|D^\beta(f-s)\|_{L_q(\Omega)} \leq C\overline{h}^{\vartheta} \max_{|\beta|=k} \|D^\beta f\|_{L_p(\Omega)}, \tag{2}$$

$$\max_{|\beta|=l} \|D^\beta(f-s)\|_{L_q(\Omega)} \leq C\overline{h}^{\vartheta} \max_{|\beta|=k} \omega(D^\beta f, \overline{h})_{L_p(\Omega)} \tag{3}$$

have been obtained, where s is a spline interpolating f at the points of Δ, ω denotes the modulus of continuity, and C and ϑ are constants depending only on m, k, l, p, q, and Ω (or C depends, moreover, on the quantity $\overline{h}/\underline{h}$, where

1991 *Mathematics Subject Classification.* Primary 41A15, 41A05, 41A63; Secondary 46E35.

$\overline{h} = \overline{h}(\Delta)$, $\underline{h} = \underline{h}(\Delta) = \inf_{x,y\in\Delta,\ x\neq y}|x-y|)$. There have also been studies of D^m-splines with boundary condition, i.e., functions minimizing the functional (1) under the conditions $u \in f + \overset{\circ}{W}{}_2^m(\Omega)$ and $u(x) = f(x)$ $(x \in \Delta)$. First of all we mention the case $n = 1$, $m = 2$, i.e., cubic splines of a single variable (Birkhoff and de Boor [7]; Ahlberg, Nilson, and Walsh [2]; Marsden [8]; Lucas [9]; Stechkin and Subbotin [10]; Zav′yalov, Kvasov, and Miroshnichenko [11]). Splines of a single variable and arbitrary order m have been considered by Ahlberg, Nilson, and Walsh [2], [12]–[14], Ziegler [15], Subbotin [16]–[18], de Boor [19], and others. For the case of several variables D^m-splines have been studied by Duchon [20], [21] and Bezhaev [22], [23], who obtained estimates of the form (2) for $p = 2$ and $k = m$ (and also for certain other cases reducing to this). Imamov [24] and Vasilenko [6] have considered convergence of splines in Hilbert spaces.

A basic object of this investigation is to get estimates of the form (2) and (3) for the error in interpolation by splines, and to use the estimates to construct bases in Sobolev spaces. These questions were studied earlier by various methods. In many papers relating to the one-dimensional case use was made of the circumstances that for $n = 1$ (in contrast to the case $n \geq 2$) the explicit form of a spline can be found for certain classes of sets Δ. The property of decrease of fundamental splines for $n = 1$ has been used by de Boor [19] and Subbotin [18]; this fact holds also in the multidimensional case (see the corollary in §1). Duchon [21] and Bezhaev [22], [23] used an approach based directly on the variational principle in the multidimensional case for $p = 2$ and $k = m$ (see Remark 5). However, this approach does not work for arbitrary p and k and does not lead to estimates of the form (3). The derivation of inequalities of the form (2) and (3) for arbitrary n, p, and k requires the use of special methods and of a considerably more complicated apparatus.

In this paper the concept of a D^m-spline is defined in a more general situation than in the foregoing work of Atteia [1], Duchon [20], and Bezhaev [23]. Suppose that the domain Ω satisfies the strong cone condition, $\overline{h} < h^*$, where $h^* = h^*(m, \Omega) > 0$ is some constant, and $\widetilde{\mathscr{D}}(\Omega)$ is the set of restrictions of functions in $\mathscr{D}(\mathbf{R}^n)$ to Ω (where $\mathscr{D}(\mathbf{R}^n)$ is the set of infinitely differentiable functions on $\mathbf{R}^n$ with compact support). We show that the linear operator $f \to s$ from $\widetilde{\mathscr{D}}(\Omega)$ to $W_2^m(\Omega)$ defined above can be *extended* to a continuous operator acting from $W_p^k(\Omega)$ to $W_2^m(\Omega, \text{fin})$ (see §1) in two cases: (i) when $\underline{h} > 0$ and $W_p^k(B)$ is imbedded in $C(B)$, where B is a ball; (ii) when $W_p^k(B)$ is imbedded in $W_2^m(B)$. However, here it is no longer possible to use the spline construction given above (if Ω is unbounded and $p > 2$). In this article we give a construction, working in cases (i) and (ii), of a spline that interpolates a function $f \in W_p^k(\Omega)$.

Assume the imbeddings $W_p^k(B) \to W_q^l(B)$ and $W_2^m(B) \to W_q^l(B)$ (k and l are nonnegative integers, $1 \leq p \leq \infty$, $1 \leq q \leq \infty$), $0 < h < h^*$, $0 \leq \alpha < \alpha^*$, where $\alpha^* = \alpha^*(\Omega) > 0$ is a constant. Let

$$K_1 = \left\{ f \in W_p^k(\Omega) : \max_{|\beta|=k} \|D^\beta f\|_{L_p(\Omega)} \leq 1 \right\},$$

$$K_2 = K_2(h) = \left\{ f \in W_p^k(\Omega) : \max_{|\beta|=k} \omega(D^\beta f, h)_{L_p(\Omega)} \leq 1 \right\}$$

and denote by $\Delta(\alpha, h)$ $(\Delta'(h))$ the class of sets $\Delta \subset \Omega$ such that $\overline{h}(\Delta) = h$ and $\underline{h}(\Delta) > \alpha h$ $(\overline{h}(\Delta) = h$, respectively), where $\overline{h}$ and $\underline{h}$ are defined above. The problem of obtaining estimates of the form (2), (3) can be formulated as the problem of estimating the quantities

$$\eta_i(h) = \sup\left\{\max_{|\beta|=l}\|D^\beta(f-s)\|_{L_p(\Omega)} : f\in K_i,\ \Delta\in\Delta(\alpha, h)\right\},$$

$$\zeta_i(h) = \sup\left\{\max_{|\beta|=l}\|D^\beta(f-s)\|_{L_q(\Omega)} : f\in K_i,\ \Delta\in\Delta'(h)\right\}, \qquad i=1, 2.$$

Obviously, $0\le\eta_i\le\zeta_i\le\infty$.

We present some results. In the cases (i) and (ii), $\eta_1=\eta_2=\infty$ if Ω is unbounded and $p>q$ (Remark 3). It will be assumed below that Ω is bounded when $p>q$. Suppose that the conditions (i) hold. Assume first that $\alpha>0$. Then $\eta_1<\infty$ for $k\le m$, and $\eta_2<\infty$ for $k<m$. If Ω is bounded, then $\eta_1=\infty$ for $k>m$, and $\eta_2=\infty$ for $k\ge m$ (Theorems 1 and 2, Remark 3). On the other hand, if $\Omega=\mathbf{R}^n$, then $\eta_1, \eta_2<\infty$ for $k<2m$ (Theorem 6). For $\alpha=0$ we have the following negative result: if Ω is a cube, $k\ne m$, and $p<\infty$ when $k=m-1$, then $\eta_1=\eta_2=\infty$ (Remarks 3 and 8).

Let us now assume conditions (ii). If $k=m$, then $\zeta_1<\infty$, and if $k\ne m$ and Ω is bounded, then $\zeta_1=\infty$; $\zeta_2=\infty$ for bounded Ω (Theorem 5, Remark 3). If $\Omega=\mathbf{R}^n$, then $\zeta_1, \zeta_2<\infty$ for $k<2m$ (Theorem 6).

Further, if $\eta_i<\infty$ or $\zeta_i<\infty$, one can pose the problem of the *order* of $\eta_i(h)$ and $\zeta_i(h)$ as $h\to 0$. In all the cases above, when one of η_i or ζ_i is finite, it turns out to have the precise order t with respect to h, where

$$t=\begin{cases} k-l+n/q-n/p & \text{if } p\le q,\\ k-l & \text{if } p>q\end{cases}$$

(Theorems 1, 2, 5, and 6, and Remark 7).

For $p=2$ and $k=m$ the result $\zeta_1(h)=O(h^t)$ was obtained by Bezhaev [23].

For splines with boundary conditions analogous results are obtained that generalize results in [22] and [23]. Some of the results obtained here were also known earlier for the one-dimensional case (Ahlberg, Nilson, and Walsh [2], Subbotin [16]–[18], de Boor [19]).

In this paper we mainly employ tools from the theory of functions of several variables.

The author expresses gratitude to Yu. N. Subbotin for suggesting the topic and for his attention to this work, as well as to N. I. Chernykh for useful discussions.

1. Some auxiliary assertions

The symbol Ω will always stand for a domain in $\mathbf{R}^n$ satisfying the *strong cone condition* ([25], §8.1). A set of the form $B(a, r)=\{x\in\mathbf{R}^n : |x-a|<r\}$ will be called a *ball*. Denote the Hausdorff distance between sets F_1 and F_2 by $\rho(F_1, F_2)$:

$$\rho(F_1, F_2)=\max\left(\sup_{x\in F_1}\inf_{y\in F_2}|x-y|,\ \sup_{x\in F_2}\inf_{y\in F_1}|x-y|\right).$$

We begin with a lemma on covering the set Ω by starlike domains.

Lemma 1. *There exist an* $\varepsilon_0>0$ *and* M *and* N *depending on* Ω *such that for any* $\varepsilon\in(0, \varepsilon_0]$ *there are domains* $D_j\subset\Omega$, $j\in J$, *satisfying the following conditions*:

a) $\operatorname{diam} D_j\le M\varepsilon$;

b) D_j *is starlike with respect to some ball of radius* ε (*i.e., starlike with respect to each point of this ball*);

c) *for any* $x\in\mathbf{R}^n$ *there exists a* $j\in J$ *with* $\Omega\cap B(x, \varepsilon)\subset D_j$;

d) *any ball of radius* ε *intersects at most* N *of the sets* D_j.

Proof $(n > 1)$. We consider open cubes in $\mathbf{R}^n$ with edges of length 4ε parallel to the coordinate axes and with vertices of the form $2\varepsilon(\zeta_1, \dots, \zeta_n)$, where $\zeta_1, \dots, \zeta_n$ are integers. Let $\{K_j : j \in J\}$ be the set of all such cubes intersecting Ω.

In the case $K_j \subset \Omega$ we take $D_j = K_j$. Suppose now that $K_j \not\subset \Omega$. Choose a point $y_j \in \partial\Omega \cap K_j$. Since Ω satisfies the strong cone condition, there exist $r > 0$, $h > 0$, and $L > 1$ (depending only on Ω) and a rectangular system of coordinates $\Gamma_j = (\xi_1, \dots, \xi_{n-1}, \eta)$ with origin at y_j such that

$$T_j(x) \subset \Omega \qquad \forall x \in \Omega \cap B(y_j, r), \tag{4}$$

where the set $T_j(x)$ is defined by

$$L|\xi - \xi^{(0)}| \le \eta - \eta^{(0)} \le h,$$

for $x = (\xi^{(0)}, \eta^{(0)})$ $(\xi = (\xi_1, \dots, \xi_{n-1}),\ |\xi| = (\sum_{i=1}^{n-1} \xi_i^2)^{1/2})$.

Denote by R_j the set

$$|\xi| < 4n\varepsilon, \quad |\eta| < 11n\varepsilon$$

and by B_j the ball

$$|\xi|^2 + |\eta - 10Ln\varepsilon|^2 < \varepsilon^2,$$

and let $D_j = \Omega \cap R_j$. Then $\Omega \cap K_j \subset D_j$ for any j. The assertion a) with $M = 24Ln$ is obvious; c) follows from the fact that if $\Omega \cap B(x, \varepsilon) \ne \varnothing$, then for the corresponding j

$$B(x, \varepsilon) \subset K_j,$$
$$\Omega \cap B(x, \varepsilon) \subset \Omega \cap K_j \subset D_j.$$

If B is a ball of radius ε, then for all $j \in J$ with $B \cap D_j \ne \varnothing$

$$\rho(B, K_j) \le (M + 4n + 2)\varepsilon.$$

Since the number of cubes K_j with this condition is bounded, we get d) from this.

We show for the proof of b) that if

$$K_j \not\subset \Omega, \qquad \varepsilon \le \varepsilon_0 = \min\left(\frac{h}{22Ln}, \frac{r}{M}\right),$$

then D_j is starlike with respect to the ball B_j (and hence is a domain). Suppose that a point $a = (\xi', \eta') \in D_j$ is given (in the system of coordinates Γ_j); then

$$\mathrm{co}(\{a\} \cup B_j) \subset \Omega \tag{5}$$

(co denotes the convex hull). Indeed, in the case $\eta' \ge 4Ln\varepsilon$ the set $\mathrm{co}(\{a\} \cup B_j)$ is contained in the set

$$|\xi| < 4n\varepsilon, \quad 4Ln\varepsilon \le \eta < 11Ln\varepsilon,$$

which is contained in $\operatorname{Int} T_j(y_j)$. Since $\operatorname{Int} T_j(y_j) \subset \Omega$ in view of (4), we get (5). If $\eta' < 4Ln\varepsilon$ and $(\xi, \eta) \in B_j$, then

$$|\xi - \xi'| < 5n\varepsilon, \quad 5Ln\varepsilon < \eta - \eta' < 22Ln\varepsilon \le h,$$

which implies that $B_j \subset T_j(a)$; from this and from (4), taking into account that $a \in B(y_j, M\varepsilon) \subset B(y_j, r)$, we get that

$$\mathrm{co}(\{a\} \cup B_j) \subset T_j(a) \subset \Omega.$$

Finally, since $B_j \subset R_j$,

$$\mathrm{co}(\{a\} \cup B_j) \subset \Omega \cap R_j = D_j.$$

The lemma is proved.

Remark 1. If the domains $D_j \subset \Omega$ satisfy the conditions a)–c) of Lemma 1 and $D_{j_1} \cap D_{j_2} \neq \varnothing$, then

$$B(x_i, \varepsilon/M) \subset D_{j_i} \cap D_{j_3}, \qquad i = 1, 2,$$

for some $x_1, x_2 \in \Omega$, $j_3 \in J$.

Indeed, suppose that $a \in D_{j_1} \cap D_{j_2}$, D_{j_i} is starlike with respect to the ball $B(z_i, \varepsilon)$, $x_i = a + M^{-1}(z_i - a)$, and j_3 is such that $\Omega \cap B(a, \varepsilon) \subset D_{j_3}$. Then for $i = 1, 2$

$$B(x_i, \varepsilon/M) \subset \operatorname{co}(\{a\} \cup B(z_i, \varepsilon)) \subset D_{j_i},$$
$$\rho(\{a\}, B(x_i, \varepsilon/M)) = \frac{1}{M}\rho(\{a\}, B(z_i, \varepsilon)) \leq \frac{1}{M} \operatorname{diam} D_{j_i} \leq \varepsilon,$$

which implies that

$$B(x_i, \varepsilon/M) \subset \Omega \cap B(a, \varepsilon) \subset D_{j_3},$$

as claimed.

We remark that a bounded domain that is starlike with respect to a ball satisfies the strong cone condition.

Everywhere below, ψ denotes a particular infinitely differentiable function supported in $B(0, 1)$ such that $\psi(x) \geq 0$, $\psi(x) > 0$ for $|x| \leq 1/2$, $\psi(0) = 1$, and $\int_{\mathbf{R}^n} \psi(x)\,dx = 1$.

We consider the Sobolev space $W_p^k(D)$ consisting of the functions $f: D \to \mathbf{R}$ for which all generalized derivatives of order $\leq k$ exist and belong to $L_p(D)$, with the norm

$$\|f\|_{W_p^k(D)} = \sum_{|\beta| \leq k} \|D^\beta f\|_{L_p(D)},$$

where D is an open set in $\mathbf{R}^n$, k is a nonnegative integer, $1 \leq p \leq \infty$, and β is a multi-index (equivalent functions are identified). We often use one and the same letter to denote a function and its restriction to some set. Denote by $W_p^k(D, \text{fin})$ the space of functions in $W_p^k(D \cap B(0, r))$ for any $r > 0$, with the locally convex topology given by the family of seminorms $\{\|\cdot\|_{W_p^k(D \cap B(0,r))} : r > 0\}$. Denote by $L_p^k(D)$ the set of functions on D for which all generalized derivatives of order k exist and belong to $L_p(D)$. Let

$$\|D^k f\|_{L_p(E)} = \begin{cases} \left(\displaystyle\int_E \sum_{|\beta|=k} \frac{k!}{\beta!} |D^\beta f|^p\,dx\right)^{1/p} & \text{if } p < \infty, \\ \displaystyle\max_{|\beta|=k} \|D^\beta f\|_{L_\infty(E)} & \text{if } p = \infty \end{cases}$$

(where $f \in L_p^k(D)$, and $E \subset D$ is a measurable set). For convenience we formulate some (well-known) assertions about functions in the classes $L_p^k(D)$; below, D denotes a bounded domain, starlike with respect to the ball $B(a, 1)$ and such that $\operatorname{diam} D \leq d$.

The following relation holds for any function f infinitely differentiable on $\mathbf{R}^n$ and any multi-index γ with $|\gamma| < k$ ([26], formulas (7.12) and (10.8)):

$$D^\gamma f(x) = D^\gamma \sum_{|\beta|<k} (x-a)^\beta \int_{B(a, 1/2)} \Phi_\beta(y-a) f(y)\,dy + R(x), \tag{6}$$

where

$$R(x) = \sum_{|\beta|=k} \int_D \frac{w_{\beta\gamma}(x-a, y-a)}{|x-y|^{n-k+|\gamma|}} D^\beta f(y)\,dy, \tag{7}$$

$x^\beta = x_1^{\beta_1} \cdots x_n^{\beta_n}$, and the functions $\Phi_\beta\colon \mathbf{R}^n \to \mathbf{R}$ and $w_{\beta\gamma}\colon \mathbf{R}^{2n} \to \mathbf{R}$ are locally bounded. Here Φ_β and $w_{\beta\gamma}$ are independent of f and D. For any $b \in \mathbf{R}^n$ let U_b be the operator

$$U_b f(x) = \sum_{|\beta|<k} (x-b)^\beta \int_{B(b, 1/2)} \Phi_\beta(y-b) f(y)\,dy, \tag{8}$$

acting from $L_1(B(b, 1/2))$ into the space of polynomials of degree less than k.

Proposition 1. *If $f \in L_p^k(D)$, then $f \in W_p^k(D)$ and*

$$\|f - U_a f\|_{W_p^k(D)} \le c_0 \|D^k f\|_{L_p(D)},$$

where c_0 depends only on d, k, and n.

To verify this assertion we must represent f as the limit of $f(x) * r^{-n}\psi(x/r)$ as $r \to 0$ (on the domain $\alpha(D-a)+a$, where $0<\alpha<1$), and then estimate $\|R\|_{L_p}$ by Young's inequality.

It follows from (6) and (7) that $U_b P(x) = P(x)$ for any polynomial $P(x)$ of degree less than k; therefore, $U_a f = U_b U_a f$. This and Proposition 1 give us

Proposition 2. *Let $B(b, 1) \subset D$ and $f \in W_p^k(D)$. Then*

$$\|U_a f - U_b f\|_{L_\infty(D)} \le c_1 \|D^k f\|_{L_p(D)},$$

where $c_1 = c_1(d, k, n)$.

From Proposition 1 and from imbedding theorems ([25], Theorem 10.4) it is easy to get

Proposition 3. *Let $f \in W_2^m(D)$ and $n < 2m$. There exist constants $r' = r'(m) \in (0, 1)$ and $c_2(d, m)$ such that if*

$$\rho(\{x \in B(a, 1) : f(x) = 0\},\ B(a, 1)) \le r',$$

then

$$\|f\|_{W_2^m(D)} \le c_2 \|D^m f\|_{L_2(D)}.$$

We now assume that $\alpha > 0$ is given and $F \subset \alpha\Omega = \{\alpha x : x \in \Omega\}$, and either F or $\alpha\Omega \backslash F$ is a bounded set. For $r>0$ denote by $\Pi_a(r)$ the component of $\alpha\Omega \cap B(a, r)$ containing the point a $(a \in \alpha\Omega)$, and for $r \le 0$ let $\Pi_a(r) = \varnothing$. We write

$$\Pi_F(r) = \bigcup_{a\in F} \Pi_a(r), \qquad \Sigma_F(r) = \alpha\Omega \backslash \overline{\Pi_F(r)}.$$

The constants ε_0, M, N, and r' in Lemma 1 and Proposition 3 are used below without reference to these statements.

Lemma 2. *If $0 < r_0 \le \alpha\varepsilon_0$ and $a \in \Pi_F(r)$, then*

$$\alpha\Omega \cap B(a, r_0) \subset \Pi_F(r + M r_0).$$

Proof. We consider a covering $\{D_j : j \in J\}$ of Ω satisfying the conditions of Lemma 1 with $\varepsilon = r_0/\alpha$. By the conditions a) and c), there exists a j_0 such that

$$a \in \alpha\Omega \cap B(a, \alpha\varepsilon) \subset \alpha D_{j_0} \subset \alpha\Omega \cap B(a, \alpha M\varepsilon). \tag{9}$$

If $a \in \Pi_b(r)$, where $b \in F$, then in view of the connectedness of D_{j_0} it follows from (9) that $\alpha\Omega \cap B(a, r_0) \subset \alpha D_{j_0} \subset \Pi_a(Mr_0) \subset \Pi_b(r + Mr_0) \subset \Pi_F(r + Mr_0)$, which is what was required.

For a function $f \in L_p^k(\Omega)$ we consider the modulus of continuity of the kth-order derivatives,

$$\omega(D^k f, \delta)_{L_p(\Omega)} = \max_{|\beta|=k} \omega(D^\beta f, \delta)_{L_p(\Omega)},$$

where

$$\omega(g, \delta)_{L_p(\Omega)} = \sup_{|a|<\delta} \|g(x) - g(x+a)\|_{L_p(\Omega_a)},$$

$$g \in L_p(\Omega), \qquad \Omega_a = \{x \in \Omega : x + a \in \Omega\},$$

along with the quantity

$$\tilde{\omega}(D^k f, \delta)_{L_p(\Omega)} = \max_{|\beta|=k} \tilde{\omega}(D^\beta f, \delta)_{L_p(\Omega)},$$

where

$$\tilde{\omega}(g, \delta)_{L_p(\Omega)} = \|g(x) - g(y)\|_{L_p(X)},$$

$$X = \{(x, y) \in \mathbf{R}^{2n} : x, y \in \Omega, \ |x - y| < \delta\}.$$

Lemma 3. *Suppose that $\alpha \geq 1$, $a \in \alpha\Omega$, and $f \in W_p^k(\alpha\Omega, \mathrm{fin})$. There exists a polynomial $P_1(x)$ of degree at most k such that for any $r \geq 1$*

$$\|f - P_1\|_{W_p^k(\Pi_a(r))} \leq \kappa r^{5n+k} \tilde{\omega}(D^k f, \delta)_{L_p(\Pi_a(r+\tau))}.$$

If $k \geq 1$, then there exists a polynomial $P_2(x)$ of degree less than k such that

$$\|f - P_2\|_{W_p^k \Pi_a(r))} \leq \kappa r^{5n+k} \|D^k f\|_{L_p(\Pi_a(r+\tau))}$$

for any $r \geq 1$. Here δ, κ, and τ are constants depending only on k and the domain $\Omega \subset \mathbf{R}^n$.

Proof. Assume that $\alpha \geq M/\varepsilon_0$ (everything can be reduced to this case with the help of a similarity transformation). Let us use Lemma 1 with $\varepsilon = M/\alpha$, and let $D'_j = \alpha D_j$ and $\rho_j = \inf\{r : D'_j \cap \Pi_a(r) \neq \varnothing\}$. For some j_0 we have that $a \in D'_{j_0}$. Let $j \in J$ be arbitrary but fixed, and choose a path $\pi\colon [0, 1] \to \alpha\Omega \cap B(a, \rho_j + 1)$ such that $\pi(0) = a$ and $\pi(1) \in D'_j$. We use induction to construct a sequence $j_0, \dots, j_l$ of elements of J. If j_i has already been defined, then we set

$$\zeta_i = \sup\{\zeta \in [0, 1] : \pi(\zeta) \in D'_{j_i}\}, \qquad y_i = \pi(\zeta_i).$$

If $\pi(1) \notin D'_{j_i}$, then we choose j_{i+1} such that $y_i \in D'_{j_{i+1}}$. Further, $j_{i+1} \neq j_i$ (because $y_i \in \partial D'_{j_i}$); therefore, $\pi(1) \in D'_{j_{l-1}}$ for some j_{l-1} (the number of D'_j intersecting $\pi([0, 1])$ is finite).

Finally, $j_l = j$. Since $D_{j_i} \cap D_{j_{i+1}} \neq \varnothing$, it follows from Remark 1 that for the corresponding j'_i each of the sets $D_{j_i} \cap D_{j'_i}$, $D_{j_{i+1}} \cap D_{j'_i}$ contains a ball of radius ε/M. Let $\Omega_0 = D'_{j_0}$, $\Omega_1 = D'_{j'_0}$, $\Omega_2 = D'_{j_1}$, $\Omega_3 = D'_{j'_1}, \dots, \Omega_t = D'_{j_l}$. Then for some $b_i \in \mathbf{R}^n$

$$B(b_i, 1) \subset \Omega_i \subset \Omega_{i+1}, \tag{10}$$
$$\operatorname{diam} \Omega_i \le M^2, \tag{11}$$
$$\Omega_i \subset E_j \tag{12}$$

$(i = 0, \dots, t)$, where $E_j = \Pi_a(\rho_j + 2M^2 + 1)$, $\Omega_i = D'_{j(i)}$, $\Omega_0 = D'_{j_0}$, and $\Omega_t = D'_j$. Moreover, it will be assumed that $\Omega_i \ne \Omega_{i'}$ for $i < i'$ (otherwise the segment $\Omega_{i+1}, \dots, \Omega_{i'}$ is deleted from the sequence $\Omega_0, \dots, \Omega_t$).

Below we use auxiliary constants A_j $(i = 1, \dots, 9)$ depending only on k and Ω. Let

$$g_i^{(\beta)} = \frac{1}{|\Omega_i|} \int_{\Omega_i} D^\beta f \, dx,$$

where $0 \le i \le t$, $|\beta| = k$. Taking (10)–(12) into account and denoting by v the volume of the n-dimensional ball of radius 1, we write

$$\begin{aligned}
|g_0^{(\beta)} - g_t^{(\beta)}| &\le \sum_{i=0}^{t-1} g_i^{(\beta)} - g_{i+1}^{(\beta)} \\
&\le \sum_{i=0}^{t-1} \sum_{l=i}^{i+1} \left| \frac{1}{|\Omega_l|} \int_{\Omega_l} D^\beta f(x)\, dx - \frac{1}{v} \int_{B(b_i, 1)} D^\beta f(x)\, dx \right| \\
&= \sum_{i=0}^{t-1} \sum_{l=i}^{i+1} \frac{1}{|\Omega_l| v} \left| \iint_{\substack{x \in \Omega_l \\ y \in B(b_i, 1)}} D^\beta f(x)\, dx\, dy \right. \\
&\qquad \left. - \iint_{\substack{x \in \Omega_l \\ y \in B(b_i, 1)}} D^\beta f(y)\, dx\, dy \right| \\
&\le v^{-2} \sum_{i=0}^{t-1} \sum_{l=i}^{i+1} \iint_{x, y \in \Omega_l} |D^\beta f(x) - D^\beta f(y)|\, dx\, dy \\
&\le 2Nv^{-2} \iint_{\substack{x, y \in E_j \\ |x-y| \le M^2}} |D^\beta f(x) - D^\beta f(y)|\, dx\, dy \\
&\le 2Nv^{-2} |E_j|^{2(1-1/p)} \tilde{\omega}(D^\beta f, M^2)_{L_p(E_j)} \\
&\le A_1 (\rho_j + 1)^{2n(1-1/p)} \tilde{\omega}(D^\beta f, M^2)_{L_p(E_j)};
\end{aligned} \tag{13}$$

$$\begin{aligned}
\|D^\beta f - g_0^{(\beta)}\|_{L_p(D'_j)} &\le \|D^\beta f - g_t^{(\beta)}\|_{L_p(\Omega_t)} + \|g_0^{(\beta)} - g_t^{(\beta)}\|_{L_p(\Omega_t)} \\
&= \frac{1}{|\Omega_t|} \left\| \int_{\Omega_t} (D^\beta f(x) - D^\beta f(y))\, dy \right\|_{L_p(\Omega_t)} + |\Omega_t|^{1/p} |g_0^{(\beta)} - g_t^{(\beta)}| \\
&\le |\Omega_t|^{-1/p} \tilde{\omega}(D^\beta f, M^2)_{L_p(\Omega_t)} \\
&\quad + |\Omega_t|^{1/p} A_1 (\rho_j + 1)^{2n(1-1/p)} \tilde{\omega}(D^\beta f, M^2)_{L_p(E_j)} \\
&\le A_2 (\rho_j + 1)^{2n(1-1/p)} \tilde{\omega}(D^\beta f, M^2)_{L_p(E_j)}.
\end{aligned}$$

Let $k \ge 1$. Taking into account that Ω_i is starlike with respect to some ball $B(q_i, 1)$, we set $Q_i = U_{a_i} f$, where U_{a_i} is an operator of the form (8). Since any ball of radius $r \ge 1$ intersects at most $A_3 r^n$ of the sets D'_j, it follows from (12) that

$t+1 \le A_3(\rho_j + 2M^2 + 1)^n$. From this, (10)–(12), and Propositions 1 and 2,

$$
\begin{aligned}
\|Q_0 - Q_t\|_{W_p^k(\Omega_t)} &\le \sum_{i=0}^{t-1} \|Q_i - Q_{i+1}\|_{W_p^k(\Omega_t)} \\
&\le \sum_{i=0}^{t-1} A_4 \rho(\{b_i\}, \Omega_t)^{k-1} \|Q_i - Q_{i+1}\|_{L_\infty(B(b_i, 1))} \\
&\le \sum_{i=0}^{t-1} A_4 (2\rho_j + 4M^2 + 2)^{k-1} (\|U_{a_i} f - U_{b_i} f\|_{L_\infty(B(b_i,1))} \\
&\qquad\qquad + \|U_{a_{i+1}} f - U_{b_i} f\|_{L_\infty(B(b_i,1))}) \\
&\le 2A_4 c_1 (2\rho_j + 4M^2 + 2)^{k-1} \sum_{i=0}^{t} \|D^k f\|_{L_p(\Omega_i)} \\
&\le A_5 (\rho_j + 1)^{n+k-1} \|D^k f\|_{L_p(E_j)}, \\
\|f - Q_0\|_{W_p^k(D_j')} &\le \|f - Q_t\|_{W_p^k(\Omega_t)} + \|Q_0 - Q_t\|_{W_p^k(\Omega_t)} \\
&\le (A_5 + c_0)(\rho_j + 1)^{n+k-1} \|D^k f\|_{L_p(E_j)},
\end{aligned} \tag{14}
$$

where $c_i = c_i(M^2, k, n)$, $i = 0, 1$.

Suppose now that $r \ge 1$, $J' = \{j \in J : \rho_j < r\}$, and $E = \Pi_a(r + 2M^2 + 1)$. Since the cardinality of J' does not exceed $A_3 r^n$ and $\Pi_a(r) \subset \bigcup_{j \in J'} D_j'$, by (13) and (14) we have

$$
\|D^\beta f - g_0^{(\beta)}\|_{L_p(\Pi_a(r))} \le A_3 r^n \max_{j \in J'} \|D^\beta f - g_0^{(\beta)}\|_{L_p(D_j')} \le A_6 r^{3n} \tilde{\omega}(D^\beta f, M^2)_{L_p(E)}, \tag{15}
$$

and for $k \ge 1$

$$
\|f - Q_0\|_{W_p^k(\Pi_a(r))} \le A_3 r^3 \max_{j \in J'} \|f - Q_0\|_{W_p^k(D_j')} \le A_7 r^{2n+k} \|D^k f\|_{L_p(E)} \tag{16}
$$

($b_0^{(\beta)}$ and Q_0 do not depend on j). Setting $P_0(x) = \sum_{|\beta|=k} (g_0^{(\beta)}/\beta!) x^\beta$, from (15) we get

$$
\|D^k (f - P_0)\|_{L_p(\Pi_a(r))} \le A_8 r^{3n} \tilde{\omega}(D^k f, M^2)_{L_p(E)}. \tag{17}
$$

The last inequality leads to the necessary result in the case $k = 0$ if we set $P_1 = P_0$. In the case $k \ge 1$ we use the polynomials $P_0 = P_0(f)$ and $Q_0 = Q_0(f)$ to construct the polynomials P_1 and P_2:

$$
P_1 = P_0(f) + Q_0(f - P_0(f)), \qquad P_2 = Q_0(f).
$$

According to (16) and (17),

$$
\begin{aligned}
\|f - P_1\|_{W_p^k(\Pi_a(r))} &= \|(f - P_0(f)) - Q_0(f - P_0(f))\|_{W_p^k(\Pi_a(r))} \\
&\le A_7 r^{2n+k} \|D^k (f - P_0(f))\|_{L_p(E)} \le A_9 r^{5n+k} \tilde{\omega}(D^k f, M^2)_{L_p(\Pi_a(r + 4M^2 + 2))}.
\end{aligned}
$$

The required result follows from this and (16). The lemma is proved.

We introduce in $L_p^k(\Omega)$ the norm

$$
\|f\|_{L_p^k(\Omega)} = \|f\|_{L_p(B)} + \|D^k f\|_{L_p(\Omega)}, \tag{18}
$$

where B is some ball, $B \subset \Omega$.

Lemma 4. *Suppose that* $f \in L_p^k(\Omega)$, $a \in \Omega$, $\zeta > 1$, $r > 0$, *and* Ω' *is a bounded open subset of* Ω. *Then*

$$\|f\|_{W_p^k(\Pi_a(r))} \leq c_3 \zeta^r \|f\|_{L_p^k(\Omega)},$$
$$\|f\|_{W_p^k(\Omega')} \leq c_3 \|f\|_{L_p^k(\Omega)},$$

where c_3 *depends only on* k, ζ, a, Ω, Ω', *and* B $(\alpha = 1)$.

The proof follows easily from Lemmas 2 and 3.

It follows from Lemma 4 that the topology of $L_p^k(\Omega)$ given by the norm (18) does not depend on B and that $L_p^k(\Omega)$ is a complete space. Denote by $C^k(\Omega)$ the subspace of $L_\infty^k(\Omega)$ consisting of the functions with uniformly continuous kth-order derivatives on Ω, and let $C(\Omega) = C^0(\Omega)$.

Lemma 5. *Assume that* $G \subset L_2^m(\alpha\Omega)$, $n < 2m$, $r_1 > 0$, *and the following conditions hold*:

a) $\rho(\{x \in \alpha\Omega : g(x) = 0\}, \alpha\Omega) \leq r_1$ *for any* $g \in G$.

b) *If* $g_0 \in W_\infty^m(\alpha\Omega)$ *is an infinitely differentiable function,* $g_0(x) = 1$ *on* F, *and* $g_0(x)$ *is constant on* $\alpha\Omega \backslash T$, *where* T *is a bounded set, then* $g g_0 \in G$ *for any* $g \in G$.

c) *There exists a* $u \in G$ *such that*

$$\|D^m u\|_{L_2(\alpha\Omega)} = \inf_{g \in G} \|D^m g\|_{L_2(\alpha\Omega)}.$$

Then there exist α_0, μ, *and* ν *(depending only on* m *and* Ω*) such that* $0 < \nu < 1$, *and in the case* $\alpha \geq \alpha_0 r_1$

$$\|u\|_{W_2^m(\Sigma_F(r))} \leq \mu \nu^{r/r_1} (1 + r_1)^m \|D^m u\|_{L_2(\alpha\Omega)}, \tag{19}$$

for any $r \in \mathbf{R}$.

Proof. We carry out the proof for $r_1 = r'/2$ (the general case can be reduced to this one by using the transformation $x \to (r'/2r_1)x$). The auxiliary constants A_i used below depend only on m and Ω $(i = 1, \dots, 4)$. Let $\alpha_0 = 2/\varepsilon_0 r'$ and $\alpha \geq \varepsilon_0^{-1}$. We fix an integer $t > 0$ and let

$$F_1 = \Pi_F(3Mt - 2M), \qquad F_2 = \alpha\Omega \cap \overline{\Pi_F(3Mt - M)},$$
$$F_3 = \Sigma_F(3Mt - M).$$

If $x \in F_1$, then by Lemma 2 $(r_0 = 1)$, $F_3 \cap B(x, 1) = \varnothing$. Therefore, if $\chi(x)$ is the characteristic function of the set of points with distance $< 1/2$ from F_1 and if $\varphi(x) = 2^n \psi(2x) * \chi(x)$ (the function ψ was defined above), then

$$\varphi(x) = \begin{cases} 1 & \text{for } x \in F_1, \\ 0 & \text{for } x \in F_3. \end{cases} \tag{20}$$

Suppose that $\{D_j : j \in J\}$ is a covering of Ω satisfying the conditions of Lemma 1 with $\varepsilon = \alpha^{-1}$, $D'_j = \alpha D_j$, B_j is a ball of radius 1 with respect to which D'_j is starlike, and

$$H_t = \Pi_F(3Mt) \backslash \overline{\Pi_F(3Mt - 3M)},$$
$$J_1 = \{j \in J : D'_j \not\subset F_1, \ D'_j \cap F_2 \neq \varnothing\},$$
$$J_2 = \{j \in J : D'_j \cap \Sigma_F(3Mt + M) \neq \varnothing\}.$$

Since $\operatorname{diam} D'_j \leq M$ and D'_j is connected, it follows that

$$\bigcup_{j\in J_1} D'_j \subset H_t, \tag{21}$$

$$\bigcup_{j\in J_2} D'_j \subset \Sigma_F(3Mt). \tag{22}$$

By the condition a) of Lemma 5 ($r_1 = r'/2$, $r' < 1$),

$$\rho(\{x \in B_j : u(x) = 0\}, B_j) \leq r'.$$

Since $\|\varphi\|_{W^m_\infty(\mathbf{R}^n)} \leq A_1$, Propositions 1 and 3 give us

$$\|D^m(\varphi u)\|_{L_2(D'_j)} \leq A_2\|u\|_{W^m_2(D'_j)} \leq A_2c_2\|D^m u\|_{L_2(D'_j)}, \tag{23}$$

where $c_2 = c_2(M, m)$.

From (20) and condition b) of Lemma 5 it follows that $\varphi u \in G$ (one of the sets F and $\alpha\Omega\backslash F$ is bounded). By condition c), $\|D^m u\|_{L_2(\alpha\Omega)} \leq \|D^m(\varphi u)\varepsilon\|_{L_2(\alpha\Omega)}$. Using (20), (21), (23), and the fact that each point is covered by at most N sets D'_j, we write

$$\begin{aligned}\|D^m u\|^2_{L_2(\alpha\Omega)} &\leq \|D^m(\varphi u)\|^2_{L_2(\alpha\Omega)} = \|D^m(\varphi u)\|^2_{L_2(F_1)} + \|D^m(\varphi u)\|^2_{L_2(F_2\backslash F_1)} \\ &\leq \|D^m u\|^2_{L_2(F_1)} + \sum_{j\in J_1} \|D^m(\varphi u)\|^2_{L_2(D'_j)} \\ &\leq \|D^m u\|^2_{L_2(F_1)} + A_2^2c_2^2 \sum_{j\in J_1} \|D^m u\|^2_{L_2(D'_j)} \\ &\leq \|D^m u\|^2_{L_2(\Pi_F(3Mt))} + A_3\|D^m u\|^2_{L_2(H_t)}.\end{aligned}$$

From this,

$$\|D^m u\|^2_{L_2(\Sigma_F(3Mt))} \leq \|D^m u\|^2_{L_2(\alpha\Omega)} - \|D^m u\|^2_{L_2(\Pi_F(3Mt))} \leq A_3\|D^m u\|^2_{L_2(H_t)};$$

letting $v_i = \|D^m u\|^2_{L_2(\Sigma_F(3Mi))}$ and using the fact that $H_t \subset \Sigma_F(3Mt-3M)\backslash\Sigma_F(3Mt)$, we obtain

$$\begin{gathered} v_t \leq A_3(v_{t-1} - v_t), \qquad v_t \leq \frac{A_3}{A_3+1} v_{t-1}, \\ v_t \leq \left(\frac{A_3}{A_3+1}\right)^t v_0 = \left(\frac{A_3}{A_3+1}\right)^t \|D^m u\|^2_{L_2(\alpha\Omega)}. \end{gathered} \tag{24}$$

In view of (22)–(24)

$$\begin{aligned}\|u\|^2_{W^m_2(\Sigma_F(3Mt+M))} &\leq A_4 \sum_{j\in J_2} \|u\|^2_{W^m_2(D'_j)} \leq A_4c_2^2 \sum_{j\in J_2} \|D^m u\|^2_{L_2(D'_j)} \\ &\leq A_4c_2^2 N v_t \leq A_4c_2^2 N \left(\frac{A_3}{A_3+1}\right)^t \|D^m u\|^2_{L_2(\alpha\Omega)}.\end{aligned} \tag{25}$$

According to (23),

$$\|u\|^2_{W^m_2(\alpha\Omega)} \leq A_4 \sum_{j\in J} \|u\|^2_{W^m_2(D'_j)} \leq A_4c_2^2 \sum_{j\in J} \|D^m u\|^2_{L_2(D'_j)} \leq A_4c_2^2 N\|D^m u\|^2_{L_2(\alpha\Omega)}. \tag{26}$$

An inequality of the form (19) with $r_1 = r'/2$ follows from (25) and (26). The lemma is proved.

Now let $\Delta \subset \alpha\Omega$ and $a \in \Delta$.

Corollary. *Assume that* $\rho(\Delta, \alpha\Omega) = r_2$, $\alpha \geq \alpha_1 r_2$, $\Delta \cap B(a, r_3) = \{a\}$, *and*

$$u = \arg\min\{\|D^m g\|_{L_2(\alpha\Omega)} : g \in L_2^m(\alpha\Omega),\ \ g(a) = 1,\ g(x) = 0\ \forall x \in \Delta\backslash\{a\}\}.$$

Then for any $r \in \mathbf{R}$

$$\|u\|_{W_2^m(\Sigma_a(r))} \leq \mu_1 \nu_1^{r/r_2}(1 + r_2)^m r_3^{n/2-m},$$

where $0 < \nu_1 < 1$. *Here* α_1, μ_1, *and* ν_1 *are constants depending only on* m *and* Ω.

Proof. Let $\alpha_1 = \max(4/\varepsilon_0, 6\alpha_0)$. It is not hard to show that $\rho(\Delta\backslash\{a\}, \alpha\Omega) \leq 6r_2$. The required assertion follows from Lemma 4, Lemma 5 ($F = \{a\}$, $r_1 = 6r_2$) and the obvious inequality $\|D^m u\|_{L_2(\alpha\Omega)} \leq A_1 r_3^{n/2-m}$, where $A_1 = A_1(m)$.

2. Estimates of the error of interpolation by D^m-splines

Suppose that $\Delta \subset \Omega$, $n < 2m$ (here the functions in $L_2^m(\Omega)$ are continuous (Theorem 10.4 in [25])), and

$$H = \{g \in L_2^m(\Omega) : g(x) = 0\ \forall x \in \Delta\}, \qquad \overline{h} = \overline{h}(\Delta) = \rho(\Delta, \Omega),$$
$$\underline{h} = \underline{h}(\Delta) = \inf\{|x - y| : x, y \in \Delta,\ x \neq y\}.$$

Definition 1. If $f \in L_2^m(\Omega)$, then an *interpolation* D^m*-spline* $s = s(f, \Delta, m, \Omega)$ is defined to be an element of the set $G_f = f + H$ minimizing the functional $\|D^m g\|_{L_2(\Omega)}$ (on this set).

We show that a minimizing element exists. It follows from Proposition 1 that the topology of the quotient space $\mathscr{H} = L_2^m(\Omega)/\mathscr{P}_{m-1}$ (where $\mathscr{P}_{m-1}$ is the space of polynomials of degree less than m) is determined by the norm $\|D^m g\|_{L_2(\Omega)}$. Since G_f is closed in $L_2^m(\Omega)$, it follows that $\pi(G_f)$ is closed in $\mathscr{H}$, where $\pi: L_2^m(\Omega) \to \mathscr{H}$ is the natural homomorphism. If $\overline{s}$ is an element of the manifold $\pi(G_f)$ with minimal norm $\|D^m g\|_{L_2(\Omega)}$ (in the Hilbert space $\mathscr{H}$), then the minimum of $\|D^m g\|_{L_2(\Omega)}$ for $g \in G_f$ is attained on the set $\pi^{-1}(\overline{s}) \cap G_f$. Obviously, the minimizing element in Definition 1 is unique if there does not exist a nonzero polynomial of degree $< m$ that equals zero on Δ. Thus, the spline s exists and is unique if $\overline{h} < h^*$, where $h^* = h^*(m, \Omega) > 0$ is some constant.

The definition of a spline in this form was given by Atteia [1], Duchon [20], and Bezhaev [23].

It will be assumed everywhere below that $\overline{h} < h^*$.

Remark 2. The foregoing gives us that the spline s is determined by the conditions

$$s \in G_f, \qquad \int_\Omega \sum_{|\beta|=m} \frac{m!}{\beta!} D^\beta s D^\beta g\, dx = 0 \quad \forall g \in H,$$

and that the operator $S: f \to s(f, \Delta, m, \Omega)$ from $L_2^m(\Omega)$ to $L_2^m(\Omega)$ is linear.

In this section we define continuous linear operators $S_{k_p}: L_p^k(\Omega) \to W_2^m(\Omega, \text{fin})$ in two cases:

(i) $\underline{h} > 0$, and $W_p^k(B(0, 1))$ is imbedded in $C(B(0, 1))$;

(ii) $W_p^k(B(0, 1))$ is imbedded in $W_2^m(B(0, 1))$.

Here any two of the operators $\{S, S_{kp}\}$ coincide on their common domain. We remark that the condition $\underline{h} > 0$ is essential for the existence of the operators S_{kp} when $k \leq m - 2$. Indeed, suppose that Ω is the cube $(0, 1)^n$, $x = (x_1, \dots, x_n)$, $\Delta = \{x \in \Omega : x_1 < 1 - h^*/2\}$ (take $h^* < 1$), $k \leq m - 2$, $g(y) = y^{m-2}$ for $y > 0$,

$g(y) = 0$ for $y \leq 0$, $g_\varepsilon = g * \frac{1}{\varepsilon}\psi(x/\varepsilon)$, and $f_\varepsilon(x) = g_\varepsilon(x_1 - 1 + h^*/2)$. Then $g_\varepsilon^{(m-1)}(0) = (m-2)!\varepsilon^{-1}$, and for $x_1 > 1 - h^*/2$

$$s(f_\varepsilon, \Delta, m, \Omega)(x) = \sum_{j=0}^{m-1} \frac{1}{j!} g_\varepsilon^{(j)}(0) \left(x_1 - 1 + \frac{h^*}{2}\right)^j,$$

so that $\|f_\varepsilon\|_{W_p^k(\Omega)} = O(1)$, $\|s(f_\varepsilon, \Delta, m, \Omega)\|_{W_2^m(\Omega)} \to \infty$ as $\varepsilon \to 0$.

It is not hard to verify that the strong cone condition for Ω is equivalent to the existence of a $d > 0$ and an L' such that

$$\Omega \cap B(x, d) = \Lambda_x \cap B(x, d), \tag{27}$$

where Λ_x is a special Lipschitz domain ([27], Chapter VI, §3.2) whose Lipschitz constant does not exceed L'. In what follows it is assumed (in addition) that

$$h^* < \min(1, d/2, \varepsilon_0/4n, 1/6\alpha_0, \alpha_1^{-1}) \tag{28}$$

(the constants α_0 and α_1 are taken from Lemma 5 and the corollary in §1).

Definition 2. If $f \in L_p^k(\Omega)$, and one of the cases (i) or (ii) holds, then an interpolation D^m-spline will be defined to be a function $s = s(f, \Delta, m, \Omega) \in W_2^m(\Omega, \text{fin})$ satisfying the following conditions:

a) if $g \in L_2^m(\Omega)$, g has bounded support, and $g(x) = 0$ on Δ, then

$$\int_\Omega \sum_{|\beta|=m} \frac{m!}{\beta!} D^\beta s D^\beta g\, dx = 0;$$

b) $s(x) = f(x)$ on Δ;

c) if $\zeta > 1$ and $a \in \Omega$, then $\|s\|_{W_2^m(\Pi_a(r))} = O(\zeta^r)$ as $r \to \infty$ $(\alpha = 1)$.

According to Remark 2 and Lemma 4, a spline in the sense of Definition 1 satisfies the conditions a)–c). The existence of a spline s in the sense of Definition 2 is established in Theorems 1 and 4.

Lemma 6. *There cannot exist two different functions s satisfying the conditions of Definition 2.*

Proof. Suppose that $f = 0$, s satisfies the conditions of Definition 2, $a \in \Omega$, $r > 0$, $\chi(x)$ is the characteristic function of the set of points with distance $< 2\varepsilon_0/3$ from $\Pi_a(r)$, $\varphi(x) = (3/\varepsilon_0)^n \psi(3x/\varepsilon_0) * \chi(x)$ (the function ψ was defined above), and $s_r = \varphi s$. Then $s(x) = s_r(x)$ on F', where

$$F' = \Omega \cap \left(\bigcup_{x \in \Pi_a(r)} B(x, \varepsilon_0/3) \right).$$

If $g \in L_2^m(\Omega)$ and $g(x) = 0$ on the set $\Delta \cup \Sigma_a(r)$, then, since $F' \cup \Sigma_a(r) = \Omega$, by the condition a) we have

$$\int_\Omega \sum_{|\beta|=m} \frac{m!}{\beta!} D^\beta s_r D^\beta g\, dx = \int_\Omega \sum_{|\beta|=m} \frac{m!}{\beta!} D^\beta s D^\beta g\, dx = 0,$$

which implies that $\|D^m s_r\|_{L_2(\Omega)} = \inf_{g \in G} \|D^m g\|_{L_2(\Omega)}$, where

$$G = \{g \in L_2^m(\Omega) \colon g(x) = s_r(x)\ \forall x \in \Delta \cup \Sigma_a(r)\}.$$

By Lemma 2 $(\alpha = 1, r_0 = \varepsilon_0)$

$$\Omega \cap \operatorname{supp} \varphi \subset \Omega \cap \left(\bigcup_{x \in \Pi_a(r)} B(x, \varepsilon_0) \right) \subset \Pi_a(r + M\varepsilon_0). \tag{29}$$

Using Lemma 4 and Lemma 5 with $\alpha = 1$, $F = \Sigma_a(r)$, $r_1 = \overline{h}$ ($\alpha = r_1\overline{h}^{-1} > \alpha_0 r_1$ in view of (28)), we get

$$\|s\|_{W_2^m(\Pi_a(1))} \leq \|s_r\|_{W_2^m(\Sigma_F^{(r-1)})} \leq \mu\nu^{(r-1)/\overline{h}}(1+\overline{h})^m \|D^m s_r\|_{L_2(\Omega)}$$
$$\leq \mu\nu^{(r-1)/\overline{h}}(1+\overline{h})^m A_1 \|s\|_{W_2^m(\Pi_a(r+M\varepsilon_0))} \to 0$$

as $r \to \infty$ by virtue of (29) and the condition c) (here $A_1 = A_1(m, \Omega)$). Thus, $s = 0$. The lemma is proved.

Let us proceed to the estimates of the interpolation error. It will be assumed that: $\underline{h} > 0$; either $f \in L_p^k(\Omega)$ and $W_p^k(B(0, 1))$ is imbedded in $C(B(0, 1))$, or $f \in C(\Omega)$, $k = 0$, and $p = \infty$; we have the imbeddings

$$W_p^k(B(0, 1)) \to W_q^l(B(0, 1)), \qquad W_2^m(B(0, 1)) \to W_q^l(B(0, 1));$$

and if Ω is unbounded, then $p \leq q$ (see Remark 3 about the last condition). Let

$$\vartheta = \vartheta(n, k, l, p, q) = \begin{cases} k - l + n/q - n/p & \text{if } p \leq q, \\ k - l & \text{if } p > q. \end{cases} \tag{30}$$

Theorem 1. *There exists a unique spline s satisfying the conditions of Definition 2, and the linear operator $f \to s(f, \Delta, m, \Omega)$ acting from $L_p^k(\Omega)$ (or $C(\Omega)$) to $W_2^m(\Omega, \text{fin})$ is continuous. If $k < m$, then*

$$\|D^l(f - s(f, \Delta, m, \Omega))\|_{L_q(\Omega)} \leq C_1 \lambda^{-18m} \overline{h}^{\vartheta} \omega(D^k f, \lambda\overline{h})_{L_p(\Omega)}$$

for any λ with $0 < \lambda \leq \underline{h}\overline{h}^{-1}$, where $C_1 = C_1(m, p, q, \Omega)$ is a constant.

Proof. The auxiliarly constants A_i used below depend only on m, p, q, and Ω; δ is the constant in Lemma 3, and we assume that $\delta > 3$. Let $\alpha = \delta/\lambda\overline{h}$, and let Z be the set of points of the form $\frac{1}{n}(\zeta_1, \ldots, \zeta_n)$ (where the ζ_i are integers) with distance < 1 from $\alpha\Omega$. We show that if

$$I_z = \{x \in Z : |x - z| < 2\}, \qquad V_z = \alpha\Omega \cap B(z, 1), \qquad V_z' = \bigcup_{x \in I_z} V_x,$$

where $z \in Z$, then

$$\|g\|_{W_q^l(V_z')} \leq A_1 \|g\|_{W_p^k(V_z')} \quad \forall g \in W_p^k(V_z'),$$
$$\|g\|_{W_q^l(V_z')} \leq A_1 \|g\|_{W_2^m(V_z')} \quad \forall g \in W_2^m(V_z'),$$
$$\|g\|_{L_\infty(V_z)} \leq A_1 \|g\|_{W_p^k(V_z')} \quad \forall g \in W_p^k(V_z')$$

(recall that $A_1 = A_1(m, p, q, \Omega)$). We establish the first of these inequalities (the other two can be verified similarly). Let

$$\varphi_z(x) = \psi(x - z) \left(\sum_{z_1 \in Z} \psi(x - z_1) \right)^{-1}$$

(ψ is defined in §1), $g \in W_p^k(V_z')$, $b \in I_z$, and $g_b = g\varphi_b$. Note that

$$\lambda \leq \underline{h}\overline{h}^{-1} \leq 2 \tag{31}$$

(according to (28) and Lemma 1), and $B(x, 4\overline{h}) \subset B(x, \varepsilon_0) \subset \Omega$ for some x. We choose points y and y' such that $y \in \Delta \cap B(x, h)$ and $|y - y'| = (1 + \varepsilon)\overline{h}$, where

$0 < \varepsilon < 1$. Then $y' \in \Omega$; therefore, there exists a $y'' \in \Delta \cap \overline{B(y', \bar{h})}$. We have that $y \neq y''$ and $|y - y''| \leq (2+\varepsilon)\bar{h}$, from which $\underline{h} \leq 2\bar{h}$. Assume that

$$B(b, 1) \not\subset \alpha\Omega, \tag{32}$$

and let $\alpha a \in \partial(\alpha\Omega) \cap B(b, 1)$. By (28) and (31), $\alpha d = \delta d/\lambda \bar{h} > 2$, which implies that $B(b, 1) \subset \alpha B(a, d)$. From this and (27),

$$\alpha\Lambda_a \cap B(b, 1) = V_b .$$

Setting $g_b(x) = 0$ on the set $\alpha\Lambda_a \backslash B(b, 1)$, we get a function $g_b \in W_p^k(\alpha\Lambda_a)$. We then extend g_b to $\mathbf{R}^n$ by the method of Stein ([27], Chapter 6, Theorem 5′):

$$\|\tilde{g}_b\|_{W_p^k(\mathbf{R}^n)} \leq A_2 \|g_b\|_{W_p^k(\alpha\Lambda_a)}, \tag{33}$$

where $\tilde{g}_b(x) = g_b(x)$ on $\alpha\Lambda_a$, and A_2 depends on m $(> n/2, \geq k)$ and L'. Since $W_p^k(\mathbf{R}^n)$ is imbedded in $W_{q'}^l(\mathbf{R}^n)$, where $q' = \max(p, q)$, we get from (33) that

$$\|g_b\|_{W_q^l(V_b)} \leq A_3 \|g_b\|_{W_p^k(V_b)} .$$

But this inequality is valid if (32) does not hold. Finally,

$$\begin{aligned} \|g\|_{W_q^l(V_z)} &\leq \Big\|\sum_{b\in I_z} g_b\Big\|_{W_q^l(V_z)} \leq \sum_{b\in I_z} \|g_b\|_{W_q^l(V_b)} \\ &\leq A_3 \sum_{b\in I_z} \|g_b\|_{W_p^k(V_b)} \leq A_4 \|g\|_{W_p^k(V_z')} . \end{aligned}$$

For any $z \in Z$ we now choose a point $z' \in V_z$ and let $\Pi_z'(r) = \Pi_{z'}(r)$, $\Sigma_z'(r) = \Sigma_{z'}(r)$, and

$$r_{yz} = \inf\{r : y \in \Pi_z'(r)\} + 5M . \tag{34}$$

Take an arbitrary $y \in \alpha\Delta$; then $y \in V_b$ for some b. By Lemma 2 ($F = \{z'\}$, $r_0 = 4$, and $\alpha\varepsilon_0 > \bar{h}^{-1}\varepsilon_0 > r_0$ according to (28)), it follows from $y \in \Pi_z'(r_{yz} - 4M)$ that

$$V_b'\alpha\Omega \cap B(y, 4) \subset \Pi_z'(r_{yz}). \tag{35}$$

Similarly

$$V_z' \subset \alpha\Omega \cap B(z', 4) \subset \Pi_z'(5M). \tag{36}$$

Since $\Pi_z'(5M) \cap \Pi_y(r_{yz}/2 - 5M) = \varnothing$ (otherwise we would have that

$$y \in \Pi_y\left(\frac{r_{yz}}{2} - 5M\right) \subset \Pi_z'\left(5M + 2\left(\frac{r_{yz}}{2} - 5M\right)\right) = \Pi_z'(r_{yz} - 5M),$$

which contradicts (34)), it follows from (36) that

$$V_z' \subset \Sigma_y(r_{yz}/2 - 5M). \tag{37}$$

Let

$$s_y = \arg\min\{\|D^m g\|_{L_2(\alpha\Omega)} : g \in L_2^m(\alpha\Omega),\ g(y) = 1,\ g(x) = 0\ \forall x \in \alpha\Delta\backslash\{y\}\},$$

where $y \in \alpha\Omega$ (Definition 1). By the corollary in §1 ($r_2 = \delta/\lambda$, $r_3 = \delta\underline{h}/\lambda\bar{h} \geq \delta$, $\alpha > \alpha_1\delta/\lambda$),

$$\|s_y\|_{W_2^m(\Sigma_y(r))} \leq \mu_1 \nu_2^{4\lambda r}(1 + \delta/\lambda)^m \delta^{n/2-m}, \tag{38}$$

where $\nu_2 = \nu_1^{1/4\delta}$. From (31), (37), and (38),

$$\|s_y\|_{W_q^l(V_z)} \le A_1 \|s_y\|_{L_2^m(V_z')} \le A_5 \lambda^{-m} \nu_2^{2\lambda r_{yz}}. \tag{39}$$

Fix some $a \in \alpha\Omega$ and $\zeta \in (1, \nu_2^{-2\lambda})$. If $z' \in \Pi_a(r)$, then $V_b' \subset \Pi_z'(r_{yz}) \subset \Pi_a(r + r_{yz})$ by (35); therefore, letting $f_1(x) = f(x/\alpha)$, we have by Lemma 4 that

$$\begin{aligned} |f_1(y)| &\le \|f_1\|_{L_\infty(V_b)} \le A_1 \|f_1\|_{W_p^k(V_b')} \\ &\le A_1 \|f_1\|_{W_p^k(\Pi_a^{(r+r_{yz})})} \le A_1' \zeta^{r+r_{yz}} \|f_1\|_{L_p^k(\alpha\Omega)}. \end{aligned} \tag{40}$$

Here and below, we denote by A_i' certain constants independent of f, y, z, and r. The number of $y \in \alpha\Delta$ with the condition $r_{yz} < r$ does not exceed $A_6 r^n$, where $r > 0$ $(\alpha \underline{h} \ge \delta)$; therefore it follows from (39) and (40) that

$$\begin{aligned} \sum_{y\in\alpha\Delta} |f_1(y)|\, \|s_y\|_{W_2^m(V_z)} &\le A_2' \sum_{i=1}^{\infty} \sum_{y\in\alpha\Delta\,:\, i-1\le r_{yz}<i} \nu_2^{2\lambda r_{yz}} \zeta^{r+r_{yz}} \|f_1\|_{L_p^k(\alpha\Omega)} \\ &\le A_2' A_6 \zeta^r \sum_{i=1}^{\infty} \nu_2^{2\lambda(i-1)} \zeta^i i^n \|f_1\|_{L_p^k(\alpha\Omega)} \le A_3' \zeta^r \|f_1\|_{L_p^k(\alpha\Omega)} \end{aligned} \tag{41}$$

(if $z' \in \Pi_a(r)$). If $V_z \cap \Pi_a(r) \neq \varnothing$, then $V_z \subset \Pi_a(r + 2M)$ by Lemma 2 $(r_0 = 2$, $\alpha\varepsilon_0 > \overline{h}^{-1}\varepsilon_0 > r_0)$, hence,

$$\Pi_a(r) \subset \bigcup_{z'\in\Pi_a(r+2M)} V_z. \tag{42}$$

From (41) and (42),

$$\|s_1\|_{W_2^m(\Pi_a(r))} \le A_4' \zeta^{r/2} \|f_1\|_{L_p^k(\alpha\Omega)},$$

where

$$s_1(x) = \sum_{y\in\alpha\Delta} f_1(y) s_y(x)$$

(if Δ is infinite, then the series converges unconditionally in $W_2^m(\Omega, \text{fin})$). Passing to functions on Ω : $s(x) = s_1(\alpha x)$ and using Remark 2, we get that s satisfies the conditions of Definition 2, and in view of (41) the operator $f \to s$ from $L_p^k(\Omega)$ (or $C(\Omega)$) to $W_2^m(\Omega, \text{fin})$ is continuous (the uniqueness of s follows from Lemma 6).

Suppose now that $k < m$. Applying Lemma 3 to the function f_1 for $a = z'$ $(\alpha > \overline{h}^{-1} > 1)$ and letting $f_0 = f_1 - P_1$, we get by (35) and (36)

$$\begin{aligned} |f_0(y)| &\le \|f_0\|_{L_\infty(V_b)} \le A_1 \|f_0\|_{W_p^k(V_b')} \le A_1 \|f_0\|_{W_p^k(\Pi_z'(r_{yz}))} \\ &\le A_7 r_{yz}^{5n+k} \tilde{\omega}(D^k f_1, \delta)_{L_P(\Pi_z'(r_{yz}+\tau))}, \end{aligned} \tag{43}$$

$$\|f_0\|_{W_q^l(V_z)} \le A_1 \|f_0\|_{W_p^k(V_z')} \le A_1 \|f_0\|_{W_p^k(\Pi_z'(5M))} \le A_8 \tilde{\omega}(D^k f_1, \delta)_{L_p(\Pi_z'(5M+\tau))}. \tag{44}$$

From (39) and (43),

(45)
$$\begin{aligned}
&\sum_{y\in\alpha\Delta} |f_0(y)|\, \|s_y\|_{W_q^l(V_z)} \\
&\le A_5A_7\lambda^{-m}\sum_{i=1}^{\infty}\ \sum_{y\in\alpha\Delta\,:\,i-1\le r_{yz}<i} r_{yz}^{5n+k}\nu_2^{2\lambda r_{yz}}\tilde{\omega}(D^k f_1\,,\,\delta)_{L_p(\Pi'_z(r_{yz}+\tau))} \\
&\le A_5A_6A_7\lambda^{-m-6n-k}\nu_2^{-2\lambda}\sum_{i=1}^{\infty}(\lambda i)^{6n+k}\nu_2^{2\lambda i}\tilde{\omega}(D^k f_1\,,\,\delta)_{L_p(\Pi'_z(i+\tau))} \\
&\le A_9\lambda^{-15m}\sum_{i=1}^{\infty}\nu_2^{\lambda i}\tilde{\omega}(D^k f_1\,,\,\delta)_{L_p(\Pi'_z(i+\tau))}\,.
\end{aligned}$$

In view of (31) and (44),

(46)
$$\|f_0\|_{W_q^l(V_z)} \le A_{10}\lambda^{-15m}\nu_2^{\lambda e}\tilde{\omega}(D^k f_1\,,\,\delta)_{L_p(\Pi'_z(e+\tau))}\,,$$

where $e=[5M]+1$. Since $\deg P_1 < m$, it follows that $s(P_1\,,\alpha\Delta\,,m\,,\alpha\Omega)=P_1$; the foregoing gives us

(47)
$$\begin{aligned}
f_1 - s_1 &= f_0 + P_1 - \sum_{y\in\alpha\Delta} f_0(y)s_y - \sum_{y\in\alpha\Delta} P_1(y)s_y \\
&= f_0 - \sum_{y\in\alpha\Delta} f_0(y)s_y\,.
\end{aligned}$$

Considering (45)–(47), we write

(48)
$$\begin{aligned}
\|f_1 - s_1\|_{W_q^l(V_z)} &\le \|f_0\|_{W_q^l(V_z)} + \sum_{y\in\alpha\Delta}|f_0(y)|\,\|s_y\|_{W_q^l(V_z)} \\
&\le A_{11}\lambda^{-15m}\sum_{i=1}^{\infty}\nu_2^{\lambda i}\tilde{\omega}(D^k f_1\,,\,\delta)_{L_p(\Pi'_z(i+\tau))}\,.
\end{aligned}$$

It is formally assumed below that $p\,,\ q<\infty$, but all the arguments carry over without difficulty to the case of infinite p or q. For $p>q$ the set Ω is bounded; therefore, the cardinality of Z does not exceed $A_{12}\alpha^n$. Setting $\alpha'=(A_{12}\alpha^n)^{1/q-1/p}$ if $p>q$ and $\alpha'=1$ otherwise, and using the inequality

$$\left(\sum_{j=1}^{\infty} d_j\zeta^j\right)^p \le (1-\zeta)^{1-p}\sum_{j=1}^{\infty} d_j^p\zeta^j\,,$$

where $0<\zeta<1$ and $d_j\ge 0$, we get from (48)

(49)
$$\begin{aligned}
\|f_1 - s_1\|_{W_q^l(\alpha\Omega)} &\le A_{13}\left(\sum_{z\in Z}\|f_1-s_1\|^q_{W_q^l(V_z)}\right)^{1/q} \le A_{13}\alpha'\left(\sum_{z\in Z}\|f_1-s_1\|^p_{W_q^l(V_z)}\right)^{1/p} \\
&\le A_{13}A_{11}\alpha'\lambda^{-15m}(1-\nu_2^{\lambda})^{1/p-1}\left(\sum_{z\in Z}\sum_{i=1}^{\infty}\nu_2^{\lambda i}\tilde{\omega}(D^k f_1\,,\,\delta)^p_{L_p(\Pi'_z(i+\tau))}\right)^{1/p} \\
&\le A_{14}\alpha'\lambda^{1/p-16m}\left(\sum_{z\in Z}\sum_{i=1}^{\infty}\sum_{|\beta|=k}\nu_2^{\lambda i}\iint_{\substack{x\,,\,y\in\Pi'_z(i+\tau)\\|x-y|<\delta}}|D^\beta f_1(x)-D^\beta f_1(y)|^p\,dx\,dy\right)^{1/p} \\
&= A_{14}\alpha'\lambda^{1/p-16m}\left(\sum_{|\beta|=k}\iint_{\substack{x\,,\,y\in\alpha\Omega\\|x-y|<\delta}}|D^\beta f_1(x)-D^\beta f_1(y)|^p\varphi(x\,,\,y)\,dx\,dy\right)^{1/p}.
\end{aligned}$$

Here

$$\varphi(x, y) = \sum_{i=1}^{\infty} \sum_{z \in Z_i} \nu_2^{\lambda i},$$

where $Z_i = Z_i(x, y) = \{z \in Z : x, y \in \Pi'_z(i+\tau)\}$. Since the cardinality of Z_i does not exceed $A_{15} i^n$, it follows that

$$\varphi(x, y) \le \sum_{i=1}^{\infty} A_{15} i^n \nu_2^{\lambda i} \le \sum_{i=1}^{\infty} A_{16} \lambda^{-n} \nu_2^{\lambda i/2} \le A_{17} \lambda^{-n-1}$$

and

$$\begin{aligned} \|f_1 - s_1\|_{W_q^l(\alpha\Omega)} &\le A_{14} A_{17}^{1/p} \alpha' \lambda^{-18m} \left(\sum_{|\beta|=k} \tilde{\omega}(D^\beta f_1, \delta)_{L_p(\alpha\Omega)}^p \right)^{1/p} \\ &\le A_{18} \alpha' \lambda^{-18m} \omega(D^k f_1, \delta)_{L_p(\alpha\Omega)}. \end{aligned}$$

Passing to functions on Ω, we obtain $(\vartheta \ge 0)$

$$\begin{aligned} \|D^l(f - s(f, \Delta, m, \Omega))\|_{L_q(\Omega)} &\le \alpha^{l-k+n/p-n/q} A_{18} \alpha' \lambda^{-18m} \omega(D^k f, \delta/\alpha)_{L_p(\Omega)} \\ &\le A_{19} \lambda^{-18m} \overline{h}^\vartheta \omega(D^k f, \lambda \overline{h})_{L_p(\Omega)}. \end{aligned}$$

The theorem is proved.

If $k < m$ and

$$\overline{h}/\underline{h} \le A \tag{50}$$

for some A, then from Theorem 1 we get the estimate $(\lambda = \frac{1}{2} \underline{h} \overline{h}^{-1} \le 1$ in view of (31))

$$\|D^l(f - s(f, \Delta, m, \Omega))\|_{L_q(\Omega)} \le C_2 \overline{h}^\vartheta \omega(D^k f, \overline{h})_{L_p(\Omega)}, \tag{51}$$

where $C_2 = C_2(A, m, p, q, \Omega)$. It will be shown below (Remark 8) that the constant C_2 in this estimate cannot be replaced by any constant $C'_2(m, p, q, \Omega)$. We also have the following result.

Theorem 2. *If $k = m$ and (50) holds, then*

$$\|D^l(f - s(f, \Delta, m, \Omega))\|_{L_q(\Omega)} \le C_3 \overline{h}^\vartheta \|D^k f\|_{L_p(\Omega)},$$

where $C_3 = C_3(A, m, p, q, \Omega)$.

The proof repeats almost literally that of Theorem 1, in which it is necessary to set $\lambda = \underline{h} \overline{h}^{-1}$. The changes are as follows. Instead of the polynomial $P_1(x)$ one must use $P_2(x)$ (Lemma 3 is to be applied to the function $f_1(x)$), with ω and $\tilde{\omega}$ replaced by the corresponding norms. Instead of the estimate (49) one should write

$$\|f_1 - s_1\|_{W_q^l(\alpha\Omega)} \le A_{14} \alpha' \lambda^{1/p-16m} \left(\sum_{|\beta|=k} \int_{\alpha\Omega} |D^\beta f_1(x)|^p \varphi_1(x)\, dx \right)^{1/p},$$

where

$$\varphi_1(x) = \sum_{i=1}^{\infty} \sum_{z \in Z'_i} \nu_2^{\lambda i}, \qquad Z'_i = \{z \in Z : x \in \Pi'_z(i+\tau)\}.$$

In the case $p \ge 2$ the constant C_3 can be taken to be independent of A (Theorem 5, part 2).

Remark 3. Estimates of the form

$$\|D^l(f-s)\|_{L_q(\Omega)} \le \begin{cases} C\omega(D^m f, \overline{h})_{L_p(\Omega)}, \\ C\|D^k f\|_{L_p(\Omega)}, \quad k > m, \end{cases}$$

do not hold in general.

We show that if Ω is bounded, $x = (x_1, \dots, x_n)$, $f(x) = x_1^m$, and $0 < h < h^*$, then there exists a set $\Delta \subset \Omega$ such that $\overline{h}(\Delta) = h$, $\underline{h}(\Delta) \ge A_1 h$, where $A_1 = A_1(\Omega) > 0$, $\|D^l(f-s)\|_{L_q(\Omega)} \ne 0$ (whereas $\omega(D^m f, \overline{h})_{L_p(\Omega)} = \|D^k f\|_{L_p(\Omega)} = 0$, $k > m$).

It will be established first that for any Ω and any $h \in (0, h^*)$ there exists a set $\Delta \subset \Omega$ such that $\overline{h}(\Delta) = h$ and $\underline{h}(\Delta) \ge A_1 h$, where $A_1 = [Mn + n]^{-1}$ (we also need this below). Let Δ' be the set of points in Ω of the form $A_1 h(\zeta_1, \dots, \zeta_n)$, where the ζ_i are integers. Since any ball of radius $A_1 hn$ lying in Ω contains a point in Δ', we have in view of Lemma 1 ($\varepsilon = A_1 hn < \varepsilon_0$) that

$$\overline{h}(\Delta') \le A_1 Mhn < h, \qquad \underline{h}(\Delta') \ge A_1 h. \tag{52}$$

There exist a ball $B(a, \varepsilon_0) \subset \Omega$ (Lemma 1) and a point $b \in \Delta'$ such that $|a-b| < h$. We assume without loss of generality that $b = 0$, and we let $K(r) = \{x \in \mathbf{R}^n : \max_{1\le i\le n} |x_i| < r\}$ and $\Delta = \Delta' \backslash K(h)$. Note that in view of (28)

$$\overline{K(2h)} \subset \overline{B(0, 2h\sqrt{n})} \subset B(a, 3h\sqrt{n}) \subset B(a, \varepsilon_0) \subset \Omega. \tag{53}$$

Since $0 \in \Omega$ and $B(0, h)\Delta = \varnothing$, it follows that $\overline{h}(\Delta) \ge h$, and $\underline{h}(\Delta) \ge \underline{h}(\Delta') \ge A_1 h$. It remains to show that $\overline{h}(\Delta) \le h$. If $x \in \Omega\backslash K(2h)$, then $|x-y| < h$ for some $y \in \Delta'$ by (52); therefore, $y \in \Delta'\backslash K(h) = \Delta$. If $x \in K(2h)\backslash K(h)$, then $|x-y| \le A_1 h\sqrt{n}/2 < h$ for some $y \in \Delta$ by (53). Let $x \in K(h)$. Assume that $\max_{1\le i\le n} |x_i| = |x_1|$ (the other cases are handled similarly). If $|x_1| < (2/n)h$, then we set $y = (h \operatorname{sgn} x_1, 0, \dots, 0)$. In this case $y \in \Delta$ by (53), and

$$|x-y|^2 = |x|^2 - 2h|x_1| + h^2 \le n|x_1|^2 - 2h|x_1| + h^2 \le h^2.$$

For $|x_1| \ge (2/n)h$ we take $y = h(\operatorname{sgn} x_1, y_2, \dots, y_n)$, where $y_i = A_1 h[x_i/(A_1 h)]$. In view of (53), $y \in \Delta$ and

$$|x-y|^2 \le \left(h - \frac{2}{n}h\right)^2 + (n-1)^2 A_1^2 h^2 \le h^2.$$

Suppose now that $e = \sup\{x_1 : x \in \Delta\}$, $g(x) = 0$ for $x_1 \le e$, and $g(x) = (x_1 - e)^m$ for $x_1 > e$. Then $g \in W_2^m(\Omega)$, $g(x) = 0$ on Δ, and

$$\int_\Omega \frac{\partial^m g}{\partial x_1^m}\, dx \ne 0.$$

From this and Remark 2 it follows that the spline $s(x_1^m, \Delta, m, \Omega)$ cannot have the form $x_1^m + P(x)$, where $P(x)$ is a polynomial of degree less than m, which is what was required.

Moreover, the interpolation error estimates obtained in this section do not hold if Ω is unbounded and $p > q$. Indeed, in this case there exist disjoint balls $B(a_i, r) \subset \Omega\backslash\Delta$, $i = 1, 2, \dots$, where $r > 0$ (since Ω satisfies the strong cone condition). Then the function

$$f(x) = \sum_{i=1}^{\infty} i^{-1/q} \psi((x - a_i)/r)$$

belongs to $W_p^k(\Omega)$, but the difference $f - s(f, \Delta, m, \Omega) = f$ does not belong to $L_q^l(\Omega)$.

3. Splines with boundary conditions

The splines indicated in the heading of this section admit the interpolation error estimates (2) and (3) with larger values of the index k than the splines considered in §2, and with constants C not dependent on $\overline{h}/\underline{h}$. In this section we consider, moreover, the case (ii) in Definition 2.

Let $\Delta \subset \Omega$, $\overline{h} < h^*$, $n < 2m$, and $f \in L_p^*(\Omega)$. Denote by $\overset{\circ}{L}{}_2^m(\Omega)$ the closure of $\mathscr{D}(\Omega)$ in $L_2^m(\Omega)$, where $\mathscr{D}(\Omega)$ is the set of all differentiable functions having compact supports in Ω.

Definition 3. If $W_p^k(B(0, 1))$ is imbedded in $W_2^m(B(0, 1))$, then an *interpolation D^m-spline with boundary conditions* is defined to be a function $\sigma = \sigma(f, \Delta, m, \Omega) \in W_2^m(\Omega, \text{fin})$ such that:

a) if $g \in \overset{\circ}{L}{}_2^m(\Omega)$, g has bounded support, and $g(x) = 0$ on Δ, then

$$\int_\Omega \sum_{|\beta|=m} \frac{m!}{\beta!} D^\beta \sigma D^\beta g\, dx = 0;$$

b) $f(x) = \sigma(x)$ on Δ;

c) $(f - \sigma)\varphi \in \overset{\circ}{L}{}_2^m(\Omega)$ for any $\varphi \in \mathscr{D}(\mathbf{R}^n)$;

d) if $\zeta > 1$ and $a \in \Omega$, then $\|\sigma\|_{W_2^m(\Pi_a(r))} = O(\zeta^r)$ as $r \to \infty$.

Remark 4. In a way analogous to that for the spline s in Definition 2 (Lemma 6) it can be proved that there can exist at most one function σ satisfying the conditions of Definition 3.

Further, by arguments analogous to those in the beginning of §2 it can be established that for $f \in L_2^m(\Omega)$ there exists a unique element σ of the set $G = \{g \in f + \overset{\circ}{L}{}_2^m(\Omega) : g(x) = f(x)\ \forall x \in \Delta\}$ such that

$$\|D^m \sigma\|_{L_2(\Omega)} = \inf_{g \in G} \|D^m g\|_{L_2(\Omega)}.$$

As is not hard to see, this σ satisfies the conditions of Definition 3 (the condition d) follows from Lemma 4). Moreover, if f is a polynomial of degree less than $2m$, then $\sigma = f$, since for any $g \in \mathscr{D}(\Omega)$

$$\int_\Omega D^\beta f D^\beta g\, dx = (-1)^m \int_\Omega D^{2\beta} f \cdot g\, dx = 0,$$

where $|\beta| = m$.

Below in this section we assume that the imbeddings $W_p^k(B(0, 1)) \to W_q^l(B(0, 1))$ and $W_2^m(B(0, 1)) \to W_q^l(B(0, 1))$ are valid, and that $p \leq q$ if Ω is unbounded; ϑ is defined in (30).

Theorem 3. *Suppose that $W_p^k(B(0, 1))$ is imbedded in $W_2^m(B(0, 1))$. There exists a unique spline σ satisfying the conditions of Definition 3, and the operator $f \to \sigma(f, \Delta, m, \Omega)$, acting from $L_p^k(\Omega)$ to $W_2^m(\Omega, \text{fin})$, is continuous. For $k < 2m$ and $\lambda \in (0, 1]$*

$$\|D^l(f - \sigma(f, \Delta, m, \Omega))\|_{L_q(\Omega)} \leq C_4 \lambda^{-18m} \overline{h}^{\vartheta} \omega(D^k f, \lambda \overline{h})_{L_p(\Omega)},$$

where $C_4 = C_4(m, p, q, \Omega)$.

Proof ($\bar{h} > 0$). Let $\delta (> 3)$, α, Z, V_z, V_z', $\Pi_z'(r)$, $\Sigma_z'(r)$, $\varphi_z(x)$, $f_1(x)$ be the same as in the proof of Theorem 1. Below we use the auxiliary constants $A_i = A_i(m, p, q, \Omega)$. For y, $z \in Z$ we set $\rho_{yz} = \inf\{r : y' \in \Pi_z'(r)\} + 5M$, where y' is some point in V_y. The inequalities

$$\|g\|_{W_q^l(V_y)} \leq A_1 \|g\|_{W_p^k(V_y')} \quad \forall g \in W_p^k(V_y'),$$
$$\|g\|_{W_q^l(V_y)} \leq A_1 \|g\|_{W_2^m(V_y')} \quad \forall g \in W_2^m(V_y'),$$
$$\|g\|_{W_2^m(V_y)} \leq A_1 \|g\|_{W_p^k(V_y')} \quad \forall g \in W_p^k(V_y'),$$

are established just as in the proof of Theorem 1. It can be shown as in the case of (35)–(37) that

$$V_y' \subset \Pi_z'(\rho_{yz}), \tag{54}$$
$$V_z' \subset \Pi_z'(5M), \tag{55}$$
$$V_z' \subset \Sigma_y'(\rho_{yz}/2 - 5M). \tag{56}$$

By virtue of Lemma 3 ($a = z'$) and the above inequalities (54) and (55), there is a polynomial P_1 of degree at most k such that for any $y \in Z$

$$\|f_0\|_{W_2^m(V_y)} \leq A_2 \rho_{yz}^{5n+k} \tilde{\omega}(D^k f_1, \delta)_{L_p(\Pi_z'(\rho_{yz}+\tau))}, \tag{57}$$
$$\|f_0\|_{W_q^l(V_z)} \leq A_3 \tilde{\omega}(D^k f_1, \delta)_{L_p(\Pi_z'(5M+\tau))}, \tag{58}$$

where $f_0 = f_1 - P_1$. Let $f_{yi} = \varphi_y f_i$ and

$$H_0 = \{g \in \overset{\circ}{L}{}_2^m(\alpha\Omega) : g(x) = 0 \; \forall x \in \alpha\Delta\},$$
$$G_{yi} = f_{yi} + H_0 \subset L_2^m(\alpha\Omega), \qquad \sigma_{yi} = \arg\min\{\|D^m g\|_{L_2(\alpha\Omega)} : g \in G_{yi}\}$$

(Remark 4), where $y \in Z$, $i = 0, 1$. It is not hard to show that

$$\rho(\alpha\Delta \backslash B(y, 1), \alpha\Omega) \leq 6\delta/\lambda.$$

If $g_0 \in W_\infty^m(\alpha\Omega)$ is an infinitely differentiable function, $g_0(x) = 1$ on V_y, and $g_0(x)$ is constant on $\alpha\Omega \backslash T$, where T is a bounded set, then $g_0 f_{yi} = f_{yi}$ and $g_0 H_0 \subset H_0$ (Lemma 4), which implies that $g_0 G_{yi} \subset G_{yi}$. Let $F = V_y$. By Lemma 2 ($r_0 = 2$, $\alpha\varepsilon_0 > \bar{h}^{-1}\varepsilon_0 > r_0$ by (28)), $F \subset \alpha\Omega \cap B(y', 2) \subset \Pi_y'(3M)$, and this gives us that $\Pi_F(r) \subset \Pi_y'(r + 3M)$ and $\Sigma_y'(r + 3M) \subset \Sigma_F(r)$. By Lemma 5 ($G = G_{yi}$, $r_1 = 6\delta/\lambda$, $\alpha > 6\alpha_0\delta/\lambda = \alpha_0 r_1$),

$$\begin{aligned} \|\sigma_{yi}\|_{W_2^m(\Sigma_y'(r+3M))} &\leq \mu\nu_0^{4\lambda r}(1 + 6\delta/\lambda)^m \|D^m \sigma_{yi}\|_{L_2(\alpha\Omega)} \\ &\leq \mu\nu_0^{4\lambda r}(1 + 6\delta/\lambda)^m \|D^m f_{yi}\|_{L_2(\alpha\Omega)} \leq A_4 \lambda^{-m} \nu_0^{4\lambda r} \|f_i\|_{W_2^m(V_y)}, \end{aligned} \tag{59}$$

where $\nu_0 = \nu^{1/24\delta}$, $r \in \mathbf{R}$.

Fix some $a \in \alpha\Omega$ and $\zeta \in (1, \nu_0^{-2\lambda})$. Suppose that y, $z \in Z$ and $z' \in \Pi_a(\rho)$. The constants A_i' used below are independent of f, y, z, and ρ. According to (54), $V_y' \subset \Pi_z'(\rho_{yz}) \subset \Pi_a(\rho + \rho_{yz})$, and by Lemma 4

$$\|f_1\|_{W_2^m(V_y)} \leq A_1 \|f_1\|_{W_p^k(V_y')} \leq A_1 \|f_1\|_{W_p^k(\Pi_a(\rho+\rho_{yz}))} \leq A_1' \zeta^{\rho+\rho_{yz}} \|f_1\|_{L_p^k(\alpha\Omega)}. \tag{60}$$

From (56), (59), and (60) we get that

$$\|\sigma_{y1}\|_{W_2^m(V_z)} \leq A_2' \nu_0^{2\lambda\rho_{yz}} \zeta^{\rho+\rho_{yz}} \|f_1\|_{L_p^k(\alpha\Omega)};$$

therefore,

$$\sum_{y\in Z}\|\sigma_{y1}\|_{W_2^m(V_z)} \le \sum_{i=1}^{\infty}\sum_{i-1\le\rho_{yz}<i} A_2'\nu_0^{2\lambda\rho_{yz}}\zeta^{p+\rho_{yz}}\|f_1\|_{L_p^k(\alpha\Omega)}$$
$$\le A_2'A_5\zeta^p\sum_{i=1}^{\infty}\nu_0^{2\lambda(i-1)}\zeta^i i^n\|f_1\|_{L_p^k(\alpha\Omega)} \le A_3'\zeta^p\|f_1\|_{L_p^k(\alpha\Omega)}.$$

This inequality and (42) give us that the function $\sum_{y\in\alpha\Delta}\sigma_{y1}(\alpha x)$ is the spline $\sigma=\sigma(f, \Delta, m, \Omega)$ (the uniqueness of σ follows from Remark 4) and that the corresponding operator $f\to\sigma$ is continuous. Thus, setting $\sigma_1(x)=\sigma(x/\alpha)$, we have that

$$\sigma_1=\sum_{y\in\alpha\Delta}\sigma_{y1}. \tag{61}$$

Let $k<2m$. According to (56), (57), and (59),

$$\|\sigma_{y0}\|_{W_q^l(V_z)} \le A_1\|\sigma_{y0}\|_{W_2^m(V_z')} \le A_6\lambda^{-m}\rho_{yz}^{5n+k}\nu_0^{2\lambda\rho_{yz}}\tilde\omega(D^kf_1, \delta)_{L_p(\Pi_z'(\rho_{yz}+\tau))},$$

from which

$$\begin{aligned}\sum_{y\in Z}\|\sigma_{y0}\|_{W_q^l(V_z)} &\le A_6\lambda^{-m}\sum_{i=1}^{\infty}\sum_{y\in Z\,:\,i-1\le\rho_{yz}<i}\rho_{yz}^{5n+k}\nu_0^{2\lambda\rho_{yz}}\tilde\omega(D^kf_1, \delta)_{L_p(\Pi_z'(\rho_{yz}+\tau))}\\ &\le A_7\lambda^{-m-6n-k}\sum_{i=1}^{\infty}(\lambda i)^{6n+k}\nu_0^{2\lambda i}\tilde\omega(D^kf_1, \delta)_{L_p(\Pi_z'(i+\tau))}\\ &\le A_3\lambda^{-15m}\sum_{i=1}^{\infty}\nu_0^{\lambda i}\tilde\omega(D^kf_1, \delta)_{L_p(\Pi_z'(i+\tau))}.\end{aligned}\tag{62}$$

Since $\sigma(P_1, \alpha\Delta, m, \alpha\Omega)=P_1$ (Remark 4), it follows from (58), (61), and (62) that

$$\begin{aligned}\|f_1-\sigma_1\|_{W_q^l(V_z)} &= \left\|f_0+P_1-\sum_{y\in Z}\sigma_{y0}-\sum_{y\in Z}\sigma(P_1\sigma_y, \alpha\Delta, m, \alpha\Omega)\right\|_{W_q^l(V_z)}\\ &=\left\|f_0-\sum_{y\in Z}\sigma_{y0}\right\|_{W_q^l(V_z)} \le A_9\lambda^{-15m}\sum_{i=1}^{\infty}\nu_0^{\lambda i}\tilde\omega(D^kf_1, \delta)_{L_p(\Pi_z'(i+\tau))}.\end{aligned}\tag{63}$$

The inequality (63) is analogous to (48). Arguing further just as in the proof of Theorem 1, we arrive at the required result. The theorem is proved.

Theorem 4. *If $W_p^k(B(0, 1))$ is imbedded in $W_2^m(B(0, 1))$, then there exists a unique interpolation D^m-spline $s(f, \Delta, m, \Omega)$, and the operator $f\to s(f, \Delta, m, \Omega)$, acting from $L_p^k(\Omega)$ to $W_2^m(\Omega, \mathrm{fin})$, is continuous.*

The proof repeats the corresponding part of the proof of Theorem 3, with σ replaced by s.

Theorem 5. *Assume the imbedding*

$$W_p^k(B(0, 1))\to W_2^m(B(0, 1)).$$

1. *For* $k=2m$

$$\|D^l(f-\sigma(f, \Delta, m, \Omega))\|_{L_q(\Omega)} \le C_5\overline{h}^{\vartheta}\|D^kf\|_{L_p(\Omega)}.$$

2. *For* $k = m$

$$\|D^l(f - s(f, \Delta, m, \Omega))\|_{L_q(\Omega)} \le C_5 \overline{h}^{\vartheta} \|D^k f\|_{L_p(\Omega)}.$$

Here $C_5 = C_5(m, p, q, \Omega)$ *is a constant.*

The proof is analogous to that of Theorem 3, in which it is necessary to replace $P_1(x)$ by $P_2(x)$ (Lemma 3, applied to the function $f_1(x)$), ω and $\tilde{\omega}$ by the corresponding norms, and, in case 2, σ by s, and λ must be set equal to 1.

Remark 5. The case $p = 2$ in Theorem 5 (2) admits the following scheme of arguments (Duchon [21] and Bezhaev [23]):

$$\|D^m(f - s)\|_{L_2(\Omega)} \le 2\|D^m f\|_{L_2(\Omega)}$$

by Definition 1, and

$$\|D^l(f - s)\|_{L_q(\Omega)} \le A_1 \overline{h}^{\vartheta} \|D^m(f - s)\|_{L_2(\Omega)} \le 2A_1 \overline{h}^{\vartheta} \|D^m f\|_{L_2(\Omega)},$$

since the function $f - s$ has an $\overline{h}$-net of zeros in Ω. Here $A_i = A_i(m, p, q, \Omega)$ are constants. Inequalities of the form (2) for splines with boundary conditions can also be obtained more simply in the case when $f \in W_p^k(\Omega)$ and $W_p^k(B(0, 1))$ is imbedded in $W_2^m(B(0, 1))$, $k \le 2m$, $1 < p \le 2 \le q$ (see Bezhaev [22], [23] for $p = 2$). Indeed, using Remark 4 and integration by parts, we write

$$\begin{aligned}
\|D^m(f - \sigma)\|^2_{L_2(\Omega)} &= \int_\Omega \sum_{|\beta|=m} \frac{m!}{\beta!} D^\beta(f - \sigma) D^\beta f\, dx \\
&= \int_\Omega \sum_{|\beta|=m} \frac{m!}{\beta!} (-1)^{k-m} D^{\beta'}(f - \sigma) D^{\beta''} f\, dx,
\end{aligned}$$

where $|\beta'| = 2m - k$, $|\beta''| = k$. Since $f - \sigma$ has an $\overline{h}$-net of zeros in Ω and $W_2^m(B(0, 1))$ is imbedded in $W_t^{2m-k}(B(0, 1))$, where $t = p/(p-1)$ ([25], Theorem 10.2), we have that

$$\begin{aligned}
\|D^m(f - \sigma)\|^2_{L_2(\Omega)} &\le A_2 \|D^{2m-k}(f - \sigma)\|_{L_t(\Omega)} \|D^k f\|_{L_p(\Omega)} \\
&\le A_3 \overline{h}^{k-m+n/t-n/2} \|D^m(f - \sigma)\|_{L_2(\Omega)} \|D^k f\|_{L_p(\Omega)},
\end{aligned}$$

$$\|D^m(f - \sigma)\|_{L_2(\Omega)} \le A_3 \overline{h}^{k-m+n/t-n/2} \|D^k f\|_{L_p(\Omega)},$$

$$\|D^l(f - \sigma)\|_{L_q(\Omega)} \le A_4 \overline{h}^{m-l+n/q-n/2} \|D^m(f - \sigma)\|_{L_2(\Omega)} \le A_3 A_4 \overline{h}^{\vartheta} \|D^k f\|_{L_p(\Omega)}.$$

We now consider separately the case $\Omega = \mathbf{R}^n$. In this case, as is easy to see, the spline $s(f, \Delta, m, \mathbf{R}^n)$ in Definition 2 satisfies all the conditions in Definition 3, i.e.,

$$s(f, \Delta, m, \mathbf{R}^m) = \sigma(f, \Delta, m, \mathbf{R}^n) \tag{64}$$

(if $W_p^k(B(0, 1))$ is imbedded in $W_2^m(B(0, 1))$).

Theorem 6. *Let* $\Omega = \mathbf{R}^n$.

1. *If* $\underline{h} > 0$, $W_p^k(B(0, 1))$ *is imbedded in* $C(B(0, 1))$, *and* $k < 2m$, *then*

$$\|D^l(f - s(f, \Delta, m, \mathbf{R}^n))\|_{L_q(\mathbf{R}^n)} \le C_6 \lambda^{-18m} \overline{h}^{\vartheta} \omega(D^k f, \lambda \overline{h})_{L_p(\mathbf{R}^n)}$$

for any λ, $0 < \lambda \le \underline{h}\overline{h}^{-1}$.

2. *If* $W_p^k(B(0,1))$ *is imbedded in* $W_2^m(B(0,1))$ *and* $k<2m$, *then the preceding estimate holds for any* $\lambda\in(0,1]$.

3. *If* $k=2m$, *then*

$$\|D^l(f-s(f,\Delta,m,\Omega))\|_{L_q(\mathbf{R}^n)}\le C_6\overline{h}^{\vartheta}\|D^kf\|_{L_p(\mathbf{R}^n)}.$$

Here $C_6=C_6(m,p,q)$ *is a constant.*

Proof. It follows from (64) and Remark 4 that $s(g,\Delta,m,\mathbf{R}^n)=g$ for any polynomial g of degree less than $2m$. Therefore, if $\Omega=\mathbf{R}^n$, then in Theorem 1 the condition $k<m$ can be replaced by the condition $k<2m$ (the proof is exactly the same). This gives us assertion 1. The assertions 2 and 3 follow from (64) and Theorems 3 and 5 ($W_p^k(B(0,1))$ is imbedded in $W_2^m(B(0,1))$ for $k=2m$, because $n<2m$ ([25], Theorem 10.2)). The theorem is proved.

Remark 6. There are no estimates of the form

$$\|D^l(f-\sigma)\|_{L_q(\Omega)}\le\begin{cases}C\omega(D^{2m}f,\overline{h})_{L_p(\Omega)},\\ C\|D^kf\|_{L_p(\Omega)}, & k>2m.\end{cases}$$

Indeed, $\sigma(x)$ is a polyharmonic function of order m on the set $\Omega\backslash\overline{\Delta}$ (see the proof of Theorem 2 in §14 of [26]). Therefore, if $f(x)=f(x_1,\dots,x_n)=x_1^{2m}$, Ω is bounded, and $\overline{\Delta}\not\supset\Omega$, then $\|D^l(f-\sigma)\|_{L_q(\Omega)}\ne0$, and $\omega(D^{2m}f,\overline{h})_{L_p(\Omega)}=\|D^kf\|_{L_p(\Omega)}=0$ $(k>2m)$. As in Remark 3, it can be shown that the estimates obtained do not hold if Ω is unbounded and $p>q$.

Remark 7. The estimates obtained in §§2 and 3 for the error of interpolation by splines are sharp with respect to order.

Suppose that $1\le p\le\infty$, $1\le q\le\infty$, k and l are nonnegative integers $\le 2m$, and $n<2m$. For any $h\in(0,h^*)$ there exist a function $f\in\mathscr{D}(\Omega)$ and a set $\Delta\subset\Omega$ such that

$$\overline{h}(\Delta)=h,\qquad \underline{h}(\Delta)\ge A_1h,$$

$$\text{(65)}\qquad \min(\|D^l(f-s(f,\Delta,m,\Omega))\|_{L_q(\Omega)},\ \|D^l(f-\sigma(f,\Delta,m,\Omega))\|_{L_q(\Omega)})\ge A_2h^{\vartheta}\|D^kf\|_{L_p(\Omega)}\ge\tfrac12A_2h^{\vartheta}\omega(D^kf,h)_{L_p(\Omega)}>0$$

(here and below, $A_i>0$ are constants, $A_1=A_1(\Omega)$, $A_i=A_i(m,\Omega)$ for $i=2,\dots,7$). Indeed, if Δ satisfies the conditions (65) (Remark 3), then there exist a_i such that the balls $B(a_i,A_3h)$ are pairwise disjoint and lie in $\Omega\backslash\Delta$ (for $i=1,\dots,t$, where $A_4\le th^n\le A_5$). Let

$$f_1(x)=\sum_{i=1}^t\psi\left(\frac{x-a_i}{A_3h}\right),\qquad f_2(x)=\psi\left(\frac{x-a_1}{A_3h}\right).$$

The functions f_1 and f_2 have compact supports in $\Omega\backslash\Delta$; therefore, the corresponding interpolation splines are zero. We have

$$\|D^lf_1\|_{L_q(\Omega)}\ge A_6h^{k-l}\|D^kf_1\|_{L_p(\Omega)},$$
$$\|D^lf_2\|_{L_q(\Omega)}\ge A_7h^{k-l+n/q-n/p}\|D^kf_2\|_{L_p(\Omega)},$$

which is what was required.

4. Convergence of the interpolation process

Assume that $n<2m$ and that the imbedding

$$W_2^m(B(0,1))\to W_p^k(B(0,1))\to L_\infty(B(0,1))$$

is valid for some p and k. Let $f \in L_p^k(\Omega)$ if $p < \infty$, and let $f \in C^k(\Omega)$ (see §1) if $p = \infty$; $\Delta_i \subset \Omega$ $(i \in \mathbf{N})$, $\overline{h}_i = \overline{h}(\Delta_i) < h^*$, $\underline{h}_i = \underline{h}(\Delta_i) > 0$, and $s_i = s(f, \Delta_i, m, \Omega)$.

Theorem 7. *If $\overline{h}_i = o(1)$ and $\overline{h}_i = O(\underline{h}_i)$ as $i \to \infty$, and if Ω is bounded in the case $k = m$, then $f - s_i \to 0$ in $W_p^k(\Omega)$, and for $k < m$*

$$\|f - s_i\|_{W_p^k(\Omega)} = O(\omega(D^k f, \overline{h}_i)_{L_p(\Omega)}).$$

Proof. For $k < m$ the result follows immediately from the estimate (51). The case $k = m$, $p = 2$ was treated by Bezhaev (Theorem 1.3 in [23]). It remains to see that for $k = m$ and $1 \le p < 2$ the space $W_2^m(\Omega)$ is dense in $W_p^k(\Omega)$ and that the norms of the operators $f \to s(f, \Delta_i, m, \Omega)$ from $W_p^k(\Omega)$ to $W_p^k(\Omega)$ are bounded uniformly with respect to i in view of Theorem 2. The theorem is proved.

Remark 8. In Theorem 7 the condition $\overline{h}_i = O(\underline{h}_i)$ (which means that the sets Δ_i must condense uniformly in Ω) is essential. We show that if Ω is the cube $(-1, 1)^n$, $k < m$, and $p < \infty$ in the case $k = m - 1$, then there exist a function f satisfying the conditions above and finite sets $\Delta_i \subset \Omega$ such that $\overline{h}(\Delta_i) \to 0$ and $\|s(f, \Delta_i, m, \Omega)\|_{L_1(\Omega)} \to \infty$ as $i \to \infty$. Suppose first that $n = 1$. Assume $\eta \in \mathbf{N}$, $0 < h < 1/\eta m$, and $k - 1/p < \alpha < m - 1$, and let

$$f(x) = \begin{cases} 0 & \text{for } x \le 0, \\ x^\alpha & \text{for } x > 0, \end{cases} \qquad P_h(x) = \prod_{i=0}^{m-2}(x + ih),$$

$$\Delta = \{-ih : i = -1, \dots, m-2\}, \qquad \Delta' = \{\pm i/\eta : i = 0, \dots, \eta - 1\},$$

$$\Delta_h = \Delta \cup \Delta',$$

$$s_h = \arg\min\{\|D^m g\|_{L_2(\Omega)} : g \in W_2^m(\Omega),\ g(x) = 0\ \forall x \in \Delta,$$

$$g(x) = h^{-\alpha} P_h(h) f(x) - P_h(x)\ \forall x \in \Delta'\}$$

(Definition 1). Obviously, $\|s_h^{(m)}\|_{L_2(\Omega)} \le c$, where c does not depend on h; since $s_h(x)$ has m zeros on the interval $[(2-m)h, h]$, we have $s_h^{(j)}(\zeta_j) = 0$ for some $\zeta_j \in [(2-m)h, h]$, $j = 0, \dots, m-1$, and for $x \in (h, 1)$

$$|s_h(x)| = \left| \int_{\zeta_0}^{x} \int_{\zeta_1}^{z_1} \cdots \int_{z_{m-1}}^{z_{m-1}} s_h^{(m)}(z_m)\, dz_1 \cdots dz_m \right|$$

$$\le (x + (m-2)h)^{m-1} \int_{(2-m)h}^{x} |s_h^{(m)}(z)|\, dz$$

$$\le (x + (m-2)h)^{m-1/2} \|s_h^{(m)}\|_{L_2(\Omega)} \le c m^m x^{m-1/2}.$$

This implies that $\underline{\lim}_{h \to 0} \|P_h + s_h\|_{L_1(\Omega)} > 0$. Let $s_{\eta h} = s(f, \Delta_h, m, \Omega)$. Using the fact that $P_h(h) = O(h^{m-1})$ and (by Remark 2)

$$s_{\eta h} = h^\alpha P_h(h)^{-1}(P_h + s_h),$$

we get $\|s_{\eta h}\|_{L_1(\Omega)} \to \infty$ as $h \to 0$ for any η, and $\lim_{\eta \to \infty} \|s_\eta h_\eta\|_{L_1(\Omega)} = \infty$ for the corresponding h_η.

Suppose now that $n > 1$, $x = (x_1, \dots, x_n)$, $\Omega = (-1, 1)^n$, $E \subset (-1, 1)$, $F_i \subset F_{i+1} \subset (-1, 1)^{n-1}$ $(i \ge 1)$, E, F_i are finite sets, $\rho(F_i(-1, 1)^{n-1}) \to 0$ as $i \to \infty$, $\Delta = E \times (-1, 1)^{n-1}$, $\Delta_i = E \times F_i$, $\overline{h}(\Delta_1) < h^*$, $s = s(f, \Delta, m, \Omega)$, and $s_i = s(f, \Delta_i, m, \Omega)$. The multidimensional case will be reduced to the one-dimensional case if we show that: 1) if $\tilde{f} \in W_2^m((-1, 1))$, $f(x) = \tilde{f}(x_1)$, then

$s(x) = s(\tilde{f}, E, m, (-1, 1))(x_1)$, and 2) $s_i \to s$ in $W_2^m(\Omega)$ as $i \to \infty$. The first follows from Remark 2. To prove 2) we note that in view of Proposition 1 the norms

$$\|g\| = \left(\sum_{x \in \Delta_1} g(x)^2 + \|D^m g\|_{L_2(\Omega)} \right)^{1/2}$$

and $\|g\|_{W_2^m(\Omega)}$ are equivalent. Since

$$\|s_i\| \leq \|s\|, \tag{66}$$

some subsequence s_{i_j} converges to some s^* weakly in $W_2^m(\Omega)$, and

$$\|s^*\| \leq \|s\|. \tag{67}$$

The set $G_i = \{g \in W_2^m(\Omega) : g(x) = f(x)\ \forall x \in \Delta_i\}$ is weakly closed in $W_2^m(\Omega)$; therefore,

$$s^* \in \bigcap_{j=1}^{\infty} G_{i_j} = \{g \in W_2^m(\Omega) : g(x) = f(x)\ \forall x \in \Delta\}.$$

By the meaning of s, this and (67) imply that $s = s^*$. Then $s_{i_j} \to s$ strongly in $W_2^m(\Omega)$ in view of (66), and from this it follows that $s_i \to s$ in $W_2^m(\Omega)$ by the choice of F_i. This is what was required.

Using Remark 3 and analogous arguments, we can show that if Ω is a cube and $0 < h < h^*$, then there exist a function $f \in W_p^k(\Omega)$ and finite sets $\Delta_i \subset \Omega$ such that $\overline{h}(\Delta_i) = h$ and $\|D^l s(f, \Delta_i, m, \Omega)\|_{L_1(\Omega)} \to \infty$ as $i \to \infty$ $(l = 0, \dots, m)$.

Assume that Ω is bounded, and let $\{a_j : j \in \mathbf{N}\}$ be a dense subset of Ω and $\Delta_i = \{a_j : j \leq i\}$. Assume that $\overline{h}_i = O(\underline{h}_i)$. For any $i \in \mathbf{N}$ there exists a $u_i \in G_i = \{g \in W_2^m(\Omega) : g(a_i) = 1,\ g(a_j) = 0\ \forall j < i\}$ such that $\|D^m u_i\|_{L_2(\Omega)} = \inf\{\|D^m g\|_{L_2(\Omega)} : g \in G_i\}$ (see the beginning of §2).

Theorem 8. *The functions* u_i, $i \in \mathbf{N}$, *form a basis in the space* $W_p^k(\Omega)$ *if* $p < \infty$, *or in* $C^k(\Omega)$ *if* $p = \infty$.

Proof. The functions u_i belong to the indicated spaces, since they can be extended to functions $\tilde{u}_i \in W_2^m(\mathbf{R}^n)$ (Theorem 5 in Chapter 6 of [27]). Let $f \in W_p^k(\Omega)$ if $p < \infty$, and $f \in C^k(\Omega)$ if $p = \infty$, and let

$$\alpha_1 = f(a_1), \qquad \alpha_i = f(a_i) - \sum_{j=1}^{i-1} \alpha_j u_j(a_i) \quad (i > 1). \tag{68}$$

Then $\sum_{j=1}^{i} \alpha_j u_j = s(f, \Delta_i, m, \Omega)$ for sufficiently large i by Remark 2, and by Theorem 7,

$$\sum_{i=1}^{\infty} \alpha_i u_i = f, \tag{69}$$

where the series converges in $W_p^k(\Omega)$. Conversely, (69) implies (68). The theorem is proved.

We remark that bases in Sobolev spaces have been studied in a number of papers (see, for example, [28] and [29] and the literature cited there).

Bibliography

1. Marc Atteia, *Existence et détermination des fonctions "spline" à plusieurs variables*, C. R. Acad. Sci. Paris Sér. A **262** (1966), 575–578.
2. J. H. Ahlberg, E. N. Nilson, and J. L. Walsh, *The theory of splines and their application*, Academic Press, New York, 1967.
3. N. P. Korneĭchuk, *Splines and approximation theory*, "Nauka", Moscow, 1984. (Russian)
4. Marc Atteia, *Généralisation de la définition et des propriétés des "spline fonctions"*, C. R. Acad. Sci. Paris **260** (1965), 3550–3553.
5. Pierre-Jean Laurent, *Approximation et optimisation*, Hermann, Paris, 1972.
6. V. A. Vasilenko, *Spline functions: theory, algorithms, and programs*, "Nauka", Novosibirsk, 1983. (Russian)
7. Garrett Birkhoff and Carl de Boor, *Error bounds for spline interpolation*, J. Math. Mech. **13** (1964), 827–835.
8. Martin Marsden, *Cubic spline interpolation of continuous functions*, J. Approx. Theory **10** (1974), 103–111.
9. Thomas R. Lucas, *Error bounds for interpolating cubic splines under various end conditions*, SIAM J. Numer. Anal. **11** (1974), 569–584.
10. S. B. Stechkin and Yu. N. Subbotin, *Splines in numerical mathematics*, "Nauka", Moscow, 1976. (Russian)
11. Yu. S. Zav'yalov, B. I. Kvasov, and V. L. Miroshnichenko, *Methods of spline functions*, "Nauka", Moscow, 1980. (Russian)
12. J. L. Walsh, J. H. Ahlberg, and E. N. Nilson, *Best approximation and convergence properties of higher-order spline fits*, Notices Amer. Math. Soc. **10** (1963), 202 (Abstract No. 63T-103).
13. J. H. Ahlberg, E. N. Nilson, and J. L. Walsh, *Best approximation and convergence properties of higher-order spline approximations*, J. Math. Mech. **14** (1965), 231–243.
14. J. H. Ahlberg and E. N. Nilson, *Polynomial splines on the real line*, J. Approx. Theory **3** (1970), 398–409.
15. Zvi Ziegler, *On the convergence of* n*th order spline functions*, Notices Amer. Math. Soc. **12** (1965), 620 (Abstract No. 65T-354).
16. Yu. N. Subbotin, *Approximation of functions of class* $W^k H^p_\omega$ *by splines of order* m, Dokl. Akad. Nauk SSSR **195** (1970), 1039–1041; English transl. in Soviet Math. Dokl. **11** (1970).
17. ———, *Approximation by "spline" functions, and estimates of widths*, Trudy Mat. Inst. Steklov. **109** (1971), 35–60; English transl. in Proc. Steklov Inst. Math. **109** (1971).
18. ———, *Spline approximation*, Theory of Functions and Approximations, Part 1 (Proc. Saratov Winter School, 1982; S. M. Nilol'skiĭ, editor), Saratov. Gos. Univ., Saratov, 1983, pp. 81–90. (Russian)
19. Carl de Boor, *Odd-degree spline interpolation at a biinfinite knot sequence*, Approximation Theory (Proc. Internat. Colloq., Bonn, 1976), Lecture Notes in Math., vol. 556, Springer-Verlag, Berlin, 1976, pp. 30–53.
20. Jean Duchon, *Interpolation des fonctions de deux variables suivant le principe de la flexion des plaques minces*, Rev. Française Automat. Informat. Recherche Opérationnelle Sér. Rouge Anal. Numér. **10** (1976), no. R-3, 5–12.
21. ———, *Sur l'erreur d'interpolation des fonctions de plusieurs variables par les* D^m*-splines*, RAIRO Anal. Numér. **12** (1978), 325–334.
22. A. Yu. Bezhaev, *On the convergence of spline interpolations with boundary conditions in a bounded domain*, The Finite Element Method in Some Problems of Numerical Analysis (V. A. Vasilenko, editor), Vychisl. Tsentr Sibirsk. Otdel. Akad. Nauk SSSR, Novosibirsk, 1984, pp. 4–20.
23. ———, D^m*-splines in problems of approximation of functions on chaotic grids*, Candidate's dissertation, Novosibirsk, 1985. (Russian)
24. A. Imamov, *Convergence of an interpolation process in a Hilbert space, and its applications*, Chisl. Metody Mekh. Sploshn. Sredy **8** (1977), no. 1, 49–58. (Russian)
25. O. V. Besov, V. P. Il'in, and S. M. Nikol'skiĭ, *Integral representations of functions and imbedding theorems*, "Nauka", Moscow, 1975; English transl., Vols. 1, 2, Wiley, New York, 1979.

26. S. L. Sobolev, *Some applications of functional analysis in mathematical physics*, Izdat. Sibirsk. Otdel Akad. Nauk SSSR, Novosibirsk, 1962; English transl., Amer. Math. Soc., Providence, RI, 1963.

27. Elias M. Stein, *Singular integrals and differentiability properties of functions*, Princeton Univ. Press, Princeton, NJ, 1970.

28. Z. Chesel′skiĭ [Ciesielski], *Bases and K-functionals for Sobolev spaces over compact manifolds of class C^∞*, Trudy Mat. Inst. Steklov. **164** (1983), 197–202; English transl. in Proc. Steklov Inst. Math. **1985**, no. 2(164).

29. ———, *Spline bases in classical function spaces on compact C^∞ manifolds*. III, Constructive Theory of Functions (Proc. Internat. Conf., Varna, 1984; Bl. Sendov et al., editors), Publ. House Bulgarian Acad. Sci., Sofia, 1984, pp. 214–223.

Translated by H. H. McFADEN

Труды Матем. Инст. Стеклова
198 (1992)

Proceedings of the Steklov
Institute of Mathematics
1994, Issue 1

ON ERRORS OF PARALLELEPIPED CUBATURE FORMULAS ON CLASSES OF DIFFERENTIABLE FUNCTIONS

UDC 517.5

A. V. REZTSOV

INTRODUCTION

We consider classes F_p^r $(r \in \mathbf{N}, 1 \leq p \leq \infty)$ of differentiable functions of m $(m \in \mathbf{N})$ variables given by restrictions on the $\mathscr{L}_p$-norm of the rth differential. The functions $f \in F_p^r$ are periodic, and the corresponding m-dimensional tori are arbitrary and nondegenerate. For these classes F_p^r we study the errors of parallelepiped cubature formulas (defined to be cubature formulas that are exact on constants and have equal weights and nodes lying on an m-dimensional lattice compatible with the periodicity of the functions in F_p^r).

The case $m = 1$ and $1 \leq p \leq \infty$ (here F_p^r is the Sobolev class $\widetilde{W}^r\mathscr{L}_p$) has been studied by many authors. See the survey [1] of Zhensykbaev and the monograph [2] of Nikol'skiĭ. In the case $m = 2$, $p = \infty$, $r = 1$ it follows from results of V. F. Babenko that parallelepiped cubature formulas with a special grid of nodes (below we call these formulas triangle formulas) or formulas close to them (see [2], §E.4) are optimal for the classes F_∞^1.

The case $p = \infty$, $m \geq 2$, $r \in \mathbf{N}$ $(r \geq 2)$ has been studied by the author [3]–[5]. The main result in [3]–[5] is the assertion that for $m = 2$ and $r = 2, 3$, or $r \geq 7$ the triangle parallelepiped cubature formulas are better than the square parallelepiped cubature formulas (i.e., those for which the nodes lie on a square lattice in $\mathbf{R}^2$; see below for the exact definition).

In the present article it is proved that the triangle parallelepiped cubature formulas are better than the square parallelepiped cubature formulas when $m = 2$, $1 \leq p \leq \infty$, $r \geq 7$.

1. CLASSES OF FUNCTIONS, AND CUBATURE FORMULAS

For an arbitrary basis $A = \{a^{(1)}, \dots, a^{(m)}\}$ in the space $\mathbf{R}^m$ ($a^{(j)}$ is a basis vector $(j = 1, \dots, m)$) the set

$$\Gamma(A) = \{x = z_1 \cdot a^{(1)} + \cdots + z_m \cdot a^{(m)} : z_j \in \mathbf{Z}\ (j = 1, \dots, m)\}$$

is called an *m-dimensional lattice.* The *fundamental parallelepiped* of the lattice $\Gamma(A)$ is defined to be

$$D(A) = \{x = t_1 \cdot a^{(1)} + \cdots + t_m \cdot a^{(m)} : 0 \leq t_j < 1\ (j = 1, \dots, m)\}.$$

The group of all shifts (parallel translations) of the space $\mathbf{R}^m$ carrying $\Gamma(A)$ into itself is denoted by $\mathscr{T}(A)$. The group $\mathscr{T}(A)$ contains, in particular, shifts T_j $(j = 1, \dots, m)$ such that $T_j x = x + a^{(j)}$ for any $x \in \mathbf{R}^m$.

1991 *Mathematics Subject Classification.* Primary 41A55, 65D32.

A function $f\colon \mathbf{R}^m \to \mathbf{R}$ is called a *function on the torus* $\mathbf{T}^m(A)$ if $f(Tx) = f(x)$ for any $T \in \mathscr{T}(A)$ and any $x \in \mathbf{R}^m$.

Denote by $d^S(f; x, h)$ $(S = 0, 1, 2, \dots)$ the usual Sth differential of the function f at the point x (if it exists), where $h \in \mathbf{R}^m$ is a vector incrementing the argument of the function.

The *norm of the differential* $d^S(f; x, h)$ at the point x (if this differential exists) is defined to be the quantity

$$d_*^S(f; x) = \sup\{|d^S(f; x, h)|/|h|^S : h \in \mathbf{R}^m \setminus 0\},$$

where $|h|$ is the length of the vector h.

We say that a function $f\colon \mathbf{R}^m \to \mathbf{R}$ *belongs to the class* $MF_p^r = MF_p^r(A)$ $(M > 0, r \in \mathbf{N}, 1 \le p \le \infty)$, if it satisfies the following conditions:

1) f is a function on the torus $\mathbf{T}^m(A)$.

2) The differential $d^{r-1}(f; x, h)$ exists and is an absolutely continuous function with respect to x along any line in $\mathbf{R}^m$.

3) The differential $d^r(f; x, h)$ exists almost everywhere with respect to x, and its norm $d_*^r(f; x)$ belongs to $\mathscr{L}_p(D(A))$ as a function of x and satisfies the inequality

$$\|d_*^r(f; \cdot)\|_{\mathscr{L}_p(D(A))} \le M.$$

The class $1F_p^r(A) = 1F_p^r$ is denoted by F_p^r.

A lattice $\Gamma(B)$ is called a *multiple* of the lattice $\Gamma(A)$ if any vector of the basis A is a linear combination, with integer coefficients, of the vectors $b^{(1)}, \dots, b^{(m)}$ in the basis B.

A *parallelepiped cubature formula* (determined by a lattice $\Gamma(B)$ that is a multiple of the lattice $\Gamma(A)$) is defined to be the functional

$$L_N f = L_{N,B} f = \frac{\operatorname{meas} D(A)}{N} \sum_{x^* \in D(A) \cap \Gamma(B)} f(x^*),$$

where N (the number of nodes of the cubature formula) is the number of points in $D(A) \cap \Gamma(B)$, which is equal to the number of vectors of the form

$$z_1 \cdot b^{(1)} + \dots + z_m \cdot b^{(m)}, \qquad z_j \in \mathbf{Z} \quad (j = 1, \dots, m),$$

belonging to the fundamental parallelepiped $D(A)$.

It is clear that for any lattice $\Gamma(A)$ there exists an infinite set of multiple lattices $\Gamma(B)$, and hence also an infinite set of parallelepiped cubature formulas (see also [4], §1.3).

2. Errors of parallelepiped cubature formulas on classes of functions

The quantity

$$E(L_N, F_p^r(A)) = \sup\left\{\left|\int \cdots \int_{D(A)} f(x)\,dx - L_N f\right| : f \in MF_p^r(A)\right\}$$

is called the *error of the cubature formula* L_N on the class $MF_p^r(A)$.

Let $G(B)$ be the group of all transformations of $\mathbf{R}^m$ that map $\Gamma(B)$ into itself. For a lattice $\Gamma(B)$ that is a multiple of $\Gamma(A)$ we denote by $MF_p^r(A|B)$ the set of all functions $f^* \in MF_p^r(A)$ such that f^* is defined on the torus $\mathbf{T}^m(B)$, $f^*(gx) = f^*(x)$ for any $x \in \mathbf{R}^m$ and $g \in G(B)$, and $f^*(x^*) = 0$ for any $x \in D(A) \cap \Gamma(B)$.

It is easy to see (since $MF_p^r(A|B) \subset MF_p^r(A)$) that the errors of the parallelepiped cubature formula $L_{N,B}$ on the classes $MF_p^r(A)$ and $MF_p^r(A|B)$ satisfy the inequality

$$E(L_{N,B}, MF_p^r(A|B)) \le E(L_{N,B}, MF_p^r(A)).$$

For any $r, m \in \mathbf{N}$, $1 \le p \le \infty$, and any lattices $\Gamma(A)$ and $\Gamma(B)$ such that $\Gamma(B)$ is a multiple of $\Gamma(A)$ we have

Theorem 1.

$$E(L_{N,B}, MF_p^r(A|B)) = E(L_{N,B}, MF_p^r(A)).$$

Theorem 1 was proved in §3 of [4] for $p = \infty$. For $1 \le p < \infty$ the proof is exactly the same as for $p = \infty$. We give a

Sketch of the proof of Theorem 1. Denote by $\mathcal{T}(B)$ the group of all shifts (parallel translations) carrying $\Gamma(B)$ into itself. It was proved in Proposition 3.1 of [4] that $\mathcal{T}(B)$ is a normal subgroup of the group $G(B)$. Therefore, we can consider the quotient group $G(B)/\mathcal{T}(B)$, whose elements we denote by $[g]$.

The quotient group $\mathcal{T}(B)/\mathcal{T}(A)$ is finite, and its order $|\mathcal{T}(B)/\mathcal{T}(A)| = N$ is the number of nodes of the corresponding parallelepiped cubature formula $L_{N,B}$. For an arbitrary $x^* \in D(A) \cap \Gamma(B)$ let $T_{\langle x^* \rangle}$ be the shift in $\mathcal{T}(B)$ such that for any $x \in \mathbf{R}^m$

$$T_{\langle x^* \rangle} x = x + x^*.$$

It is clear that for different $x^* \in D(A) \cap \Gamma(B)$ the shifts $T_{\langle x^* \rangle}$ belong to different cosets $[T_*] \in \mathcal{T}(B)/\mathcal{T}(A)$.

For any $[g] \in G(B)/\mathcal{T}(B)$ and any $f \in MF_p^r$ the $[g]$-*symmetrization of* f is defined to be the function

$$f_{[g]}(x) = \frac{1}{N} \sum_{x^* \in D(A) \cap \Gamma(B)} f(T_{\langle x^* \rangle} g x),$$

where g is an arbitrary representative of the coset $[g]$.

The definition of $[g]$-symmetrization is unambiguous, that is, it does not depend on the choice of $g \in [g]$, as was proved in [4], Proposition 3.3.

For any $f \in MF_p^r(A)$ the $G(B)$-*symmetrization of* f is defined to be the function

$$\tilde{f}(x) = \frac{1}{|G(B)/\mathcal{T}(B)|} \sum_{[g] \in G(B)/\mathcal{T}(B)} f_{[g]}(x),$$

where $|G(B)/\mathcal{T}(B)|$ is the order of the finite group $G(B)/\mathcal{T}(B)$.

The finiteness of the group $G(B)/\mathcal{T}(B)$ was proved in [4], Proposition 3.4.

According to §3.3 in [4], the $G(B)$-symmetrization of an arbitrary function $f \in MF_p^r(A)$ has the following properties:

a) $\tilde{f}(x^*) = f(0)$ for any $x^* \in D(A) \cap \Gamma(B)$;

b) $\tilde{f} \in MF_p^r(A)$;

c) $\tilde{f}$ is a function on the torus $\mathbf{T}^m(B)$, and $\tilde{f}(gx) = \tilde{f}(x)$ for any $x \in \mathbf{R}^m$ and $g \in G(B)$;

d) $\int \cdots \int_{D(A)} \tilde{f}(x)\, dx = \int \cdots \int_{D(A)} f(x)\, dx$;

e) $L_{N,B} \tilde{f} = L_{N,B} f$.

The formula $L_{N,B}$ is exact on constants; therefore, for the function $f^*(x) = \tilde{f}(x) - f(0)$ the equality

$$\int \cdots \int_{D(A)} \tilde{f}(x)\, dx - L_{N,B}\tilde{f} = \int \cdots \int_{D(A)} f^*(x)\, dx - L_{N,B} f^* = \int \cdots \int_{D(A)} f^*(x)\, dx$$

is satisfied, and the properties a)–e) are preserved.

The proof of Theorem 1 is complete.

Corollary 1. *For a cubature formula $\widetilde{L}_B$ with the same nodes as for the parallelepiped cubature formula $L_{N,B}$ and with arbitrary weights the error on the class $MF_p^r(A)$ satisfies*

$$E(\widetilde{L}_B, MF_p^r(A)) \geq E(L_{N,B}, MF_p^r(A)).$$

To prove Corollary 1 it suffices to consider the functions $f^{**}(x) = f^*(x) + C$, $f^* \in MF_p^r(A|B)$, $C \in \mathbf{R}$.

Corollary 2.

$$E(L_{N,B}, MF_p^r(A)) = \sup\left\{\int \cdots \int_{D(A)} f^*(x)dx : f^* \in MF_p^r(A|B)\right\}.$$

This is obvious, since $L_{N,B}f^* = 0$ for any function $f^* \in MF_p^r(A|B)$.

If for a given torus $\mathbf{T}^2(A_\square)$ $(m = 2)$ there exists a parallelepiped cubature formula $L_{N',B_\square}$ such that $|b_\square^{(1)}| = |b_\square^{(2)}|$ and $b_\square^{(1)} \perp b_\square^{(2)}$, then such a formula is called a *square formula* and is denoted by $L_{N',B_\square}^\square$.

If for a given torus $\mathbf{T}^2(A_\triangle)$ there exists a parallelepiped cubature formula $L_{N'',B_\triangle}$ such that $|b_\triangle^{(1)}| = |b_\triangle^{(2)}|$ and (the angle) $\widehat{b_\triangle^{(1)}, b_\triangle^{(2)}} = \pi/3$, then such a formula is called a *triangle formula* and is denoted by $L_{N'',B_\triangle}^\triangle$.

It is clear (see [4]) that the expressions for the errors of the cubature formulas $L_{N',B_\square}^\square$ and $L_{N'',B_\triangle}^\triangle$ on the respective classes $M'F_p^r(A_\square)$ and $M''F_p^r(A_\triangle)$ have the forms

$$E(L_{N',B_\square}^\square, M'F_p^r(A_\square)) = c_\square \frac{M'}{(\operatorname{meas} D(A_\square))^{1/p}} \frac{(\operatorname{meas} D(A_\square))^{r/2+1}}{(N')^{r/2}},$$

$$E(L_{N'',B_\triangle}^\triangle, M''F_p^r(A_\triangle)) = c_\triangle \frac{M''}{(\operatorname{meas} D(A_\triangle))^{1/p}} \frac{(\operatorname{meas} D(A_\triangle))^{r/2+1}}{(N'')^{r/2}}.$$

Here $c_\square = c_\square(r)_p$ and $c_\triangle = c_\triangle(r)_p$ depend only on r and p.

3. The main theorem

Theorem 2. *For any $r \in \mathbf{N}$ with $r \geq 2$*

$$c_\triangle(r) \leq c_\square(r) \frac{3^{-r/4} + 3^{r/4}}{2^{r/2}} \left(\frac{3}{2} + \frac{1}{2^{r/2+1}}\right).$$

Corollary 3. *For any $1 \leq p \leq \infty$ and $r \geq 7$ the triangle parallelepiped cubature formulas are better than the square parallelepiped cubature formulas.*

We need the following two lemmas.

Lemma 1. *Let $L_{N,B}^\square$ be a square parallelepiped cubature formula, and suppose that the lattice $\Gamma(B)$ is a multiple of the lattice $\Gamma(A)$. Then the quantity*

$$H = \sup\{f((b^{(1)} + b^{(2)})/2) : f \in MF_p^r(A|B)\}$$

satisfies the inequality

$$H \leq 2E(L_{N,B}^\square, MF_p^r(A|B))/\operatorname{meas} D(A).$$

Proof of Lemma 1. For an arbitrary $f \in MF_p^r(A|B)$ we construct the function $g(x) = f((b^{(1)}+b^{(2)})/2) - f(x-(b^{(1)}+b^{(2)})/2)$. It is easy to see that $g \in MF_p^r(A|B)$, and

$$\int \cdots \int_{D(A)} f(x)\,dx + \int \cdots \int_{D(A)} g(x)\,dx = (\operatorname{meas} D(A)) f((b^{(1)}+b^{(2)})/2).$$

The statement of Lemma 1 (see Corollary 2) is obtained by passing to the supremum over $f \in MF_p^r(A|B)$ on the left-hand and right-hand sides of the equality.

Lemma 2. *Let $L^{\square}_{N,B}$ be a square parallelepiped cubature formula, let the lattice $\Gamma(B)$ be a multiple of the lattice $\Gamma(A)$, and let $[MF_p^r(A|B)]$ be the set of all $f \in MF_p^r(A|B)$ such that the values of f at the points $b^{(1)}/2$, $b^{(2)}/2$, $(b^{(1)}+b^{(2)})/2$, $b^{(1)}+b^{(2)}/2$, $b^{(1)}/2+b^{(2)}$ coincide. Then the quantity*

$$H^* = \sup\{f((b^{(1)}+b^{(2)})/2 : f \in [MF_p^r(A|B)]\}$$

satisfies the inequality

$$H^* \le \frac{H}{2^{r/2-1}} \le (E(L^{\square}_{N,B}, MF_p^r(A|B))/\operatorname{meas} D(A))\frac{1}{2^{r/2-2}}.$$

Proof of Lemma 2. Let $\Gamma(B^*)$ be the square lattice with basis $B^* = \{b_*^{(1)} = (b^{(1)}+b^{(2)})/2,\ b_*^{(2)} = (b^{(2)}-b^{(1)})/2\}$. It is clear that $\Gamma(B^*)$ is a multiple of the lattice $\Gamma(B)$, and the corresponding square parallelepiped cubature formula $L^{\square}_{N^*,B^*}$ has $N^* = 2N$ nodes. Denote by f^* the $G(B^*)$-symmetrization of the function $\frac{1}{2}(f(x)+f(x-b_*^{(1)})) - \frac{1}{2}f(b_*^{(1)})$. Here $f^* \in MF_p^r(A|B)$; therefore $f^*((b_*^{(1)}+b_*^{(2)})/2)$, which is equal to $f((b^{(1)}+b^{(2)})/2)/2$, satisfies the inequality (see Lemma 1)

$$f^*((b_*^{(1)}+b_*^{(2)})/2) \le 2E(L^{\square}_{2N,B^*}, MF_p^r(A))/\operatorname{meas} D(A).$$

The statement of Lemma 2 is obtained by passing on the left-hand side to the supremum over f^* (and hence over $f \in [MF_p^r(A|B)]$).

Proof of Theorem 2. Suppose that $f \in M''F_p^r(A_\triangle|B_\triangle)$, and $\underline{f}$ is the dilation of the function f by a factor of $\sqrt{3}$ along the vector $b^{(2)}-b^{(1)}$. Under this dilation the triangle lattice $\Gamma(B_\triangle)$ passes into the square lattice $\Gamma(B_\square)$, where

$$B_\square = \left\{ b_\square^{(1)} = \frac{1-\sqrt{3}}{2}b_\triangle^{(1)} + \frac{11+\sqrt{3}}{2}b_\triangle^{(2)},\ b_\square^{(2)} = \frac{11+\sqrt{3}}{2}b_\triangle^{(1)} + \frac{1-\sqrt{3}}{2}b_\triangle^{(2)} \right\}.$$

Denote by $f_\square$ the $G(B_\square)$-symmetrization of $\underline{f}$. Clearly $f_\square \in M'F_p^r(A_\square|B_\square)$, where

$$M' = M''((3^{-r/2}+1)3^{1/2p}/2),$$

and $\operatorname{meas} D(A_\square) = (\operatorname{meas} D(A_\triangle))\sqrt{3}$. By property c) of the $G(B_\triangle)$-symmetrization, the values of f at the points $b_\triangle^{(1)}/2$, $b_\triangle^{(2)}/2$, $(b_\triangle^{(1)}+b_\triangle^{(2)})/2$, $b_\triangle^{(1)}+b_\triangle^{(2)}/2$, $b_\triangle^{(1)}/2+b_\triangle^{(2)}$ coincide and are equal to the values of $f_\square$ at the points $b_\square^{(1)}/2$, $b_\square^{(2)}/2$, $(b_\square^{(1)}+b_\square^{(2)})/2$, $b_\square^{(1)}+b_\square^{(2)}/2$, $b_\square^{(1)}/2+b_\square^{(2)}$. Consequently, $f_\square \in [M'F_p^r(A_\square|B_\square)]$, and from this and the equality

$$\sqrt{3}\int \cdots \int_{D(A_\triangle)} f(x)\,dx = \int \cdots \int_{D(A_\square)} f(x)\,dx = \int \cdots \int_{D(A_\square)} f_\square(x)\,dx$$

we get the inequality

$$\begin{aligned}&\sqrt{3}E(L^{\triangle}_{N'',B_\triangle}, M''F^r_p(A_\triangle|B_\triangle))\\&\qquad\le \sup\left\{\int\cdots\int_{D(A)} f_\square(x)\,dx : f_\square\in[M'F^r_p(A_\square|B_\square)]\right\}.\end{aligned}$$

For any $f_\square\in[M'F^r_p(A_\square|B_\square)]$ we construct the function

$$\begin{aligned}&\tfrac14(f_\square(x)+f_\square(x-b^{(1)}_\square/2)+f_\square(x-b^{(2)}_\square/2)\\&\qquad+f_\square(x-(b^{(1)}_\square+b^{(2)}_\square)/2))-\tfrac34 f_\square(b^{(1)}_\square/2)\end{aligned}$$

and denote by $f^*_\square$ its $G(\underline{B})$-symmetrization, where $\underline{B}=\{b^{(1)}_\square/2, b^{(2)}_\square/2\}$. The lattice $\Gamma(\underline{B})$ is square and is a multiple of the lattice $\Gamma(A_\square)$, and the cubature formula $L^{\square}_{\underline{N},B}$ has $\underline{N}=4N''=4N'$ nodes. Since $f^*_\square\in M'F^r_p(A_\square|\underline{B})$, and

$$\int\cdots\int_{D(A_\square)} f^*_\square(x)\,dx=\int\cdots\int_{D(A_\square)} f_\square(x)\,dx-\tfrac34 f_\square(b^{(1)}_\square/2)\operatorname{meas}D(A_\square),$$

we conclude by passing to the supremum over $f_\square\in[M'F^r_p(A_\square|B_\square)]$ (with Lemma 2 taken into account) that

$$\begin{aligned}&\sup\left\{\int\cdots\int_{D(A_\square)} f_\square(x)\,dx : f_\square\in[M'F^r_p(A_\square|B_\square)]\right\}\\&\qquad\le E(L^{\square}_{\underline{N},B}, M'F^r_p(A_\square|\underline{B}))+\tfrac34 H^*(\operatorname{meas}D(A_\square)).\end{aligned}$$

In summary, we obtain

$$\begin{aligned}&\sqrt{3}c_\triangle\frac{M''}{(\operatorname{meas}D(A_\triangle))^{1/p}}\frac{(\operatorname{meas}D(A\triangle))^{r/2+1}}{(N'')^{r/2}}\\&\quad\le c_\square M''(3^{-r/2}+1)\cdot\frac12\frac{1}{(\operatorname{meas}D(A_\triangle))^{1/p}}\frac{(\operatorname{meas}D(A_\triangle))^{r/2+1}}{(N'')^{r/2}\cdot 2^r}\cdot 3^{r/4+1/2}\\&\quad+\frac34(\operatorname{meas}D(A_\square))\frac{1}{(\operatorname{meas}D(A_\square))}\frac{1}{2^{r/2-2}}c_\square\frac{M''}{2}(3^{-r/2}+1)\\&\quad\times\frac{1}{(\operatorname{meas}D(A_\triangle))^{1/p}}\frac{(\operatorname{meas}D(A_\triangle))^{r/2+1}}{(N'')^{r/2}}3^{r/4+1/2}.\end{aligned}$$

Therefore,

$$c_\triangle\le c_\square(3^{-r/2}+1)\cdot 3^{r/4}\left(\frac{1}{2^{r+1}}+\frac34\frac{1}{2^{r/2-1}}\right)=c_\square(3^{-r/4}+3^{r/4})\frac{1}{2^{r/2+1}}\left(\frac{1}{2^{r/2}}+3\right).$$

The proof of Theorem 2 is complete.

Corollary 3 can be proved by verifying that $(3^{-r/4}+3^{r/4})(3/2+1/2^{r/2+1})/2^{r/2}$ is less than 1 for $r\ge 7$.

The author thanks S. B. Stechkin for posing the problem of studying the classes MF^r_p and for his undivided attention to this work.

BIBLIOGRAPHY

1. A. A. Zhensykbeav, *Monosplines of minimal norm and best quadrature formulas*, Uspekhi Mat. Nauk **36** (1981), no. 4 (220), 107–159; English transl. in Russian Math. Surveys **36** (1981).
2. S. M. Nikol′skiĭ, *Quadrature formulas*, 3rd ed., with a supplement by N. P. Korneĭchuk, "Nauka", Moscow, 1979; English transl. of 1st ed., Hindustan, Delhi, 1964; Spanish transl. of 4th ed., "Mir", Moscow, 1990.

3. A. V. Reztsov, *Optimal cubature formulas on classes of differentiable functions*, Mat. Zametki **45** (1989), no. 4, 121–124. (Russian)

4. ——, *Optimal cubature formulas on classes of differentiable functions*, Manuscript No. 4227-B-89, deposited at VINITI, 1989. (Russian)

5. ——, *On parallelepiped cubature formulas and perfect splines*, Mat. Zametki **48** (1990), no. 4, 79–87; English transl. in Math Notes **48** (1990).

Translated by H. H. McFADEN

Труды Матем. Инст. Стеклова
198 (1992)

Proceedings of the Steklov
Institute of Mathematics
1994, Issue 1

ON DIVERGENCE OF FOURIER SERIES IN REARRANGED PRICE SYSTEMS

UDC 517.52

V. V. TARKAEV

1. Introduction

It is well known that there is a definite connection between the Lebesgue constants and the convergence of Fourier series. Thus, for example, for Fourier series in the Walsh system in the Paley arrangement (referred to henceforth as the Walsh-Paley system), as for Fourier series in the trigonometric system, the following estimate holds almost everywhere (cf. [1]):

$$|S_n(f, x) - f(x)| = o_x(L_n), \qquad f \in L.$$

However, the behavior of the Lebesgue constants for the Walsh-Paley system differs significantly from that of the Lebesgue constants for the trigonometric system, primarily in the existence of increasing sequences $\{n_k\}$ such that $\sup_k L_{n_k} < \infty$.

Thus it follows in particular from the result of Schneider mentioned above [1] that if an increasing sequence of natural numbers $\{n_k\}_{k=1}^{\infty}$ is such that $\sup_{k\geq 1} L_{n_k} < \infty$, then $S_{n_k}(f, x) \to f(x)$ almost everywhere for any integrable function f, where S_n and L_n are respectively the partial sums of the Fourier series in the Walsh-Paley system and the Lebesgue constants of the system .

On the other hand, the author has remarked (cf. [2]) that if $\{w_n^{\mathrm{K}}(x)\}_{n=0}^{\infty}$ is the Walsh system in the Kaczmarz arrangement, and L_n^{K} its Lebesgue constants, there exist an integrable function f and an increasing sequence of natural numbers $\{n_k\}_{k=1}^{\infty}$ such that $\sup_{k\geq 1} L_{n_k}^{\mathrm{K}} < \infty$, while $\sup_{k\geq 1} |S_{n_k}^{\mathrm{K}}(f, x)| = \infty$ almost everywhere; here S_n^{K} are the partial sums of the Fourier series in the Walsh-Kaczmarz system.

In connection with the discovery of this difference in properties of the Fourier series in the Walsh-Paley and Walsh-Kaczmarz systems, the problem naturally arose of finding the reason for the difference, i.e., the problem of finding some property of the Dirichlet kernels that is the reason for the existence of an integrable function the partial sums of whose Fourier series in some class of orthogonal systems diverge almost everywhere over a subsequence for which the Lebesgue constants remain bounded.

Two hypotheses were suggested by the proofs of the above-mentioned results. The first derives from the fact that $\sup_n |D_n(x)| = \infty$ on a set of positive measure. The second was that it is a matter of the absence of a localization principle. Only the first of these has been partially confirmed.

The early studies focused on a certain class of rearrangements of the Walsh system. The corresponding results were partly published in [2] and [3]. Then the class of systems was enlarged, and analogous questions were studied for a certain class

1991 *Mathematics Subject Classification.* Primary 42C20, Secondary 42C10.

Key words and phrases. Walsh system, Lebesgue constants, rearranged Price system, Kaczmarz arrangement, Paley arrangement, localization principle, Dirichlet kernels.

of rearrangements of Price systems (precise definitions are given in the following section).

In order to give a compact exposition of the results we make the following definitions:

Definition 1. A Price system $\{\varphi_n(x)\}_{n=0}^{\infty}$ that has been rearranged in some manner *possesses property* $*$ if there exist an integrable function f and an increasing sequence of natural numbers $\{n_k\}$ such that $\sup_k L_{n_k} < \infty$ while $\sup_k |S_{n_k}(f, x)| = \infty$ almost everywhere.

Definition 2. The system $\{\varphi_n(x)\}_{n=0}^{\infty}$ *possesses property* $**$ if there exists a set E of positive measure such that $\sup_n |D_n(x)| = \infty$ for every $x \in E$.

We use the symbols D_n, L_n, and S_n respectively to denote the Dirichlet kernels, the Lebesgue constants of the system $\{\varphi_n(x)\}$, and the partial sums of a Fourier series with respect to this system.

Section 3 of this paper is devoted to the study of property $**$. Three theorems are proved there. Theorem 1: Among the rearrangements of a Price system we are considering there exist some that possess property $**$. Theorem 2: Suppose a Price system is generated by a bounded sequence. Then if some rearrangement from the class under consideration possesses property $**$, it also possesses property $*$. Theorem 3: For rearrangements of Price systems generated by unbounded sequences property $*$ does not in general follow from property $**$. Section 3 also contains a proposition according to which property $**$ entails the absence of a localization principle. No other cause-and-effect connections between the three properties ($*$, $**$, and the localization principle) exist. This is confirmed by examples constructed in §4 (Theorems 4, 5, 6).

This paper also contains three propositions due to the author: the first is an estimate of the Lebesgue constants of a Price system; the second is an estimate of the Dirichlet kernels; the third is the fact that a Price system in the Paley arrangement does not possess property $*$.

2. Definitions and preliminary information

Here we shall give the definitions of Price systems and a certain class of rearrangements of them, after which we shall give two propositions that will be used below.

The modified interval $[0; 1]$. Let us fix some sequence $\{p_j\}_{j=0}^{\infty}$ of natural numbers larger than one. We introduce the following notation:

$$P_j = \begin{cases} 1 & \text{if } j = 0, \\ \prod_{k=0}^{j-1} p_k & \text{if } j = 1, 2, \ldots. \end{cases} \tag{2.1}$$

Then any number x with $0 \le x < 1$ can be written as

$$x = \sum_{j=0}^{\infty} x_j / P_j, \tag{2.2}$$

where the integers x_j satisfy the condition $0 \le x_j < p_j$ for each $j = 0, 1, \ldots$.

Such a representation is in general not unique. A countable set of numbers called p-rationals have two representations of the form (2.2), one of which contains an infinite number of nonzero terms and the other only a finite number. For uniqueness we agree to ascribe to a p-rational number only the expansion with an infinite number of nonzero terms.

Sometimes, however, we shall consider instead of the usual interval $[0; 1]$ the so-called modified interval $[0; 1]$, which differs from the usual one in that the p-rational numbers in it are "doubled", i.e., the elements of the modified interval are not numbers but formal series of the form (2.2), or, more precisely, their sequences of coefficients $\{x_j\}_{j=0}^{\infty}$, $0 \le x_j < p_j$, $j = 0, 1, \dots$.

Addition on the modified interval, denoted by $\dot{+}$, differs significantly from the usual addition: by definition, $z = x \dot{+} y$ if for each $j = 0, 1, \dots$ the number z_j is the sum of x_j and y_j modulo p_j. The operation inverse to $\dot{+}$ is denoted $\dot{-}$. Thus the modified interval is the group $G = \prod_{j=0}^{\infty} Z_{p_j}$, where Z_p is the cyclic group of order p. A natural topology can be introduced on G, which turns it into a zero-dimensional compact Abelian group satisfying the second axiom of countability. Such groups have been studied in detail in the monograph [4]. A normalized measure that is invariant with respect to the operation $\dot{+}$ can also be defined on this group (cf. [5]).

There is a natural correspondence between the elements of the usual interval and the modified interval; it is given by (2.2). We emphasize that the measures of two measurable subsets that correspond to each other are the same. In what follows we shall not distinguish a number x belonging to $[0; 1)$ and the sequence of integers $\{x_j\}_{j=0}^{\infty}$ corresponding to it.

The investigations below will be mainly focused on the usual interval; however, to obtain a convenient expression for the partial sums of a Fourier series in terms of the Dirichlet kernels, the group structure of the modified interval is essential. The transition from the usual interval to the modified interval and back is effected using the mapping (2.2) and will not be mentioned explicitly in the sequel.

The Price system. On the interval $[0; 1]$ we define a system of generalized Rademacher functions $\{\Phi_j(x)\}_{j=0}^{\infty}$ by setting

$$\Phi_j(x) = \exp(2\pi i x_j / p_j), \quad j = 0, 1, \dots . \tag{2.3}$$

If $p_j = 2$ for any j, the system so defined is the classical Rademacher system.

We note that the functions $\Phi_j(x)$ are piecewise constant, left continuous (when regarded as functions of a numerical argument), and periodic with period $1/P_j$. Moreover, they are pairwise independent.

The Price functions are finite products of integral powers of generalized Rademacher functions. We order them as follows. Any integer $n \ge 0$ can be written in a natural way in the number system generated by the sequence $\{p_j\}_{j=0}^{\infty}$, namely

$$n = \sum_{j=0}^{\infty} a_j(n) P_j, \tag{2.4}$$

where the coefficients $a_j(n)$ are integers such that $0 \le a_j(n) < p_j$, $j = 0, 1, \dots$ (for the definition of the quantities P_j cf. (2.1)). The sum on the right-hand side is in fact finite.

The Price system in the Paley arrangement will be denoted $\{\varphi_n^{\Pi}(x)\}$,

$$\varphi_n^{\Pi}(x) = \prod_{j=0}^{\infty} (\Phi_j(x))^{a_j(n)} \tag{2.5}$$

(n and the coefficients $a_j(n)$, $j = 0, 1, \dots$, are connected by relation (2.4)). In the case $p_j = 2$ for any j the system so obtained is the Walsh system in the Paley arrangement.

The set of Price functions $\{\varphi_k^{\Pi} : P_N \le k < P_{N+1}\}$, $N \ge 0$, will be called the packet with index N. The subset of it given by $\{\varphi_k^{\Pi}(x) : \nu P_N \le k < (\nu+1)P_N\}$, $1 \le \nu < p_N$, $N \ge 0$, is the νth subpacket of the Nth packet.

The Price system is a complete orthonormal system in $L_2[0;1]$. We note further that it is a system of characters of the modified interval. Locally compact Abelian groups satisfying the second axiom of countability and the system of characters of them are studied in [4].

The class of rearrangements. We now give the definition of the class of rearrangements of the Price system introduced by Gosselin and Young [6]. In the particular case of the Walsh system it is a subclass of the class of piecewise-linear rearrangements introduced by Schipp [7].

We denote the rearranged Price system by $\{\varphi_n^{\Sigma}(x)\}$. All rearrangements Σ permute functions within subpackets. Moreover, inside the νth subpacket of the Nth packet, Σ is defined by a permutation $\sigma_{N,\nu}$ of the nonnegative integers such that

$$\sigma_{N,\nu}(j) = j, \qquad \forall j \ge N,\ 1 \le \nu < p_N,\ N = 0, 1, \dots . \tag{2.6}$$

We introduce the notation

$$\tilde{p}_j = p_{\sigma_{N,\nu}(j)}, \qquad \widetilde{P}_j = \begin{cases} 1 & \text{when } j = 0, \\ \prod_{k=0}^{j-1} \tilde{p}_k & \text{when } j \ge 1, \end{cases} \qquad \tilde{\Phi}_j(x) = \Phi_{\sigma_{N,\nu}(j)}(x).$$

The subsequent procedure of defining the arrangements in the νth subpacket of the Nth packet is analogous to the definition of the Price system in the Paley arrangement, except that the generating sequence $\{p_j\}_{j=0}^{\infty}$ must be replaced by the rearranged sequence $\{\tilde{p}_j\}_{j=0}^{\infty}$, and the generalized Rademacher functions $\Phi_j(x)$ must be replaced by the renumbered functions $\tilde{\Phi}_j(x)$, $j = 0, 1, \dots$. To be specific, let

$$\nu P_N \le n < (\nu+1)P_N, \qquad 1 \le \nu < p_N,\ N \ge 0. \tag{2.7}$$

We now write n in the number system generated by the sequence $\{\tilde{p}_j\}$:

$$n = \sum_{j=0}^{\infty} \tilde{a}_j(n)\widetilde{P}_j = \nu P_N + \sum_{j=0}^{N-1} \tilde{a}_j(n)\widetilde{P}_j, \tag{2.8}$$

where $0 \le \tilde{a}_j(n) < \tilde{p}_j$, $j = 0, 1, \dots$ (using (2.6) and (2.7)). For indices n satisfying condition (2.7) we set

$$\varphi_n^{\Sigma}(x) = \prod_{j=0}^{\infty} (\tilde{\Phi}_j(x))^{\tilde{a}_j(n)} = (\Phi_N(x))^{\nu} \prod_{j=0}^{N-1} (\tilde{\Phi}_j(x))^{\tilde{a}_j(n)}, \tag{2.9}$$

where n is connected with the coefficients $\tilde{a}_j(n)$, $j = 0, 1, \dots$, by the relation (2.8). (Because of (2.6), $\tilde{\Phi}_j(x) = \Phi_j(x)$ when $j \ge N$, and $\varphi_0^{\Sigma}(x) = \varphi_0^{\Pi}(x) = 1$.)

We emphasize that functions are in general permuted differently in different subpackets of the same packet.

The Paley arrangement also belongs to the class of arrangements just defined; it is obtained with $\sigma_{N,\nu}(j) = j$ for all N, ν, and j.

Remark 1. If an arrangement of a Price system belongs to the class under consideration, then, knowing the representation of some Price function as a product of generalized Rademacher functions, one can determine the packet and subpacket to which the function belongs. To do this one must find the function having largest index among those occurring in the product of generalized Rademacher functions.

That index will be the index of the packet, and the index of the subpacket will be the exponent the function bears.

Let Σ be some rearrangement of a Price system of the class defined above. Then we shall denote by $D_n^{\Sigma}(x)$ the Dirichlet kernels of the rearranged Price system $D_n^{\Sigma}(x) = \sum_{k=0}^{n-1} \varphi_k^{\Sigma}(x)$, $n \geq 1$.

Remark 2. Since all the rearrangements of this class preserve subpackets, the Dirichlet kernels with indices of the form νP_N are independent of the arrangement for $1 \leq \nu < p_N$. Consequently, for the Dirichlet kernels with these indices the superscript that indicates the arrangement can be omitted.

We now give three properties of the Dirichlet kernels with indices of this form. First of all (cf. [4])

$$D_{P_j}(x) = \begin{cases} P_j & \text{when } x \in [0; 1/P_j], \\ 0 & \text{when } x \notin [0; 1/P_j], \end{cases} \qquad j = 1, 2. \tag{2.10}$$

Second,

$$\int_0^1 |D_{P_j}(x)|\, dx = 1, \quad j = 0, 1, \ldots . \tag{2.11}$$

Third,

$$\operatorname{supp} D_{\nu P_j}(x) \subset [0; 1/P_j], \qquad j = 0, 1, \ldots, \quad 1 \leq \nu < p_j. \tag{2.12}$$

This last relation follows directly from the first and the fact that for the Price system in any arrangement of this class the following obviously holds: if $\nu P_N \leq k < (\nu+1)P_N$, $1 \leq \nu < p_N$, then there exists k', $0 \leq k' < P_N$, such that

$$\varphi_k^{\Sigma}(x) \equiv (\Phi_N(x))^{\nu} \cdot \varphi_{k'}^{\Sigma}(x).$$

We now pass to the discussion of the results to be used in the following section.

The idea of the following lemma, which makes it possible to replace the rearranged Price system by the unaltered system on a "rearranged" interval in the study of certain questions, is borrowed from a paper of Skvortsov [8].

Lemma. *Let $\{\varphi_n^{\Sigma}(x)\}_{n=0}^{\infty}$ be the Price system generated by a sequence $\{p_j\}_{j=0}^{\infty}$ and arranged using permutations $\sigma_{N,\nu}$ according to the procedure described above. Then if $n = \nu P_N + s$, where $1 \leq \nu < p_N$ and $1 \leq s < P_N$, it follows that*

$$D_n^{\Sigma}(x) - D_{\nu P_N}(x) = (\Xi_N(\tilde{x}))^{\nu} \cdot D_s^{\xi}(\tilde{x}), \tag{2.13}$$

where $\{\xi_k(\tilde{x})\}_{k=0}^{\infty}$ is the Price system in the Paley arrangement generated by the sequence $\{\tilde{p}_j\}_{j=0}^{\infty}$, $\tilde{p}_j = p_{\sigma_{N,\nu}(j)}$, $D_k^{\xi}(\tilde{x})$ are its Dirichlet kernels, and $\{\Xi_j(\tilde{x})\}_{j=0}^{\infty}$ is the corresponding system of generalized Rademacher functions. Here $\tilde{x} = F_{N,\nu}(x)$, where $F_{N,\nu}$ is the measure-preserving mapping of the interval $[0; 1]$ onto itself given by the relation

$$\tilde{x} = F_{N,\nu}(x) \leftrightarrow \tilde{x}_j = x_{\sigma_{N,\nu}(j)}. \tag{2.14}$$

This mapping is an isomorphism of the groups generated by the sequences $\{p_j\}_{j=0}^{\infty}$ and $\{\tilde{p}_j\}_{j=0}^{\infty}$.

Proof of the lemma. The identity (2.13) will be proved if the equality

$$\varphi_{\nu P_N + k}^{\Sigma}(x) = (\Xi_N(\tilde{x}))^{\nu} \cdot \xi_k(\tilde{x}) \tag{2.15}$$

holds for any k, $1 \leq k < P_N$.

Thus we fix some k such that $1 \le k < P_N$. Then (cf. (2.9)) $\varphi^{\Sigma}_{\nu P_N+k}(x) = (\Phi_N(x))^{\nu} \cdot \prod_{j=0}^{N-1}(\tilde{\Phi}_j(x))^{\tilde{a}_j(k)}$, where $\tilde{a}_j(k)$ are the coefficients used in writing the number k in the number system generated by the sequence $\{\tilde{p}_j\}_{j=0}^{\infty}$.

But $\{\xi_i(x)\}_{i=0}^{\infty}$ is generated by the same sequence $\{\tilde{p}_j\}_{j=0}^{\infty}$ and is arranged in the Paley order. Consequently (cf. (2.5))

$$\xi_{\nu P_N+k}(\tilde{x}) = (\Xi_N(\tilde{x}))^{\nu} \cdot \prod_{j=0}^{N-1}(\Xi_j(\tilde{x}))^{\tilde{a}_j(k)}.$$

Thus (2.15) will be proved if $\tilde{\Phi}_j(x) = \Xi_j(\tilde{x})$ for any $j \ge 0$. But this is true, since

$$\tilde{\Phi}_j(x) = \Phi_{\sigma_{N,\nu}(j)}(x) = \exp\frac{2\pi i x_{\sigma_{N,\nu}(j)}}{p_{\sigma_{N,\nu}(j)}} = \exp\frac{2\pi i \tilde{x}_j}{\tilde{p}_j} = \Xi_j(\tilde{x}),$$

where the second equality is due to (2.3), and the third to (2.14).

The fact that the mapping $F_{N,\nu}$ preserves measure follows from the fact that it reduces to a rearrangement of intervals of the form $(l/P_N; (l+1)/P_N]$, $l = 0, 1, \dots, P_N - 1$. And this results from the fact that $\tilde{x}_j = x_j$ for all $j \ge N$.

The lemma is now proved.

Remark. A slightly stronger result also holds ([9], p. 53): under the hypotheses of the lemma $D^{\Sigma}_n(x) = D^{\xi}_n(\tilde{x})$.

We now introduce certain notation. Let $n = \sum_{j=0}^{\infty} a_j(n)P_j$, where $a_j(n)$ are the coefficients used in writing n in the number system generated by the sequence $\{p_j\}_{j=0}^{\infty}$. We then set

$$(2.16)\qquad \begin{gathered} q_j(n) = \begin{cases} 0 & \text{when } j = 0, \\ \frac{1}{P_j}\sum_{i=0}^{j-1} a_i(n)P_i & \text{when } j \ge 1, \end{cases} \\ b_j(n) = \begin{cases} a_j(n) & \text{when } q_j(n) \le 1/2, \\ a_j(n)+1 & \text{when } q_j(n) > 1/2, \end{cases} \\ c_j(n) = \min(b_j(n), p_j - b_j(n)). \end{gathered}$$

The author ([9], p. 57) has proved the following proposition, which is a generalization to Price systems of a result obtained by Schneider [1] for the Walsh system in the Paley arrangement. An equivalent result was obtained independently of Schneider by Fine [10].

Proposition 1. *Let L^{Π}_n, $n \ge 1$, be the Lebesgue constants for the Price system in the Paley arrangement. Then there exist positive constants $C_2 > C_1$ such that for any $n \ge 1$ the inequalities $C_1\mathrm{H}(n) \le L^{\Pi}_n < C_2\mathrm{H}(n)$ hold, where $\mathrm{H}(n) = \sum_{j=0}^{\infty}\eta_j(n)$, and*

$$\eta_j(n) = \begin{cases} c_j(n) & \textit{if } c_j(n) = 0, 1, \\ 2\max(1, \ln c_j(n)) & \textit{if } c_j(n) \ge 2. \end{cases}$$

(We recall that $L^{\Pi}_n = \int_0^1 |D^{\Pi}_n(x)|\,dx$).

The remark after the proof of the lemma makes it possible to generalize this result from the Price system in the Paley arrangement to Price systems in any arrangement of the class under consideration. To do this it is necessary to take $\tilde{a}_j(n)$—the coefficients of the expansion of n in the corresponding number system—in place of $a_j(n)$.

Corollary 1. *Let the increasing sequence* $\{n_k\}_{k=1}^{\infty}$ *of natural numbers be such that* $\sup_k L_{n_k}^{\Pi} < \infty$. *Then there exists a positive constant* C_3 *such that* $c_j(n_k) < C_3$ *for all* $j \geq 0$ *and* $k \geq 1$.

Corollary 2. *Let the increasing sequence of natural numbers* $\{n_k\}_{k=1}^{\infty}$ *be such that* $\sum_{j=0}^{\infty} a_j(n_k) < C_4$. *Then* $\sup_{k\geq 1} L_{n_k}^{\Pi} < \infty$.

In conclusion we note the following: If $P_j \leq n < P_{j+1}$, then $c_{j+1}(n)$ can assume only the values 0 and 1, while $c_l(n) = 0$ for $l > j+1$. Consequently,

$$\sum_{l=j+1}^{\infty} \eta_l(n) \leq 1. \tag{2.17}$$

3. On property **

We begin by proving that among the rearrangements of the Price system generated by any sequence $\{p_j\}_{j=0}^{\infty}$ ($p_j \geq 2$ for all j) there exists a rearrangement having property $**$.

The Price system in the Kaczmarz arrangement. We set

$$\sigma_{N,\nu}(j) = \sigma_N(j) = \begin{cases} j & \text{when } j \geq N, \\ N-j-1 & \text{when } 0 \leq j < N, \end{cases} \qquad N = 1, 2, \dots. \tag{3.1}$$

Definition. A *Price system in the Kaczmarz arrangement* is a Price system numbered according to the procedure discussed in §2 with the permutations (3.1) taken as the defining permutations.

In the case $p_j = 2$ for all j the resulting system is the Walsh-Kaczmarz system.

Theorem 1. *Let* $\{\varphi_n^{\mathrm{K}}(x)\}_{n=0}^{\infty}$ *be some Price system in the Kaczmarz arrangement, and* $D_n^{\mathrm{K}}(x)$, $n = 1, 2, \dots$, *its Dirichlet kernels. Then*

$$\sup_{n\geq 1} |D_n^{\mathrm{K}}(x)| = \infty \quad \textit{almost everywhere.}$$

In the special case of the Walsh system Theorem 1 is a corollary of the results of Schneider [1] and Balashov [11], who studied the rate at which the quantity $s_k(x) = \sup_{1\leq n<k} |D_n^{\mathrm{K}}(x)|$ tends to infinity in the case of the Walsh-Kaczmarz system.

Proof of Theorem 1. We fix some positive constant A, and we shall prove that

$$\sup_{n\geq 1} |D_n^{\mathrm{K}}(x)| > A \quad \text{almost everywhere}, \tag{3.2}$$

whence by the arbitrariness of A the assertion of the theorem will follow.

In the proof it seems reasonable to consider two cases separately.

Case 1. Suppose the sequence that generates the Price system is bounded:

$$\sup_{j\geq 0} p_j = P < \infty. \tag{3.3}$$

We use the notation

$$\widetilde{P}_j(N) = \begin{cases} 1 & \text{when } j = 0, \\ \prod_{k=0}^{j-1} p_{N-k-1} & \text{when } 1 \leq j < N, \end{cases} \qquad N = 2, 3, \dots, \tag{3.4}$$

i.e., $\widetilde{P}_j(N)$ is the product of the first j terms of the sequence $\{\tilde{p}_j(N)\}_{j=0}^{\infty}$, where $\tilde{p}_j(N) = p_{\sigma_N(j)}$ (cf. (3.1)).

Let C be some positive constant. Then for any sufficiently large N there exists $j(N)$ such that

$$\widetilde{P}_{j(N)-1}(N) \leq C < \widetilde{P}_{j(N)}(N) < P_N. \tag{3.5}$$

We use the notation

$$\widetilde{P}(N, C) = \widetilde{P}_{j(N)}(N) \tag{3.6}$$

(we recall that $P_j = \prod_{k=0}^{j-1} p_j$).

Applying the lemma from §2, we obtain

$$D^{\mathrm{K}}_{P_N+\tilde{P}(N, C)}(x) - D_{P_N}(x) = \Xi_N(\tilde{x}) \cdot D^{\xi}_{\tilde{P}(N, C)}(\tilde{x})$$

(the lemma is applicable due to the rightmost inequality of (3.5)), where

$$\tilde{x} = F_N(x) \leftrightarrow \tilde{x}_j = \begin{cases} x_{N-j-1} & \text{when } 0 \leq j < N, \\ x_j & \text{when } j \geq N. \end{cases} \tag{3.7}$$

Here $D^{\xi}_k(x)$ are the Dirichlet kernels of the Price system in the Paley arrangement generated by the sequence $\{p_j\}_{j=0}^{\infty}$ rearranged using the permutation σ_N, and $\{\Xi_j(\tilde{x})\}_{j=0}^{\infty}$ is the corresponding system of generalized Rademacher functions.

The quantity $\widetilde{P}(N, C)$ is (cf. (3.4), (3.6)) the product of the first $j(N)$ terms of the sequence that generates the system $\{\xi_n(x)\}_{n=0}^{\infty}$; consequently, applying the identity (2.10), we obtain

$$D^{\xi}_{\widetilde{P}(N, C)}(\tilde{x}) = \widetilde{P}(N, C) \qquad \text{when } x \in [0; 1/\widetilde{P}(N, C)].$$

Therefore, taking account of the fact that the absolute value of the generalized Rademacher functions is one, we obtain

$$|D^{\mathrm{K}}_{P_N+\tilde{P}(N, C)}(x) - D_{P_N}(x)| = \widetilde{P}(N, C)$$

for any x belonging to E_N, where

$$E_N = F_N^{-1}([0; 1/\widetilde{P}(N, C)]), \tag{3.8}$$

whence, using one of the inequalities (3.5), we obtain

$$\max(|D_{P_N}(x)|, |D^{\mathrm{K}}_{P_N+\tilde{P}(N, C)}(x)|) > C/2 \qquad \text{for any } x \in E_N. \tag{3.9}$$

We now verify that $\bigcup_N E_N$ is of full measure. Indeed, first of all, $\tilde{x} \in [0; 1/\widetilde{P}_l] \leftrightarrow x_j = 0$ for $0 \leq j < l$, whence, using the definition of the set E_N in (3.8) and the mapping F_N of (3.7), we obtain $x \in E_N \leftrightarrow x_j = 0$ for $N - j \leq j < N$. Thus if we take the N not in order but such that the inequality $N_k < N_{k+1} - j(N_{k+1})$ holds, the sets E_{N_k} will be pairwise independent. It is possible to do so because the sequence $\{j(N)\}$ is bounded (indeed, $\widetilde{P}(N, C) = 2^{j(N)}$ since $p_j \geq 2$, and so by the definition (3.6) we have $j(N) \leq \log_2 C + 1$).

Second, according to the lemma of §2, the mapping F_N preserves measure. Consequently (cf. (3.8)), we have $\mu E_N = 1/\widetilde{P}(N, C)$. Using the assumption that the generating sequence is bounded (cf. (3.3)), the definition of $\widetilde{P}(N, C)$, and the leftmost inequality of (3.5), we obtain $\widetilde{P}(N, C) = P_{N-j(N)-1}\tilde{p}_{j(N)-1}(N) \leq PC$. Hence $\mu E_N \geq 1/PC$.

Thus, the measures of all the sets E_N cannot be smaller than $1/PC$, and among them one can select infinitely many pairwise independent sets. Consequently, by the Borel-Cantelli lemma, the set $\bigcup_N E_N$ has full measure, which together with (3.9) yields (3.2) with $A = C/2$.

Case 2. Let the sequence that generates the Price system be unbounded, i.e.,

$$\sup_{j\geq 0} p_j = \infty. \tag{3.10}$$

Let us fix some $C > 0$ and impose the additional condition

$$p_j \geq 100C \quad \forall j \geq 0 \tag{3.11}$$

on the generating sequence $\{p_j\}$. This condition is inessential, since the reasoning given below also goes through for any subsequence of the generating sequence.

Now, as in Case 1, we select a subsequence of indices such that the corresponding Dirichlet kernels will be large on sets that are independent and have measures larger than some positive constant, making it possible to establish (3.2) by relying once again on the Borel-Cantelli lemma. In the sequel we shall need an estimate for the Dirichlet kernels of the Price system in the Paley arrangement proved by the author in [9], p. 43. The remark after the proof of the lemma in §2 makes it possible to extend this estimate to the Dirichlet kernels of the Price system in any arrangement belonging to the class we are considering.

Proposition 2. *Let* $P_N \leq n < P_{N+1}$ *and* $x \in (l/P_j;\,(l+1)/P_j]$ *for some* l *and* j *such that* $1 \leq l < P_j$, $1 \leq j \leq N$. *Then*

$$\begin{aligned} P_j\left(-\frac{1}{2}+\frac{p_j}{\pi}\frac{|\sin\frac{\pi l c_j(n)}{p_j}|}{\min(l,\,p_j-l)}\right) &\leq |D_n^{\Pi}(x)| \\ &\leq P_j\left(\frac{1}{2}+\frac{p_j}{2}\frac{|\sin\frac{\pi l c_j(n)}{p_j}|}{\min(l,\,p_j-l)}\right) \qquad \textit{when } c_j(n)\neq 0 \end{aligned} \tag{3.12}$$

and

$$D_n^{\Pi}(x) \equiv P_j \min(q_j(n),\,1-q_j(n)) \qquad \textit{when } c_j(n) = 0.$$

(For the definition of the quantities $c_j(n)$ and $q_j(n)$ cf. (2.16).)

We now return to the proof of the theorem.

We shall assume from now on that the constant C is an integer. We fix some $N \geq 0$. Applying the lemma of §2, as in Case 1, we obtain $|D^{\mathrm{K}}_{P_N+C}(x) - D_{P_N}(x)| = |D^{\xi}_{2C}(\tilde{x})|$, where $\tilde{x} = F_N(x)$. (For the definition of the mapping F_N cf. (3.7).) But $D^{\xi}_{2C}(\tilde{x})$ is the Dirichlet kernel in some Price system in the Paley arrangement, and consequently estimates (3.12) apply to it with $j=0$, $\tilde{p}_0 = p_{N-1}$ in place of p_0, $\widetilde{P}_0 = 1$, $\widetilde{P}_1 = \tilde{p}_0$, $\tilde{c}_0(2C) = 2C$, since (cf. (3.11)) $2C < \tilde{p}_0/2$. (For the definition of the quantity $c_j(n)$ cf. (2.16).)

We set

$$V_N = \bigcup_{l,\,1\leq l\leq \tilde{p}_0/4C} (l/\widetilde{P}_1;\,(l+1)/\widetilde{P}_1];$$

then for any $\tilde{x} \in V_N$ the leftmost of inequalities (3.12) yields

$$|D^{\xi}_{2C}(\tilde{x})| \geq -\frac{1}{2} + \frac{\tilde{p}_0}{\pi}\frac{4C}{\tilde{p}_0} \geq C$$

(in the sine function of the numerator we have substituted $\tilde{p}_0/4C$ for l, i.e., the largest possible value, and we have used the fact that the function $(\sin \pi x)/x$ is monotone decreasing on $[0,1]$). We use the notation $U_N = F_N^{-1}(V_N)$. (For the definition of the mapping F_N cf. (3.7).)

We remark that by the definition of the set V_N

$$\mu V_N = ([\tilde{p}_0/4C] - 1)/\tilde{p}_0 > 1/5C,$$

since $\tilde{p}_0$ is large in comparison with C (cf. (3.11)) and $\widetilde{P}_1 = \tilde{p}_0$. But the mapping F_N preserves measure, and consequently $\mu U_N \geq 1/5C$ for all N.

Moreover, by definition

$$V_N = \{\tilde{x} : 1 \leq \tilde{x}_0 < \tilde{p}_0/4C\}.$$

Consequently,

$$U_N = \{\tilde{x} : 1 \leq x_{N-1} < p_{N-1}/4C\},$$

from which one can see that the sets U_N are independent for different N. Thus, according to the Borel-Cantelli lemma, $\bigcup_N U_N$ has full measure.

Thus, we have proved that

$$\sup_N |D^{\mathrm{K}}_{P_N+2C}(x) - D_{P_N}(x)| \geq C \quad \text{almost everywhere},$$

and so

$$\sup_N \max(|D_{P_N}(x)|, |D^{\mathrm{K}}_{P_N+2C}(x)|) \geq C/2 \quad \text{almost everywhere},$$

which yields (3.2) with A replaced by $C/2$. Since the generating sequence is unbounded (cf. (3.10)), analogous constructions can be carried out for any C, no matter how large. Moreover, the condition (3.11) is inessential, since for any C there exists a subsequence of indices for which it holds, and this suffices for our reasoning.

Thus, inequality (3.2) holds almost everywhere for any A, which implies the assertion of the theorem.

Theorem 1 is now proved completely.

Remark. In the course of the proof of Theorem 1 the quantity $|D^{\mathrm{K}}_{P_N+k}(x) - D_{P_N}(x)|$ was estimated for a specially chosen k using the lemma of §2. The same results could have been obtained using the same reasoning, if we estimated $|D^{\mathrm{K}}_{\nu P_N+k}(x) - D_{\nu P_N}(x)|$ for the same k and any ν, $1 \leq \nu < p_N$. Thus, to prove (3.2) it would suffice to have just one subpacket arranged in the Kaczmarz order in each packet.

We shall now prove that the Price system generated by a bounded sequence, numbered using the procedure discussed in §2, and possessing property $**$, also possesses property $*$.

We recall our notation: $\{\varphi^{\Sigma}_n(x)\}_{n=0}^{\infty}$ is a Price system with an arrangement from the class defined in §2; $D^{\Sigma}_n(x)$, $n \geq 1$, are its Dirichlet kernels; L^{Σ}_n, $n \geq 1$, are its Lebesgue constants; and $S^{\Sigma}_n(f, x)$ are the partial sums of the Fourier series of an integrable function f in the system $\{\varphi^{\Sigma}_n(x)\}$.

Theorem 2. *Let $\{\varphi^{\Sigma}_n(x)\}$ be the Price system generated by a bounded sequence $\{p_j\}_{j=0}^{\infty}$ in some arrangement belonging to the class defined in §2, and suppose there exists a set E with $\mu E > 0$ such that*

$$\sup_{n\geq 1} |D^{\Sigma}_n(x)| = \infty \qquad \text{for any } x \in E. \tag{3.13}$$

Then there exist an integrable function f and an increasing sequence $\{n_k\}$ of natural numbers such that

$$\sup_k L^{\Sigma}_{n_k} = L < \infty, \tag{3.14}$$

yet

$$\sup_k |S^{\Sigma}_{n_k}(f, x)| = \infty \quad \text{almost everywhere.} \tag{3.15}$$

A special case of this proposition relating to the Walsh system was published by the author in [3].

Proof of Theorem 2. First of all we construct an increasing sequence $\{n_k\}_{k=1}^{\infty}$ of natural numbers such that condition (3.14) holds and

$$\sup_{k\geq 1}|D_{n_k}^{\Sigma}(x)| = \infty \qquad \text{for any } x\in E. \tag{3.16}$$

After this is done, the construction of an integrable function satisfying (3.15) will present no difficulty.

For a natural number k there exist unique $N(k)>0$ and $\nu(k)$, $1\leq \nu(k) < p_{N(k)}$, such that $\nu(k)P_{N(k)}\leq k < (\nu(k)+1)P_{N(k)}$. We use the notation $R(k)=\nu(k)P_{N(k)}$.

We fix some positive constant C. According to (3.13), there exists an increasing sequence of indices $\{s_i(C)\}_{i=1}^{\infty}$ such that

$$\sup_{i\geq 1}|D_{s_i(C)}^{\Sigma}(x)| > C \qquad \text{for any } x\in E. \tag{3.17}$$

From this subsequence we construct a second one: we use the notation

$$q_i(C) = R(s_i(C)) + P^*(s_i(C)), \tag{3.18}$$

where the quantity $P^*(s)$ is chosen as follows. Let $P=\sup_{j\geq 0} p_j$. If $N(s)$ is sufficiently large, for example if $P_{N(s)} > CP$, there exists $j(s, C)$ such that, in the notation

$$P^*(s) = \prod_{j=0}^{j(s,C)-1} p_{\sigma_{N(s),\nu(s)}(j)}, \tag{3.19}$$

the following inequalities hold:

$$C/P^2 \leq P^*(s) < 2C/P < P_{N(s)}. \tag{3.20}$$

We first verify that if

$$|D_s^{\Sigma}(x)| \geq C, \tag{3.21}$$

then

$$\max(|D_{R(s)}(x)|, |D_{R(s)+P^*(s)}^{\Sigma}(x)|) \geq C/P^2. \tag{3.22}$$

Indeed,

$$D_s^{\Sigma}(x) = D_{R(s)}(x) + (D_s^{\Sigma}(x) - D_{R(s)}(x)).$$

It follows from (3.21) that the absolute value of at least one of the terms is larger than $C/2$. If that term is the first, then (3.22) is proved; if it is the second, then applying to it the lemma of §2, we obtain

$$|D_s^{\Sigma}(x) - D_{R(s)}(x)| \equiv |D_{s-R(s)}^{\xi}(\tilde{x})| \geq C/2,$$

where $D_n^{\xi}(x)$ are the Dirichlet kernels of the Price system in the Paley arrangement generated by the sequence $\{\tilde{p}_j\}_{j=0}^{\infty}$ with $\tilde{p}_j = p_{\sigma_{N(s),\nu(s)}(j)}$.

From results of Vilenkin [12], [13] and Shcherbakov [14], who studied the majorants of the Dirichlet kernels of Price systems in the Paley arrangement, it follows in particular that $|D_n^{\xi}(\tilde{x})| \leq P/\tilde{x}$ for any $n\geq 1$ and for any $\tilde{x}\neq 0$, whence $\tilde{x} < 2P/C$.

On the other hand, we obtain similarly

$$|D_{R(s)+P^*(s)}^{\Sigma}(x) - D_{R(s)}(x)| \equiv |D_{P^*(s)}^{\xi}(\tilde{x})|.$$

But $P^*(s)$ is (cf. (3.19)) the product of a certain number of initial terms of the generating sequence. Consequently, the equality (2.10) is applicable to $D^{\xi}_{P^*(s)}(\tilde{x})$, and according to that equality $|D^{\xi}_{P^*(s)}(\tilde{x})| = P^*(s)$ for any $x \in [0; 1/P^*(s)]$. Consequently (cf. (3.20)) $D^{\xi}_{P^*(s)}(\tilde{x}) \geq C/P^2$ for any $x \in [0; 1/P^*(s)]$. But the leftmost inequality of (3.20) yields $2P/C \leq 1/P^*(s)$. This completes the proof that (3.22) follows from (3.21).

The proposition just proved, together with (3.17) and the definition of the sequence $\{q_i(C)\}_{i=1}^{\infty}$, yields

$$\sup_{i\geq 1} \max(|D_{R(s_i(C))}(x)|, |D^{\Sigma}_{q_i(C)}(x)|) \geq \frac{C}{2P^2} \tag{3.23}$$

for any $x \in E$.

We now verify that the Lebesgue constants with the indices of the Dirichlet kernels occurring in (3.23) are bounded. To do this we remark that in the expression of $R(s)$ in the number system generated by the sequence $\{\tilde{p}_j\}_{j=0}^{\infty}$, only one coefficient is nonzero, while in the expression for $q_i(C)$ there are two nonzero coefficients. But since the generating sequence is bounded, the coefficients cannot exceed $P-1$. Hence the hypotheses of Corollary 2 of Proposition 1 in §2 hold, and accordingly the Lebesgue constants with these indices are bounded:

$$\sup_{i} \max(L_{R(s_i(C))}, L^{\Sigma}_{q_i(C)}) < \infty, \tag{3.24}$$

and the bound is independent of C.

As the sequence $\{n_k\}_{k=1}^{\infty}$, one can take, for example, the set of terms of the sequences $\{R(s_i(2^r))\}_{i,r=1}^{\infty}$ and $\{q_i(2^r)\}_{i,r=1}^{\infty}$ arranged in increasing order. With this definition, (3.14) follows from (3.24) and (3.16) from (3.23).

We now turn to the definition of the integrable function $f(x)$ satisfying (3.15).

By hypothesis the set E has positive measure, and consequently, moving it over the modified interval, one can obtain a set of full measure (we recall that the Haar measure on the modified interval is invariant under the operation $\dot{+}$), i.e., there exists a sequence $\{x_l\}_{l=1}^{\infty}$, $x_l \in [0; 1]$, such that $\mu \bigcup_{l=1}^{\infty}(E \dot{+} x_l) = 1$.

Because of (3.16) we can choose a strictly increasing sequence $\{N_r\}_{r=1}^{\infty}$ such that for some set $H_r \subset E$ the inequality $\sup_k |D^{\Sigma}_{n_k}(x)| > 3^r$ holds for any $x \in H_r$, $P_{N_r} < n_k < P_{N_{r+1}}$, and $\mu(E \setminus H_r) < 1/r$. The function $f(x)$ can be taken as

$$f(x) = \sum_{r=1}^{\infty} 2^{-r}(D_{Q_{r+1}}(x \dot{-} x_{l(r)}) - D_{Q_r}(x \dot{-} x_{l(r)}),$$

where $Q_r = P_{N_r}$ and $l(r) = k$ if $r = 2^j + k$, $1 \leq k \leq 2^j$, $j = 0, 1, \dots$, $l(1) = 1$.

The integrability of the function f follows from the fact (cf. (2.11)) that

$$\int_0^1 |D_{P_N}(x)|\,dx = 1 \qquad \text{for any } N \geq 0.$$

It is not difficult to verify that (3.15) holds, for example, by choosing the terms with the same translate (the lth) from the sum that defines f. In doing this we obtain a function f_l with Fourier series that diverges unboundedly almost everywhere on $E \dot{+} x_l$ over the sequence $\{n_k\}$. But the union of these translates has full measure. The different terms of the sum that defines f do not "interfere" with one another, since they are disjoint polynomials. Relation (3.14) has already been verified.

Theorem 2 is now proved completely.

The theorem given below shows that the condition of boundedness of the generating sequence in Theorem 2 is essential, i.e., for Price systems generated by unbounded sequences, property $*$ does not in general follow from property $**$.

In the proof of Theorem 3 we shall need the following proposition proved by the author ([9], p. 86), which is a generalization to Price systems in the Paley arrangement of the result of Schneider [1] mentioned in the Introduction.

Proposition 3. *Let $\{\varphi_n^{\Pi}(x)\}_{n=0}^{\infty}$ be a Price system in the Paley arrangement, and let L_n^{Π}, $n \geq 1$, and $S_n^{\Pi}(f, x)$, $n \geq 1$, respectively, be the Lebesgue constants and the partial sums of the Fourier series of an integrable function f in this system. If the increasing sequence $\{n_k\}_{k=1}^{\infty}$ of natural numbers is such that $\sup_{k\geq 1} L_{n_k}^{\Pi} < \infty$, then $S_{n_k}^{\Pi}(f, x) \to f(x)$ almost everywhere for any integrable function f.*

Theorem 3. *Let the sequence $\{p_j\}_{j=0}^{\infty}$ be unbounded. Then in the class of rearrangements of the Price system corresponding to it (cf. §2) there exists a rearrangement $\{\varphi_n^{\Sigma}(x)\}_{n=0}^{\infty}$ such that*

1. *$\sup_{n\geq 1} |D_n^{\Sigma}(x)| = \infty$ almost everywhere, where D_n^{Σ} are the Dirichlet kernels of the system $\{\varphi_n^{\Sigma}(x)\}$;*
2. *if the increasing sequence of natural numbers $\{n_k\}_{k=1}^{\infty}$ is such that $\sup_{k\geq 1} L_{n_k}^{\Sigma} < \infty$, then for any integrable function f*

$$S_{n_k}^{\Sigma}(f, x) \to f(x) \quad \textit{almost everywhere},$$

where L_n^{Σ} and $S_n^{\Sigma}(f, x)$ are, respectively, the Lebesgue constants in the system $\{\varphi_n^{\Sigma}(x)\}_{n=0}^{\infty}$ and the partial sums of the Fourier series in the system.

Proof of Theorem 3. We assume at first that the sequence $\{p_j\}_{j=0}^{\infty}$ that generates the Price system is such that

$$p_{j+1} > p_j, \qquad j = 0, 1, \dots . \tag{3.25}$$

We use the notation $\nu_j = [p_j/2]$ (where the brackets denote the greatest integer function).

We now define the permutations that determine the arrangement. In one subpacket of each packet we set up the Kaczmarz arrangement:

$$\sigma_{N,\nu_N}(j) = \begin{cases} N - j - 1 & \text{when } 0 \leq j < N, \\ j & \text{when } j \geq N, \end{cases}$$

and in all other subpackets we keep the Paley ordering:

$$\sigma_{N,\nu}(j) = j \qquad \text{for any } j,\ N = 0, 1, \dots, \nu \neq \nu_N.$$

We denote the Price system numbered using these arrangements by $\{\varphi_n^{\Sigma}(x)\}_{n=0}^{\infty}$.

The fact that $\sup_{n\geq 1} |D_n^{\Sigma}(x)| = \infty$ almost everywhere follows from the remark to Theorem 1, since there is an infinite set of subpackets arranged in the Kaczmarz order.

If n belongs to the Kaczmarz subpacket of the jth packet, we set $a_j(n) = \nu_j$, where $a_j(n)$ is the leading coefficient of n in the corresponding number system. Moreover, $c_j(n) \geq \nu_j - 1$ (for the definition of the coefficients $c_j(n)$ cf. (2.16)). But ν_j grows unboundedly together with p_j. Consequently, if infinitely many terms of the sequence $\{n_k\}_{k=1}^{\infty}$ fall into Kaczmarz subpackets, then $\overline{\lim}_{k\to\infty} c_{j_k}(n_k) = \infty$, where $j(k)$ denotes the index of the packet containing n_k. But this implies, according to Corollary 1 of Proposition 1 of §2, that the sequence $\{L_{n_k}^{\Sigma}\}_{k=1}^{\infty}$ is unbounded.

Thus, if the Lebesgue constants $L_{n_k}^{\Sigma}$ are bounded, then all n_k from some k_0 on belong to the Paley subpackets. But in this case $D_{n_k}^{\Sigma}(x) \equiv D_{n_k}^{\Pi}(x)$ for $k > k_0$, since the permutations of the class under consideration preserve subpackets. This means that

$$S^{\Sigma}_{n_k}(f, x) = S^{\Pi}_{n_k}(f, x) \qquad \text{for any } k > k_0.$$

But for partial sums of the Fourier series in the Price system in the Paley arrangement, according to Proposition 3 given before Theorem 3, the boundedness of the Lebesgue constants implies that the corresponding sequence of partial sums converges almost everywhere.

Condition (3.25) that the generating sequence be increasing is inessential, since all the constructions carried out above go through for an increasing subsequence also, and such a subsequence exists if the Paley arrangement is kept in the other packets, since $\{p_j\}_{j=0}^{\infty}$ is unbounded.

Theorem 3 is now proved.

To complete the picture we exhibit one more cause-effect connection: Property $**$ (in fact, a significantly weaker property) implies the absence of a localization principle.

Proposition. *Let the rearrangement from the class defined in §2 for some Price system be such that there exists $x_0 \neq 0$ such that*

$$\sup_{n\geq 1} |D^{\Sigma}_n(x_0)| = \infty,$$

where D^{Σ}_n, $n \geq 1$, are the Dirichlet kernels for the rearranged system. Then there is no localization principle for Fourier series in this system.

Indeed, let N_0 be such that $1/P_{N_0} < x_0$. Then according to (2.10)

$$\operatorname{supp} D_{P_N}(x) \subset [0; 1/P_{N_0}]$$

when $N \geq N_0$. Hence one can select an increasing sequence $\{N_k\}_{k=1}^{\infty}$, $N_k > N_0$ for any k, such that the Fourier series of the function $f(x) = \sum_{k=1}^{\infty} 2^{-k} D_{Q_k}(x)$, where $Q_k = P_{N_k}$, will diverge at x_0. The integrability of f follows from the fact (cf. (2.11)) that $\int_0^1 |D_{P_j}(x)|\,dx = 1$ for any $j \geq 0$. At the same time the numbers N_k are chosen so that

$$\operatorname{supp} f \subset [0; 1/P_{N_0}] \not\ni x_0.$$

It follows from this that no localization principle can hold for a Fourier series in this system.

4. On the property $*$ and the localization principle

In this section we give examples of four arrangements of any Price system that realize all possible combinations of presence (or absence) of property $*$ and the localization principle for Fourier series. This shows that there is no connection between these two properties. Property $**$ will be absent in all four cases. Thus no combination of the presence (or absence) of a localization principle and property $*$ is sufficient for property $**$. The necessity was studied in §3.

As our first example we can take the Price system in the Paley arrangement. By Proposition 3, it does not have property $*$. The existence of a finite majorant for its Dirichlet kernels everywhere except at zero and a localization principle follow, for example, from the results of Shcherbakov [14].

In this section we shall study only those rearrangements of the class defined in §2 that permute the functions in the same way in all subpackets of a single packet, i.e., the generating permutations $\sigma_{N,\nu}(j)$ are independent of ν. Therefore, the index that shows the subscript of the subpacket will be absent below.

Lemma 1. *Suppose the system* $\{\varphi_n^\Sigma(x)\}_{n=0}^\infty$ *is a Price system numbered according to the procedure described in §2 using permutations* $\sigma_N(j)$ *such that for any natural number* r *there exists* $N(r) > 0$ *with the property that, for any* $N > N(r)$,

$$\sigma_N(j) = j \qquad \text{when } 0 \le j < r. \tag{4.1}$$

Then:

1. *The localization principle holds for Fourier series in the system* $\{\varphi_n^\Sigma(x)\}_{n=0}^\infty$.
2. $\sup_{n\ge1} |D_n^\Sigma(x)| < \infty$ *for any* $x \ne 0$, *where* D_n^Σ *are the Dirichlet kernels of the system* $\{\varphi_n^\Sigma(x)\}$.

Proof. We fix some r and N such that $N > \max(r, N(r))$.

We begin by proving that for any n with $P_N \le n < P_{N+1}$ there exists $q(n)$ such that

$$D_n^\Sigma(x) = D_{q(n)}^\Pi(x) \qquad \text{for any } x > 1/P_r, \tag{4.2}$$

where $P_N \le q(n) < P_{N+1}$ (the superscript Π denotes the Paley arrangement).

Both assertions of the lemma follow obviously from this identity and the fact that the Price system in the Paley arrangement, as was noted above, possesses both of these properties. We remark only that (4.2) remains true after x is replaced by $\dot{-}x$, and this makes it possible to assert that $S_n^\Sigma(f, 0) \equiv S_{q(n)}^\Pi(f, 0)$ if $f(x) = 0$ for any $x \le 1/P_r$, which is sufficient to establish the localization principle.

We now turn to the proof of (4.2). Let $i = kP_r + s$, where $k > 0$ is an integer, $0 \le s < P_r$, and $P_N < i < P_{N+1}$. Then

$$\varphi_i^\Sigma(x) \equiv \varphi_{kP_r}^\Sigma(x)\varphi_s^\Pi(x). \tag{4.3}$$

Indeed, by definition (cf. (2.7)–(2.9))

$$\varphi_i^\Sigma(x) = \prod_{j=0}^{\infty}(\tilde\Phi_j(x))^{\tilde a_j(i)} = \left(\prod_{j=0}^{r-1}(\tilde\Phi_j(x))^{\tilde a_j(i)}\right)\varphi_{kP_r}^\Sigma(x) = \left(\prod_{j=0}^{r-1}(\tilde\Phi_j(x))^{\tilde a_j(s)}\right)\varphi_{kP_r}^\Sigma(x),$$

but by choice $N > N(r)$. Consequently, by hypothesis (cf. (4.1)) we have $\tilde p_j = p_{\sigma_N(j)} = p_j$ for $0 \le j < r$, and so the expressions for the number s in the number systems generated by the sequences $\{p_j\}_{j=0}^\infty$ and $\{\tilde p_j\}_{j=0}^\infty$ coincide; moreover $\tilde\Phi_j(x) = \Phi_{\sigma_N(j)}(x) = \Phi_j(x)$ for $0 \le j < r$. All this, together with the definition of the Price system in the Paley arrangement (cf. (2.4) and (2.5)), yields $\prod_{j=0}^{r-1}(\Phi_j(x))^{a_j(s)} \equiv \varphi_s^\Pi(x)$, which completes the proof of (4.3).

We now remark that there exists k^* such that $\varphi_{kP_r}^\Sigma(x) \equiv \varphi_{k^*P_r}^\Pi(x)$, and $P_N \le k^*P_r < P_{N+1}$.

Indeed, all the permutations of the class under consideration preserve subpackets, and hence also preserve packets; the fact that the Paley index of this function has the form k^*P_r follows from the fact that in the product that defines it there are no generalized Rademacher functions with indices less than r.

We denote $k^*P_r + s$ by $q(i)$. Then $\varphi_i^\Sigma(x) \equiv \varphi_{q(i)}^\Pi(x)$.

We now fix some $x \ge 1/P_r$, so that $D_{kP_r}^\Sigma(x) = D_{k^*P_r}^\Pi(x) = 0$. Indeed, if $kP_r = P_N$, then (cf. (2.10)) $D_{P_N} = 0$, since $N > r$.

Now suppose $j = P_N/P_r$. Then

$$D_{kP_r}^\Sigma(x) = D_{P_N}^\Pi(x) + \sum_{l=j}^{k-1}\sum_{s=0}^{P_r-1}\varphi_{lP_r+s}^\Sigma(x) = D_{P_r}(x)\sum_{l=j}^{k-1}\varphi_{lP_r}^\Sigma(x) = 0,$$

since according to (2.10)

$$D_{P_r} = 0 \qquad \text{for any } x > 1/P_r.$$

On the other hand,

$$D^{\Pi}_{k^* P_r}(x) = D_{P_N}(x) + \sum_{l=j}^{k^*-1} \sum_{s=0}^{P_r-1} \varphi^{\Pi}_{lP_r+s}(x) = D_{P_r}(x) \sum_{l=j}^{k^*-1} \varphi^{\Pi}_{lP_r}(x) = 0.$$

Thus, let

$$n = kP_r + s, \quad 1 \le s \le P_r;$$

then

$$D^{\Sigma}_n(x) = D^{\Sigma}_{kP_r}(x) + \sum_{i=kP_r}^{n-1} \varphi^{\Sigma}_i(x) = \sum_{i=k^* P_r}^{k^* P_r+s-1} \varphi^{\Pi}_i(x)$$

$$= D^{\Pi}_{k^* P_r}(x) + \sum_{i=k^* P_r}^{k^* P_r+s-1} \varphi^{\Pi}_i(x) = D^{\Pi}_{k^* P_r+s}(x) = D^{\Pi}_{q(n)}(x).$$

The identity (4.2), and with it Lemma 1, is proved.

Theorem 4. *In the class herein considered of rearrangements of any Price system there exists* $\{\varphi^{\Sigma}_n(x)\}_{n=0}^{\infty}$ *such that:*

1. *For Fourier series in the system* $\{\varphi^{\Sigma}_n(x)\}$ *there is no localization principle.*
2. $\sup_{n\ge 1} |D^{\Sigma}_n(x)| < \infty$ *for any* $x \ne 0$, *where* D^{Σ}_n *are the Dirichlet kernels of the system* $\{\varphi^{\Sigma}_n(x)\}$.
3. *There exist an integrable function* f *and an increasing sequence of natural numbers* $\{n_k\}_{k=1}^{\infty}$ *such that* $\sup_{k\ge 1} L^{\Sigma}_{n_k} < \infty$, *yet* $\sup_{k\ge 1} |S^{\Sigma}_{n_k}(f, x)| = \infty$ *almost everywhere, where* L^{Σ}_n *and* S^{Σ}_n *are respectively the Lebesgue constants of the system* $\{\varphi^{\Sigma}_n(x)\}$ *and the partial sums of the Fourier series in the system.*

Proof. We define the permutations σ_N that determine the arrangement: For $N \le 10$ let $\sigma_N(j) = j$ for any j. If $10^s < N \le 10^{s+1}$ for some $s \ge 1$, we set

$$\sigma_N(j) = \begin{cases} j & \text{when } 0 \le j < 10^s, \\ j & \text{when } j \ge N, \\ N - l - 1 & \text{when } j = 10^s + l, \quad 0 \le l < N - 10^s. \end{cases} \tag{4.4}$$

These permutations obviously satisfy condition (4.1), which, according to Lemma 1, is sufficient for assertions 1 and 2 of the theorem.

We now introduce the notation $j(s, k) = 10^s + 2ks$, $s \ge 1$, $k \ge 1$.

To construct the function f whose existence is asserted in part 3 of the theorem, we need the following polynomials:

$$Q_{s,k,l}(x) = \begin{cases} 1 + \Phi_{j(s,k)-l}(x) - \overline{\Phi_{j(s,k)-l}(x)}/2i & \text{when } p_{j(s,k)-l} \ne 2, \\ 1 + \Phi_{j(s,k)-l}(x) & \text{when } p_{j(s,k)-l} = 2, \end{cases}$$

$$s = 1, 2, \dots, \quad k = 1, 2, \dots, 4^s, \quad l = 0, 1, \dots, s;$$

$$R_{s,k}(x) = \prod_{l=1}^{s} Q_{s,k,l}(x), \qquad s = 1, 2, \dots, k = 1, 2, \dots, 4^s; \tag{4.5}$$

$$f_s(x) = \prod_{k=1}^{4^s} Q_{s,k,0}(x) R_{s,k}(x), \qquad s = 1, 2, \dots . \tag{4.6}$$

We now establish some of their properties. The polynomials $Q_{s,k,l}(x)$ are nonnegative pairwise independent functions, and

$$\int_0^1 |Q_{s,k,l}(x)|\,dx = \int_0^1 Q_{s,k,l}(x)\,dx = 1$$

for all values of s, k, l. Moreover, choosing different values of $p_{j(s,k)-l}$, it is not difficult to verify that

$$\mu\{x : Q_{s,k,l}(x) \geq 3/2\} \geq 1/4 \quad \text{for any } s, k, l.$$

It follows from the pairwise independence of the $Q_{s,k,l}(x)$ that the $R_{s,k}(x)$ are also pairwise independent. Moreover, if we use the notation

$$E_{s,k} = \{x : |R_{s,k}(x)| \geq (3/2)^s\},$$

then the sets $E_{s,k}$ are pairwise independent, and $\mu E_{s,k} \geq 4^{-s}$ for any s, k, i.e., according to the Borel-Cantelli lemma

$$\mu\,\overline{\lim} E_{s,k} = 1. \tag{4.7}$$

We shall use the notation

$$T(s, k) = \prod_{j=0}^{10^s+s-1} P_{\sigma_{j(s,k)}(j)}.$$

Then

$$S^{\Sigma}_{P_{j(s,k)}+T(s,k)}(f_s, x) - S_{P_{j(s,k)}}(f_s, x) = \Phi_{j(s,k)}(x) R_{s,k}(x) \tag{4.8}$$

for any s and k.

Indeed the left-hand side of (4.8) contains only functions from the packet with index $j(s, k)$; but according to Remark 1 from §2, the products of the form (2.9) that determine them cannot contain functions $\Phi_j(x)$ for $j > j(s, k)$, i.e., all the factors starting from $(k+1)$st in the product defining $f_s(x)$ (cf. (4.6)) can be eliminated.

The left-hand side of (4.8) can contain only functions of the first subpacket of packet $j(s, k)$. Thus, in the product that defines these functions the function $\Phi_{j(s,k)}(x)$ must occur to the first power.

If the function $\varphi^{\Sigma}_n(x)$ belongs to the required subpacket, but the product that defines it contains at least one function $\tilde{\Phi}_j(x)$ $(\tilde{\Phi}_j(x) = \Phi_{\sigma_{j(s,k)}(j)}(x)$ for $j \geq 10^s + s)$, then it no longer belongs to this segment of the series. The reason is that

$$T(s, k) = \prod_{j=0}^{10^s+s-1} \tilde{p}_j, \qquad \tilde{p}_j = p_{\sigma_{j(s,k)}(j)}.$$

Thus this segment of the series can contain only functions that are obtained after multiplication in the right-hand side of (4.8). We note further that it follows immediately from the properties of $Q_{s,k,l}(x)$ that $\int_0^1 |f_s(x)|\,dx = 1$. Thus the function $f(x) = \sum_{s=1}^{\infty} f_s(x)/s^2$ is integrable.

It follows from (4.8) and the fact that the polynomials $f_s(x)$ are disjoint for different s that

$$\begin{aligned} &S^{\Sigma}_{P_{j(s,k)}+T(s,k)}(f, x) - S_{P_{j(s,k)}}(f, x) \\ &\quad= \frac{1}{s^2}\left(S^{\Sigma}_{P_{j(s,k)}+T(s,k)}(f_s, x) - S_{P_{j(s,k)}}(f_s, x)\right) = \frac{1}{s^2}(\Phi_{j(s,k)}(x) R_{s,k}(x)).\end{aligned}$$

Consequently (cf. (4.6)),

$$\left|S^{\Sigma}_{P_{j(s,k)}+T(s,k)}(f, x) - S_{P_{j(s,k)}}(f, x)\right| \geq \frac{(3/2)^s}{s^2}$$

for any $x \in E_{s,k}$, $s = 1, 2, \dots$, $k = 1, 2, \dots, 4^s$.

But for any integrable function

$$S_{P_N}(f, x) \to f(x) \quad \text{almost everywhere as } N \to \infty.$$

The proof of this fact is analogous to the proof that in the case of the Walsh system

$$S_{2^N}(f, x) \to f(x) \quad \text{almost everywhere as } N \to \infty.$$

Consequently,

$$\sup_{\substack{s \geq 1 \\ 1 \leq k \leq 4^s}} |S^{\Sigma}_{P_{j(s,k)}+T(s,k)}(f, x)| = \infty \qquad \text{for any } x \in \overline{\lim} E_{s,k},$$

i.e., almost everywhere (cf. (4.7)).

It remains to show that

$$\sup_{s,k} L^{\Sigma}_{P_{j(s,k)}+T(s,k)} < \infty.$$

This follows from Corollary 2 of Proposition 1 in §2, for in the expression for the numbers $P_{j(s,k)} + T(s, k)$ in the corresponding number systems there are only two nonzero coefficients and they are both equal to 1.

Theorem 4 is now proved.

Remark. If in defining the permutations $\sigma_N(j)$ that determine the arrangement (cf. (4.4)) we reject 10^s, we obtain an example that shows that the Kaczmarz arrangement of any Price system possesses property $*$. The fact that it possesses property $**$ has already been noted (cf. Theorem 1). Consequently, there is also no localization principle for it. Thus, in the given class of arrangements of any Price system the Kaczmarz arrangement is the worst in respect to the relation we are interested in. The optimal one, according to what was said at the beginning of §4, is the Paley arrangement.

In the two remaining examples the permutations $\sigma_N(j)$ that determine the arrangement of the Price system will have the following property:

$$\sigma_N(N-1) = 0 \qquad \text{for any } N \geq 1. \tag{4.9}$$

In what follows we shall need an auxiliary system $\{\omega_k(x)\}_{k=0}^{\infty}$, which can be constructed by repeating the procedure described in §2 for constructing the Price system generated by the sequence $\{q_j\}_{j=0}^{\infty}$, $q_j = p_{j+1}$, $j = 0, 1, \dots$; the role of generalized Rademacher functions here is played by the functions $\Omega_j(x) = \Phi_{j+1}(x)$, $j = 0, 1, \dots$, where $\{\Phi_j(x)\}_{j=0}^{\infty}$ is the system of generalized Rademacher functions corresponding to the sequence $\{p_j\}_{j=0}^{\infty}$. Let the arrangement of the system $\{\omega_k(x)\}_{k=0}^{\infty}$ be given using the permutations

$$\sigma_N^*(j) = \begin{cases} \sigma_{N+1}(j) & \text{when } 0 \leq j < N, \\ j & \text{when } j \geq N. \end{cases} \tag{4.10}$$

The system so defined is the Price system generated by the sequence $\{p_1, p_2, \dots\}$ with arrangement given by the permutations (4.10), "compressed" p_0 times and then extended periodically to the whole interval $[0; 1]$. It is clear that the system $\{\omega_k(x)\}_{k=0}^{\infty}$ is a subsystem of the system $\{\varphi_n^{\Sigma}(x)\}_{n=0}^{\infty}$ generated by the sequence $\{p_0, p_1, p_2, \dots\}$ and numbered using the permutations (4.9).

The Dirichlet kernels of the system $\{\omega_k(x)\}_{k=0}^{\infty}$ will be denoted $D_k^{\omega}(x)$, $k \geq 1$.

Lemma 2. *Let the arrangement of the system* $\{\varphi_n^\Sigma(x)\}$ *be given using permutations* $\sigma_N(j)$, $N = 1, 2, \dots$, *satisfying* (4.9). *Then*

1. *The localization principle does not hold for Fourier series in* $\{\varphi_n^\Sigma(x)\}_{n=0}^\infty$.
2. *The identity*

$$D_n^\Sigma(x) \equiv \sum_{s=0}^{p_0-1} (\Phi_0(x))^s D_{r_s(n)}^\omega(x) \tag{4.11}$$

holds for each $n \ge 1$, *where* D_n^Σ *are the Dirichlet kernels of the system* $\{\varphi_n^\Sigma(x)\}$, *and the quantities* $r_s(n)$ *are defined as follows: Let*

$$n = \nu P_N + \sum_{j=0}^{N-1} \tilde{a}_j(n)\widetilde{P}_j, \qquad 1 \le \nu < p_N,$$

be the expression for the number n *in the number system generated by the sequence* $\{\tilde{p}_j\}_{j=0}^\infty$, $\tilde{p}_j = p_{\sigma_N(j)}$. *Then*

$$r_s(n) = \begin{cases} (\nu+1)P_n/p_0, & \text{if } s < \tilde{a}_{N-1}(n), \\ \nu P_N/p_0 + \sum_{j=0}^{N-2} \tilde{a}_j(n)\widetilde{P}_j & \text{if } s = \tilde{a}_{N-1}(n), \\ \nu P_N/p_0, & \text{if } s > \tilde{a}_{N-1}(n). \end{cases} \tag{4.12}$$

Proof. The permutations that define the arrangement in the Nth packet of the system $\{\varphi_n^\Sigma(x)\}_{n=0}^\infty$ and packet $(N-1)$ of the system $\{\omega_k(x)\}$ (cf. (4.9) and (4.10)) are consistent, hence $p_{\sigma_N(j)} = q_{\sigma_{N-1}^*(j)}$ and $\tilde{\Phi}_j(x) = \Omega_j(x)$ when $0 \le j < N-1$. Consequently, if

$$l = \nu P_N + s\widetilde{P}_{N-1} + t, \qquad 1 \le \nu < p_N, \quad 0 \le s < p_0, \quad 0 \le t < \widetilde{P}_{N-1},$$

then

$$\varphi_l^\Sigma(x) \equiv (\Phi_0(x))^s \omega_{\nu Q_{N-1}+t}(x)$$

$(Q_{N-1} = \widetilde{P}_{N-1} = P_N/P_0)$, i.e., the system $\{\varphi_n^\Sigma(x)\}_{n=0}^\infty$ "decomposes" into p_0 subsystems

$$\{(\Phi_0(x))^s \omega_k(x)\}_{k=0}^\infty, \qquad s = 0, 1, \dots, p_0 - 1.$$

Here the functions $\varphi_n^\Sigma(x)$ such that

$$\nu P_N + s\widetilde{P}_{N-1} \le n < \nu P_N + (s+1)\widetilde{P}_{N-1}$$

form the νth subpacket of packet $(N-1)$ of the system $\{\omega_k(x)(\Phi_0(x))^s\}_{k=0}^\infty$, and the order of the functions is preserved, which guarantees that (4.11) holds.

Remark. The proof given below of the first assertion of the lemma goes through only for the modified interval, though the fact itself holds for the ordinary interval. If the generalized Rademacher functions had been defined to be continuous from the right rather than the left, this proof would also go through.

Let $x_0 = 1/p_0 = \{1, 0, 0, \dots\}$. According to (4.11)

$$|D^\Sigma_{P_N+\widetilde{P}_{N-1}}(x_0) - D_{P_N}(x_0)| = |D^\omega_{Q_{N-1}}(x_0)| = Q_{N-1},$$

since D_k^ω are the Dirichlet kernels of the Price system generated by the sequence $\{q_j\}_{j=0}^\infty$, "compressed" p_0 times and then extended periodically to the whole interval, and $Q_N \to \infty$ as $N \to \infty$. Consequently, $\sup_{n\ge1} |D_n^\Sigma(x_0)| = \infty$, which according to the proposition proved at the end of §3 implies the absence of a localization principle.

Lemma 2 is now proved.

Theorem 5. *In the class of arrangements under consideration for any Price system, there exists* $\{\varphi_n^\Sigma(x)\}_{n=0}^\infty$ *such that*

1) *the localization principle holds for Fourier series in the system;*

2) $\sup_{n\geq 1}|D_n^\Sigma(x)|<\infty$ *almost everywhere, where* D_n^Σ *are the Dirichlet kernels of the system; and*

3) *if the increasing sequence of natural numbers* $\{n_k\}_{k=0}^\infty$ *is such that* $\sup_{k\geq 1} L_{n_k}^\Sigma<\infty$, *then for any integrable function* f, $S_{n_k}^\Sigma(f, x)\to f(x)$ *almost everywhere, where* L_n^Σ *and* S_n^Σ *are, respectively, the Lebesgue constants of the system* $\{\varphi_n^\Sigma(x)\}$ *and the partial sums of the Fourier series in the system.*

Proof. We define the permutations that determine the arrangement:

$$\sigma_N(j)=\begin{cases} j+1 & \text{when } 0\leq j<N-1,\\ 0 & \text{when } j=N-1,\\ j & \text{when } j\geq N.\end{cases} \tag{4.13}$$

These permutations satisfy the condition (4.9). Consequently, the absence of a localization principle is guaranteed by Lemma 2.

If we denote the Price system generated by the sequence $\{p_j\}_{j=0}^\infty$ and arranged using the permutations (4.13) by $\{\varphi_n^\Sigma(x)\}_{n=0}^\infty$, then the system $\{\sigma_k(x)\}_{k=0}^\infty$ corresponding to it, whose construction procedure was discussed before the statement of Lemma 2, is the Price system generated by the sequence $\{p_1, p_2, \dots\}$ in the Paley arrangement, "compressed" p_0 times and then extended periodically to the whole interval (cf. (4.10)). But the Dirichlet kernels of any Price system in the Paley arrangement have almost everywhere a finite majorant. Consequently $\sup_{k\geq 1}|D_k^\omega(x)|<\infty$ almost everywhere, which together with (4.11) yields the second assertion of the theorem.

We now turn to the proof of the third assertion. We fix some integrable function f. It can be represented in the form

$$f(x)=\sum_{s=0}^{p_0-1}(\Phi_0(x))^s f_s(x),$$

where the functions $f_s(x)$, $s=0, 1, \dots, p_0-1$, are periodic with period $1/p_0$.

Applying (4.11), we obtain

$$S_n^\Sigma(f, x)\equiv\sum_{s=0}^{p_0-1}(\Phi_0(x))^s S_{r_s(n)}^\omega(f_s, x),$$

where S_k^ω are the partial sums of the Fourier series in the system $\{\omega_k(x)\}_{k=0}^\infty$.

Hence it suffices to prove that, for any increasing sequence $\{n_i\}_{i=1}^\infty$ of natural numbers, the inequality

$$\sup_{i\geq 1} L_{n_i}^\Sigma<\infty \tag{4.14}$$

implies that $S_{r_s(n_i)}^\omega(f_s, x)\to f_s(x)$ almost everywhere, $s=0, 1, \dots, p_0-1$.

This, because of Proposition 3, will be proved if

$$\sup_{i\geq 1} L_{r_s(n_i)}^\omega<\infty, \qquad s=0, 1, \dots, p_0-1, \tag{4.15}$$

where L_k^ω are the Lebesgue constants of the system $\{\omega_k(x)\}_{k=0}^\infty$, for the functions $f_s(x)$, $s=0, 1, \dots, p_0-1$, are periodic with period $1/p_0$, and the system $\{\omega_k(x)\}$ is the Price system in the Paley arrangement "compressed" p_0 times and then extended periodically to the whole interval.

Thus it remains to prove that (4.15) follows from (4.14).

We shall use $\tilde{a}_j(n)$, $\tilde{b}_j(n)$, $\tilde{c}_j(n)$, $j = 0, 1, \ldots$, to denote the coefficients that occur in the expression of the number n in the number system generated by the sequence $\{p_{\sigma_N(j)}\}_{j=0}^{\infty}$. The symbols $a_j^{\omega}(l)$, $b_j^{\omega}(l)$, and $c_j^{\omega}(l)$ are the coefficients that occur in writing the number l in the number system generated by the sequence $\{q_{\sigma_{N-1}^*(j)}\}_{j=0}^{\infty}$, where $q_j = p_{j+1}$, $j = 0, 1, \ldots$.

We fix some i and let

$$\nu P_N + s^* \widetilde{P}_{N-1} \le n_i < \nu P_N + (s^* + 1)\widetilde{P}_{N-1},$$

where $1 \le \nu < p_N$ and $0 \le s^* < p_0$. Then, by the definition (4.12),

$$\nu Q_{N-1} \le r_s(n_i) \le (\nu + 1)Q_{N-1}, \quad s = 0, 1, \ldots, p_0 - 1.$$

When $s \ne s^*$, there can be only one nonzero coefficient in the expression for $r_s(n_i)$ in the corresponding number system. This is either

$$a_{N-1}^{\omega}(r_s(n_i)) = \begin{cases} \nu & \text{when } s > s^*, \\ \nu + 1 & \text{when } s < s^*,\ \nu + 1 = p_0, \end{cases}$$

or $a_N^{\omega}(r_s(n_i)) = 1$, which is possible when $s < s^*$ and $\nu + 1 = p_0$. In the latter case $L_{r_s(n_i)}^{\omega} = 1$.

In the first case, from the fact that $\tilde{a}_N(n_i) = \nu$ and $a_j^{\omega}(r_s(n_i)) = 0$ for $j < N-1$, it follows that $\tilde{b}_N(n_i)$ and $b_{N-1}^{\omega}(r_s(n_i))$ can differ by at most one. But $p_N = q_{N-1}$, and consequently $\tilde{c}_N(n_i)$ and $c_{N-1}^{\omega}(r_s(n_i))$ also differ by at most one. Moreover, $c_j^{\omega}(r_s(n_i)) = 0$ for $j < N-1$ and for $j > N$ (cf. the last paragraph of §2), while $c_N^{\omega}(r_s(n_i)) \le 1$; consequently, with $s \ne s^*$,

$$\sum_{j=0}^{\infty} c_j^{\omega}(r_s(n_i)) \le \tilde{c}_N(n_i) + 2. \tag{4.16}$$

But according to Corollary 1 of Proposition 1 it follows from (4.14) that there exists a constant C_1 independent of i such that $\tilde{a}_N(n_i) < C_1$. Hence the sum on the left-hand side of (4.16) does not exceed $C_1 + 2$ for all i, which according to Corollary 2 of Proposition 1 implies the existence of a constant C_2 such that

$$\sup_{i \ge 1,\, s \ne s^*} L_{r_s(n_i)} < C_2;$$

this together with (4.14) and the inequality

$$L_{r_{s^*}(n_i)}^{\omega} \le L_{n_i}^{\Sigma} + \sum_{s=0,\, s \ne s^*}^{p_0 - 1} L_{r_s(n_i)}^{\omega},$$

which follows from (4.11), yields (4.15).

Theorem 5 is now completely proved.

Theorem 6. *In the class of rearrangements under consideration for any Price system, there exists* $\{\varphi_n^{\Sigma}(x)\}_{n=0}^{\infty}$ *such that*

1) *the localization principle does not hold for Fourier series in the system;*

2) $\sup_{n \ge 1} |D_n^{\Sigma}(x)| < \infty$ *almost everywhere, where* D_n^{Σ} *are the Dirichlet kernels;*

3) *there exist an integrable function* f *and an increasing sequence of natural numbers* $\{n_k\}_{k=1}^{\infty}$ *such that* $\sup_{k \ge 1} L_{n_k}^{\Sigma} < \infty$ *almost everywhere, where* L_n^{Σ} *and* S_n^{Σ} *are,*

respectively, the Lebesgue constants of the system $\{\varphi_n^{\Sigma}(x)\}_{n=0}^{\infty}$ *and the partial sums of a Fourier series in it.*

Proof. We define the permutations $\sigma_N^*(j)$ that determine the arrangement of the Price system having the desired properties as follows. Let $\sigma_N(j)$ be the permutations defined by (4.4). We set

$$\sigma_{N+1}^*(j) = \begin{cases} \sigma_N(j)+1 & \text{when } 0 \le j < N, \\ 0 & \text{when } j = N, \\ j & \text{when } j \ge N+1, \end{cases} \tag{4.17}$$

$N = 1, 2, 3, \dots$, and $\sigma_1^*(j) = j$ for any j.

Let $\{\varphi_n^{\Sigma}(x)\}$ be a Price system arranged using the permutations (4.17), and $\{\omega_k(x)\}$ the subsystem of it whose construction procedure was described just before the statement of Lemma 2. A comparison of (4.4), (4.10), and (4.17) shows that $\{\omega_k(x)\}$ is the rearrangement of the Price system that was constructed for the proof of Theorem 4, "compressed" p_0 times and then extended periodically to the whole interval.

Hence assertions 2) and 3) of Theorem 6 can be deduced from parts 2 and 3 of Theorem 4 using the identity (4.11). But the localization principle does not hold for Fourier series in the system $\{\varphi_n^{\Sigma}(x)\}$ because of Lemma 2, since the permutations (4.17) possess the property (4.9).

Theorem 6 is now proved.

Bibliography

1. A. A. Shneider [Schneider], *On convergence of Fourier series in the Walsh system*, Mat. Sb. **34 (76)** (1954), 441–472. (Russian)
2. V. V. Tarkaev, *On divergence of Fourier series in the Walsh system*, Extremal Problems, Functional Analysis, and Their Applications, Izdat. Moskov. Gos. Univ., Moscow, 1988, pp. 31–33. (Russian)
3. ——, *On divergence of Fourier-Walsh series*, Vestnik Moskov. Univ. Ser. 1 Mat. Mekh. **1988**, no. 5, 50–53; English transl. in Moscow Univ. Math. Bull. **43** (1988).
4. G. N. Agaev et al., *Multiplicative systems of functions and harmonic analysis on zero-dimensional groups*, "Èlm", Baku, 1981. (Russian)
5. P. R. Halmos, *Measure theory*, Van Nostrand, New York, 1950.
6. J. Gosselin and W. S. Young, *On rearrangements of Vilenkin-Fourier series which preserve almost-everywhere convergence*, Trans. Amer. Math. Soc. **209** (1975), 157–174.
7. F. Shipp [Schipp], *On some rearrangements of series in the Walsh system*, Mat. Zametki **18** (1975), 193–201; English transl. in Math. Notes **18** (1975).
8. V. A. Skvortsov, *On Cesàro means for Fourier series in multiplicative Walsh systems*, Vestnik Moskov. Univ. Ser. 1 Mat. Mekh. **1982**, no. 1, 7–11; English transl. in Moscow Univ. Math. Bull. **36** (1982).
9. V. V. Tarkaev, *On divergence of Fourier series in multiplicative systems*, Candidate's dissertation, Moscow, 1988. (Russian)
10. N. J. Fine, *On the Walsh functions*, Trans. Amer. Math. Soc. **65** (1949), 372–414.
11. L. A. Balashov, *Divergence of Fourier-Walsh series*, Dokl. Akad. Nauk SSSR **274** (1984), 779–780; English transl. in Soviet Math. Dokl. **29** (1984).
12. N. Ya. Vilenkin, *On a class of complete orthogonal systems*, Izv. Akad. Nauk SSSR Ser. Mat. **11** (1947), 363–400; English transl. in Amer. Math. Soc. Transl. (2) **28** (1963).
13. S. Kaczmarz and H. Steinhaus, *Theorie der Orthogonalreihen*, PWN, Warsaw, 1935; reprint, Chelsea, New York, 1951 (supplement (by N. Ya. Vilenkin) to the Russian transl., Fizmatgiz, Moscow, 1958, pp. 457–493; English transl. in Amer. Math. Soc. Transl. (2) **17** (1961)).
14. V. I. Shcherbakov, *The Dini-Lipschitz test and convergence of Fourier series in multiplicative systems*, Anal. Math. **10** (1984), 133–150. (Russian)

Translated by R. L. COOKE

Труды Матем. Инст. Стеклова
198 (1992)

Proceedings of the Steklov
Institute of Mathematics
1994, Issue 1

ON ASYMPTOTICALLY MONOGENIC FUNCTIONS

UDC 517.53

D. S. TELYAKOVSKIĬ

A function $f(z)$ of a complex variable defined in a neighborhood of a point ξ is said to be *asymptotically monogenic* at this point if $f(z)$ has at ξ a finite derivative with respect to z along some measurable set $G(\xi)$ for which ξ is a point of density (with respect to planar Lebesgue measure).

Men′shov proved the following theorem on the holomorphicity of asymptotically monogenic functions.

Men′shov's Theorem ([1], Theorem 2). *A function continuous in a domain D and asymptotically monogenic at each point of it is holomorphic in D except for an at most countable set of points.*

In the same paper Men′shov gave an example, going back to Luzin, of an unbounded function that is asymptotically monogenic in a domain but not holomorphic there, and he posed the question of whether continuity can be replaced by boundedness in his theorem. The author [2] gave a positive answer to this question.

In the present article Men′shov's theorem is generalized to functions f such that $\log^+ |f(z)|$ is integrable with respect to planar Lebesgue measure.

Theorem. *If $f(z)$ is asymptotically monogenic at each point of the domain D and $\log^+ |f(z)|$ is locally integrable in D, then $f(z)$ is holomorphic in D.*

In contrast to Men′shov's theorem, it can be assumed in this theorem that $f(z)$ fails to be asymptotically monogenic on an at most countable set e of points only if additional conditions are imposed on $f(z)$ that exclude isolated singular points. It suffices to assume, for example, that for each point $\xi \in e$ the function $f(z)$ is bounded on some measurable set having ξ as a point of density; the proof of the theorem does not change.

We introduce some notation. The closed disk about ξ of radius r is denoted by $S(\xi, r)$, and its boundary by $C(\xi, r)$. If $A(\xi)$ is a set depending on ξ, then we let $A(\xi, r) = A(\xi) \cap S(\xi, r)$.

Lemma. *Let D be some domain, $P \subset D$ a perfect set with respect to D, and $e \subset P$ an at most countable set of points. If $f(z)$ is defined in D and is asymptotically monogenic at each point $\xi \in P \setminus e$, then there exist a portion Π of P and a number σ with $1/8 > \sigma > 0$ such that, after $f(z)$ is redefined if necessary on $e \cap \Pi$, it is possible to associate with each point $\xi \in \Pi$ a set $E(\xi) \subset S(\xi, \sigma)$ satisfying the following conditions*:

1°. $\mathrm{meas}_2\, E(\xi, r) \geq (1 - \sigma)\pi r^2$ *for* $0 < r \leq \sigma$;
2°. $|f(z) - f(\xi)| \leq 1|z - \xi|/\sigma$ *for* $z \in E(\xi)$.

1991 *Mathematics Subject Classification.* Primary 30A05.

Further,

$$|f(\xi') - f(\xi'')| \le 2|\xi' - \xi''|/\sigma$$

for any $\xi', \xi'' \in \Pi$, $|\xi' - \xi''| \le \sigma$.

The proof of the lemma is based on ideas of Men′shov, who used them to prove Lemmas 1 and 2 in [1]. Here we do not present this proof, because it is analogous to that of a lemma in [2] whose basic idea was communicated to the author by E. P. Dolzhenko.

Proof of the theorem. Assume that $f(z)$ is not holomorphic in D. Denote by P the exceptional set of $f(z)$, i.e., the set of points of D at which f is not holomorphic (as usual, a function is understood to be holomorphic at a point if it is holomorphic in a sufficiently small neighborhood of this point). Since $f(z)$ is holomorphic on $D \setminus P$, the function $\log^+ |f(z)|$ is subharmonic on $D \setminus P$.

It is clear that P is closed with respect to D. We show that it is perfect with respect to D. If not, then interior to D there is an isolated singular point ξ of $f(z)$. In any neighborhood of ξ the function $f(z)$ is not bounded. It can be assumed that $f(\xi) = 0$. For a sufficiently small radius r the function $f(z)$ is holomorphic on $S(\xi, r) \setminus \xi$, at points $z \in G(\xi, r)$ we have

$$\log^+ |f(z)| = 0, \tag{1}$$

and for each radius $\rho < r$

$$\operatorname{meas}_2 G(\xi, \rho) \tfrac{44}{45} \pi \rho^2. \tag{2}$$

Recall that $G(\xi)$ is a set along which an asymptotic derivative exists at the point ξ.

Let

$$M = \sup \log^+ |f(z)| \qquad \text{for } z \in \overline{S(\xi, r) \setminus S(\xi, r/2)}.$$

Since $f(z)$ is unbounded in any neighborhood of ξ, there is a point $\nu_1 \in S(\xi, r/2)$ at which

$$\log^+ |f(\nu_1)| \ge M + 1.$$

Let $a_1 = |\xi - \nu_1|$. Since the function $\log^+ |f(z)|$ is subharmonic in $S(\nu_1, a_1/2)$, it follows that

$$\begin{aligned} \iint_{S(\nu_1, a_1/2)} \log^+ |f(z)|\, dx\, dy &\ge \pi (a_1/2)^2 \log^+ |f(\nu_1)| \\ &\ge \pi (a_1/2)^2 (M+1) = K > 0. \end{aligned} \tag{3}$$

Since $2a_1 < r$, it follows from (2) that

$$\operatorname{meas}_2 G(\xi, \tfrac{3}{2} a_1) \ge \tfrac{44}{45} \operatorname{meas}_2 S(\xi, \tfrac{3}{2} a_1). \tag{4}$$

Therefore, since $S(\nu_1, a_1/2) \subset S(\xi, \frac{3}{2} a_1)$ and

$$\operatorname{meas}_2 S(\xi, \tfrac{3}{2} a_1) = 9 \operatorname{meas}_2 S(\nu_1, a_1/2),$$

we get

$$\operatorname{meas}_2 (G(\xi, \tfrac{3}{2} a_1) \cap S(\nu_1, \tfrac{1}{2} a_1)) \ge \tfrac{4}{5} \operatorname{meas}_2 S(\nu_1, \tfrac{1}{2} a_1), \tag{5}$$

and hence by (1)

$$\operatorname{meas}_2 (z \in S(\nu_1, a_1/2) : \log^+ |f(z)| > 0) \le \tfrac{1}{5} \operatorname{meas}_2 S(\nu_1, a_1/2). \tag{6}$$

Consequently, in view of (3) the disk $S(\nu_1, a_1/2)$ contains a point ν_2 at which

$$\log^+ |f(\nu_2)| \ge 5 \log^+ |f(\nu_1)| \ge 5(M+1).$$

It is clear that $a_2 = |\xi - \nu_1| \geq a_1/2$. By the definition of the number M, the point ν_2 lies in the disk of $S(\xi, r/2)$. Hence, since the function $\log^+|f(z)|$ is subharmonic interior to the disk $S(\nu_2, a_2/2) \subset S(\xi, r)\backslash\xi$, (3) implies that

$$\iint_{S(\nu_2, a_2/2)} \log^+|f(z)|\,dx\,dy \geq \pi(a_2/2)^2 \log^+|f(\nu_2)| \geq \tfrac{5}{4}K. \tag{7}$$

We repeat the arguments, replacing ν_1 by ν_2, $S(\nu_1, a_1/2)$ by $S(\nu_2, a_2/2)$, and $G(\xi, 3a_1/2)$ by $G(\xi, 3a_2/2)$. It follows from relations analogous to (4)–(7) that there is a point $\nu_3 \in S(\nu_2, a_2/2)$, at which

$$\log^+|f(\nu_3)| \geq 5\log^+|f(\nu_2)| \geq 5^2(M+1);$$

moreover, $a_3 = |\xi - \nu_3| \geq a_2/2 \geq (1/2)^2 a_1$ and

$$\int_{S(\nu_3, a_2/2)} \log^+|f(z)|\,dx\,dy \geq (\tfrac{5}{4})^2 K.$$

Arguing further in this way, we get a sequence of points $\nu_i \in S(\xi, r/2)$ at which

$$\log^+|f(\nu_i)| \geq 5^{i-1}(M+1);$$

moreover, $a_i = |\xi - \nu_i| \geq (1/2)^{i-1} a_1$ and

$$\int_{S(\nu_i, a_i/2)} \log^+|f(z)|\,dx\,dy \geq (5/4)^{i-1} K.$$

Hence, for sufficiently large i the value of the integral $\iint_{S(\nu_i, a_i/2)} \log^+|f(z)|\,dx\,dy$ exceeds any preassigned quantity, and this contradicts the assumption that $\log^+|f(z)|$ is integrable in D. Consequently, the exceptional set P does not contain isolated points.

According to the lemma, there exist a positive number $\sigma < 1/81$ and a nonempty portion Π of P with diam $\Pi < \sigma$, such that to each point $\zeta \in \Pi$ there corresponds a set $E(\zeta)$ satisfying the conditions 1° and 2° of the lemma, and $f(z)$ satisfies a Lipschitz condition on Π. Therefore, it can be assumed that for all $\zeta \in \Pi$ and all $z \in E(\zeta)$

$$\log^+|f(z)| = 0. \tag{8}$$

For this we may need to consider the function $(f(z) - f(\omega))/C$ instead of $f(z)$, where ω is some point in Π, and the modulus of the number C is sufficiently large. It will also be assumed that the portion Π is $P \cap S(\xi, R)$, where $\xi \in P$.

In the disk $S(\xi, R/8)$ there is a point ν_1 at which

$$\log^+|f(\nu_1)| > \frac{288}{\pi R^2} \iint_{S(\xi, R)} \log^+|f(z)|\,dx\,dy. \tag{9}$$

Take a radius a_1 such that the disk $S(\nu_1, a_1)$ contains points of Π only on its boundary $C(\nu_1, a_1)$. Let m be the least positive integer with

$$(3/2)^m a_1 > R/4.$$

Let η_1 be an arbitrary point of $\Pi \cap C(\nu_1, a_1)$. Since $|\nu_1 - \eta_1| \leq |\nu_1 - \xi|$, it follows that

$$|\eta_1 - \xi| \leq |\eta_1 + \nu_1| + |\nu_1 - \xi| \leq 2|\nu_1 - \xi| < R/4. \tag{10}$$

We set

$$I_1 = \pi(a_1/2)^2 \log^+|f(\nu_1)|.$$

Since $\log^+|f(z)|$ is subharmonic in $S(\nu_1, a_1/2)$,

$$\iint_{S(\nu_1, a_1/2)} \log^+|f(z)|\,dx\,dy \geq \pi(a_1/2)^2 \log^+|f(\nu_1)| = I_1. \tag{11}$$

We have $S(\nu_1, a_1/2) \subset S(\eta_1, 3a_1/2)$, and

$$\operatorname{meas}_2 S(\eta_1, 3a_1/2) = 9 \operatorname{meas}_2 S(\nu_1, a_1/2).$$

By property $2°$ in the lemma,

$$\operatorname{meas}_2 E(\eta_1, 3a_1/2) \geq \tfrac{80}{81} \operatorname{meas}_2 S(\eta_1, 3a_1/2);$$

therefore,

$$\operatorname{meas}_2\{E(\eta_1, 3a_1/2) \cap S(\nu_1, a_1/2)\} \geq \tfrac{8}{9} \operatorname{meas}_2 S(\nu_1, a_1/2).$$

Consequently, by the condition (8),

$$\operatorname{meas}_2\{z \in S(\nu_1, a_1/2)\colon \log^+|f(z)| > 0\} \leq \tfrac{1}{9} \operatorname{meas}_2 S(\nu_1, a_1/2). \tag{12}$$

Hence, in view of the inequality (10)

$$\log^+|f(\nu_2)| = 9 \log^+|f(\nu_1)| \tag{13}$$

at some point $\nu_2 \in S(\nu_1, a_1/2)$. There is a point ν_2 at which this equality holds, because $\log^+|f(z)|$ is continuous in $S(\nu_1, a_1/2)$.

Take a disk $S(\nu_2, a_2)$ containing points of Π only on its boundary. It is clear that

$$a_1/2 \leq a_2 \leq 3a_1/2. \tag{14}$$

Therefore, if $z \in S(\nu_2, a_2)$, then

$$|z - \eta_1| \leq |z - \nu_2| + |\nu_2 - \nu_1| + |\nu_1 - \eta_1| \leq 3a_1 < 2(3/2)^2 a_1, \tag{15}$$

and $S(\nu_2, a_2) \subset S(\eta_1, 2(3/2)^2 a_1)$.

Let

$$a_2^* = 3a_1/2, \qquad a_{*2} = a_1/2,$$
$$I_2 = \pi(a_2/2)^2 \log^+|f(\nu_2)|, \qquad I_2^* = \pi(a_2^*/2)^2 \log^+|f(\nu_2)|,$$
$$I_{*2} = \pi(a_{*2}/2)^2 \log^+|f(\nu_2)|.$$

By the conditions (13) and (14),

$$\tfrac{9}{4} I_1 = I_{*2} \leq I_2 \leq I_2^* = \tfrac{81}{4} I_1.$$

Let $\eta_2 \in \Pi \cap C(\nu_2, a_2)$ be an arbitrary point. We repeat the preceding arguments, considering ν_2 instead of ν_1, η_2 instead of η_1, and $S(\nu_2, a_2)$ instead of $S(\nu_1, a_1)$. By relations analogous to (11)–(13), the disk $S(\nu_2, a_2/2)$ contains a point ν_3 at which

$$\log^+|f(\nu_3)| = 9 \log^+|f(\nu_2)|.$$

For the point ν_3 take a disk $S(\nu_3, a_3)$ containing points of Π only on its boundary. Here

$$(1/2)^2 a_1 \leq (1/2)a_2 \leq a_3 \leq (3/2)a_2 \leq (3/2)^2 a_1.$$

Therefore, since (15) holds at the point $\nu_3 \in S(\nu_2, a_2)$,

$$\begin{aligned} |z - \eta_1| &\leq |z - \nu_3| + |\nu_3 - \eta_1| \leq a_3 + |\nu_3 - \eta_1| \\ &\leq (3/2)^2 a_1 + 2(3/2)^2 a_1 = 2(3/2)^3 a_1 \end{aligned}$$

at each point $z \in S(\nu_3, a_3)$, and hence $S(\nu_3, a_3) \subset S(\eta_1, 2(3/2)^3 a_1)$.

Let

$$a_3^* = (3/2)^2 a_1, \qquad a_{*3} = (1/2)^2 a_1,$$
$$I_3 = \pi(a_3/2)^2 \log^+ |f(\nu_3)|, \qquad I_3^* = \pi(a_3^*/2)^2 \log^+ |f(\nu_3)|,$$
$$I_{*3} = \pi(a_{*3}/2)^2 \log^+ |f(\nu_3)|.$$

The quantities I_3, I_3^*, and I_{*3} satisfy the relations

$$(9/4)^2 I_1 = I_{*3} \leq I_3 \leq I_3^* = (81/4)^2 I_1.$$

We repeat these arguments as long as the disk $S(\nu_i, a_i)$ lies entirely in $S(\xi, R)$. At the ith step we take a point $\nu_i \in S(\nu_{i-1}, a_{i-1}/2)$ at which

$$\log^+ |f(\nu_i)| = 9 \log^+ |f(\nu_{i-1})| = 9^{i-1} \log^+ |f(\nu_1)|, \tag{16}$$

along with a disk $S(\nu_i, a_i)$ corresponding to ν_i that contains points of Π only on its boundary. The radius a_i satisfies the inequalities

$$(1/2)^{i-1} a_1 \leq (1/2) a_{i-1} \leq a_i \leq (3/2) a_{i-1} \leq (3/2)^{i-1} a_1. \tag{17}$$

Let

$$a_i^* = (3/2)^{i-1} a_1, \qquad a_{*i} = (1/2)^{i-1} a_1,$$
$$I_i = \pi(a_i/2)^2 \log^+ |f(\nu_i)|, \qquad I_i^* = \pi(a_i^*/2)^2 \log^+ |f(\nu_i)|,$$
$$I_{*i} = \pi(a_{*i}/2)^2 \log^+ |f(\nu_i)|.$$

It is clear that

$$(9/4)^{i-1} I_1 = I_{*i} \leq I_i \leq I_i^* = (81/4)^{i-1} I_1,$$

which implies that

$$I_i \geq 9^{1-i} I_i^*. \tag{18}$$

For $i = 2$ and 3

$$S(\nu_i, a_i) \subset S(\eta_1, 2(3/2)^i a_1).$$

We verify that if this relation holds at the $(j-1)$st step, then it holds also at the jth step. Since $\nu_j \in S(\nu_{j-1}, a_{j-1})$, it follows by assumption that

$$|\nu_j - \eta_1| \leq 2(3/2)^{j-1} a_1.$$

Therefore, by (17), at each point

$$|z - \eta_1| \leq |z - \nu_j| + |\nu_j - \eta_1| \leq (3/2)^{j-1} a_1 + 2(3/2)^{j-1} a_1 = 2(3/2)^j a_1,$$

i.e.,

$$S(\nu_j, a_j) \subset S(\eta_1, 2(3/2)^j a_1).$$

Since m is the minimal positive integer such that $(3/2)^m a_1 \geq R/4$, it follows that $(3/2)^m < (3/8)R$. Therefore,

$$S(\eta_1, 2(3/2)^m a_1) \subset S(\eta_1, 3R/4).$$

Since $\eta_1 \in S(\xi, R/4)$ by (10), we have

$$S(\eta_1, 2(3/2)^m a_1) \subset S(\xi, R),$$

and for all $i \leq m$

$$S(\eta_1, 2(3/2)^i a_1) \subset S(\xi, R).$$

At the ith step

$$S(\nu_i, a_i) \subset S(\eta_1, 2(3/2)^i a_1),$$

and the point ν_i is sought as long as the disk $S(\nu_i, a_i)$ lies in $S(\xi, R)$. Hence, there are at least m points ν_i.

According to the definition of m and the relations (16) and (17),

$$I_m^* = \pi \left(\frac{a_m^*}{2}\right)^2 \log^+ |f(\nu_m)| = \frac{\pi}{4}\left(\left(\frac{3}{2}\right)^{m-1} a_1\right)^2 9^{m-1} \log^+ |f(\nu_1)|$$
$$\geq \frac{1}{144} 9^{m-1} \pi R^2 \log^+ |f(\nu_1)|.$$

Consequently, by (9),

$$I_m^* \geq 2 \cdot 9^{m-1} \iint_{S(\xi, R)} \log^+ |f(z)|\, dx\, dy.$$

Therefore, by (18),

$$I_m \geq 2 \iint_{S(\xi, R)} \log^+ |f(z)|\, dx\, dy.$$

Since $\log^+ |f(z)|$ is subharmonic interior to the disk $S(\nu_m, a_m/2)$, it follows that

$$\iint_{S(\nu_m, a_m/2)} \log^+ |f(z)|\, dx\, dy \geq \pi (a_m/2)^2 \log^+ |f(\nu_m)| = I_m$$
$$\geq 2 \int_{S(\xi, R)} \log^+ |f(z)|\, dx\, dy > 0.$$

Since $S(\nu_m, a_m/2) \subset S(\xi, R)$ and $\log^+ |f(z)| \geq 0$, this inequality is impossible. Hence, the assumption that $f(z)$ is unbounded on any disk $S(\xi, \rho)$ is false.

Thus, $f(z)$ is bounded in some neighborhood U of the point $\xi \in \Pi$. Therefore, $f(z)$ is holomorphic in U by the theorem on bounded asymptotically monogenic functions. Consequently, the portion Π is empty, and the exceptional set P is empty. The theorem is proved.

BIBLIOGRAPHY

1. D. E. Men'shov, *On asymptotic monogenicity*, Mat. Sb. **1 (43)** (1936), 189–210. (Russian)

2. D. S. Telyakovskiĭ, *On asymptotically monogenic bounded functions*, Mat. Sb. **129 (171)** (1986), 434–439; English transl. in Math. USSR Sb. **57** (1987).

Translated by H. H. McFADEN

Труды Матем. Инст. Стеклова
198 (1992)

Proceedings of the Steklov
Institute of Mathematics
1994, Issue 1

ON APPROXIMATION OF DIFFERENTIABLE FUNCTIONS OF HIGH SMOOTHNESS BY FOURIER SUMS

UDC 517.518

S. A. TELYAKOVSKIĬ

1. Introduction

We study approximation properties of partial sums of Fourier series on classes of differentiable functions of high smoothness.

Beginning with Lebesgue's paper [1], the behavior of the suprema

$$S_n(\mathfrak{M}, X) := \sup_{f\in\mathfrak{M}} \|f - s_{n-1}(f)\|_X \tag{1.1}$$

of the deviations over function classes as $n \to \infty$ has been used as a characteristic of the approximation properties of Fourier sums. The first results on the order of the suprema (1.1) were obtained by Lebesgue himself, and on their asymptotic behavior by Kolmogorov [2].

We introduce the following function classes in order to formulate Kolmogorov's theorem and the generalizations of it connected with the present article.

Let $W_\alpha^r(C)$ be the class of functions representable as a convolution

$$f(x) = \frac{a_0}{2} + \frac{1}{\pi}\int_0^{2\pi} \psi(t-x)\varphi(t)\,dt, \tag{1.2}$$

where

$$\psi(t) := \sum_{k=1}^{\infty} k^{-r}\cos\left(kt + \frac{\alpha\pi}{2}\right), \tag{1.3}$$

$r > 0$, α is an arbitrary number, and the function φ satisfies the conditions

$$|\varphi(t)| \le 1, \qquad \int_0^{2\pi} \varphi(t)\,dt = 0 \tag{1.4}$$

(see [3], [4]). If $\alpha = r$ here, then φ is equivalent to the rth derivative of f (for nonintegral r we have in mind the derivative in the sense of Weyl). If $\alpha = r - 1$, then φ is equivalent to the function conjugate to the derivative $f^{(r)}$. These classes are denoted by $W^r(C)$ and $\overline{W}^r(C)$, respectively.

Further, $W_\alpha^r(L)$, $r > 0$, α arbitrary, is the class of functions representable as a convolution

$$f(x) = \frac{a_0}{2} + \frac{1}{\pi}\int_0^{2\pi} \psi(t-x)\,dv(t), \tag{1.5}$$

1991 *Mathematics Subject Classification*. Primary 42A10; Secondary 42A16.

where ψ is given by the formula (1.3), and v is a function of bounded variation satisfying the conditions

$$(1.6) \qquad \int_0^{2\pi} |dv(t)| \le 1, \qquad \int_0^{2\pi} dv(t) = 0$$

(see [5], Russian p. 387; [6]). The classes corresponding to $\alpha = r$ and $\alpha = r - 1$ are denoted by $W^r(L)$ and $\overline{W}^r(L)$.

We remark that if only absolutely continuous functions v are considered in the definition of $W^r_\alpha(L)$, then we get a narrower class. The functions in this class are representable as a convolution (1.2), where φ satisfies the conditions

$$\int_0^{2\pi} |\varphi(t)|\, dt \le 1, \qquad \int_0^{2\pi} \varphi(t)\, dt = 0.$$

In this case the values of the suprema (1.1) (for $X = L$) will be the same as for $W^r_\alpha(L)$, but such classes do not contain functions on which the suprema are attained (see [6]).

For brevity we write $S_n(W^r_\alpha, C)$ and $S_n(W^r_\alpha, L)$ instead of $S_n(W^r_\alpha(C), C)$ and $S_n(W^r_\alpha(L), L)$, respectively.

The investigations of Kolmogorov and of the authors who continued his work led to the asymptotic estimate

$$(1.7) \qquad S_n(W^r_\alpha, C) = \frac{4}{\pi^2} n^{-r} \log n + O(n^{-r}).$$

Kolmogorov [2] proved it for the classes $W^r(C)$ with integer r, Pinkevich [7] for the remaining $r > 0$, and Nikol′skiĭ ([8], [9]) for the classes $\overline{W}^r(C)$ with $r > 0$. In the general case the estimate (1.7) follows from results of Efimov [10] and the author [11].

In all these references the numbers r and α were taken to be fixed, and the question of how the remainder term in (1.7) depends on these parameters was not posed. This dependence was first investigated by Sokolov [12], who proved the estimate

$$S_n(W^r_\alpha, C) \le \frac{4}{\pi^2} n^{-r} \log n + c n^{-r}$$

for integral r, where c is an absolute constant. The dependence of the remainder term in (1.7) on r and α was studied by Selivanov [13] and Natanson [14].

An estimate with remainder term uniform with respect to $r > 0$ and α was obtained by the author in [15]:

$$(1.8) \qquad S_n(W^r_\alpha, C) = \frac{4}{\pi^2} n^{-r} \log \frac{n}{\min(n, r+1)} + \frac{2}{\pi r} \left| \sin \frac{\alpha\pi}{2} \right| n^{-r} + O(n^{-r}).$$

It gives the principal term in the asymptotics of the quantities $S_n(W^r_\alpha, C)$ if $r = o(n)$ as $n \to \infty$, and it shows that if this condition does not hold, then $S_n(W^r_\alpha, C) = O(n^{-r})$ uniformly with respect to all the parameters.

The principal term in the asymptotics of the suprema $S_n(W^r_\alpha, C)$ as $n \to \infty$ and $r \to \infty$ was found by Stechkin [16]. He proved that

$$(1.9) \qquad S_n(W^r_\alpha, C) = \frac{8}{\pi^2} n^{-r} K(e^{-r/n}) + O\left(\frac{1}{r} n^{-r}\right)$$

uniformly with respect to $r \ge 1$ and α, where

$$K(q) = \int_0^{\pi/2} \frac{du}{\sqrt{1 - q^2 \sin^2 u}}$$

is the complete elliptic integral of the first kind. For rapidly increasing r (when $r/n \to \infty$ as $n \to \infty$) sharper results were obtained in [16]: if $r \geq n+1$, then

$$S_n(W^r_\alpha, C) = n^{-r}\left\{\frac{4}{\pi} + O\left(\left(1+\frac{1}{n}\right)^{-r}\right)\right\}, \tag{1.10}$$

and if $n+1 \leq r \leq n^2$, then

$$S_n(W^r_\alpha, C) = n^{-r}\left\{\frac{8}{\pi^2}K(e^{-r/n}) + O(rn^{-2}e^{-r/n})\right\}; \tag{1.11}$$

moreover, both these estimates are uniform with respect to all parameters. We remark that for $r = o(n^2)$ the estimate (1.11) is sharper than (1.10).

In [17] the author improved the remainder term in (1.10). Namely,

$$S_n(W^r_\alpha, C) = n^{-r}\left\{\frac{4}{\pi} + O\left(\left(1+\frac{2}{n}\right)^{-r}\right)\right\} \tag{1.12}$$

uniformly with respect to $r \geq n+1$ and all α.

If

$$r = n\log(n/\log n), \tag{1.13}$$

then the expressions $rn^{-2}e^{-r/n}$ and $(1+2/n)^{-r}$ in the remainder terms of the estimates (1.11) and (1.12) have the same order $((\log n)/n)^2$, i.e., these estimates are equivalent. If r grows more slowly as $n \to \infty$, then the estimate (1.11) is better, but if r grows more rapidly, then (1.12) is better.

In [17] it was shown also that for the case when $r \to \infty$ and $r = o(n)$ as $n \to \infty$, the following refinement of the estimate (1.8) is a consequence of (1.9):

$$S_n(W^r_\alpha, C) = n^{-r}\left\{\frac{4}{\pi^2}\log\frac{n}{r} + \frac{12}{\pi^2}\log 2 + O\left(\frac{1}{r} + \frac{r}{n}\log\frac{n}{r}\right)\right\} \tag{1.14}$$

uniformly with respect to $r \geq 1$ and all α. But if the ratio r/n tends to zero slowly as $n \to \infty$, then (1.9) enables us to single out the subsequent terms in the asymptotics of the quantities $S_n(W^r_\alpha, C)$.

Estimates analogous to (1.7)–(1.12) and (1.14) hold also for the suprema $S_n(W^r_\alpha, L)$. An estimate of the form (1.7) for the classes $W^r(L)$ and $\overline{W}^r(L)$ with integral r was established by Nikol′skiĭ [18], who was the first to consider these problems for approximations in the metric of L. In the general case it was obtained by Stechkin and the author [6]. Estimates of $S_n(W^r_\alpha, L)$ analogous to (1.8)–(1.12) and (1.14) were proved in the same articles.

Our main results here are Theorems 1 and 2, which give a refinement of estimates of the form (1.12) for $S_n(W^r_\alpha, C)$ and $S_n(W^r_\alpha, L)$ by singling out the next terms of the asymptotics. Further, the difference in the asymptotic behavior of the suprema is revealed.

Theorem 1. *Let $r \geq 2n+1$. Then*

$$S_n(W^r_\alpha, C) = n^{-r}\left\{\frac{4}{\pi} + \frac{1}{\pi}\left(1+\frac{1}{n}\right)^{-2r} + O\left(\left(1+\frac{4}{n}\right)^{-r}\right)\right\} \tag{1.15}$$

uniformly with respect to all parameters.

We compare the remainder terms in the estimates (1.11) and (1.15). For

$$r = \tfrac{1}{3}n\log(n/\log n) \tag{1.16}$$

each of the expressions $rn^{-2}e^{-r/n}$ and $(1+4/n)^{-r}$ has order $(\log n/n)^{4/3}$, and hence the estimates are equivalent. If r grows more slowly as $n \to \infty$, then the estimate (1.11) is better, but if it grows more rapidly, then (1.15) is better.

Theorem 2. *Suppose that* $r \geq 2n+1$. *Then*

$$S_n(W_\alpha^r, L) = n^{-r}\left\{\frac{4}{\pi} + \frac{1}{\pi}\left(1+\frac{1}{n}\right)^{-2r}\cos^2\frac{\pi}{2n} + O\left(\left(1+\frac{4}{n}\right)^{-r}\right)\right\} \tag{1.17}$$

uniformly with respect to all parameters.

If r satisfies the conditions

$$2n+1 \leq r \leq n\log n, \tag{1.18}$$

then

$$(1+1/n)^{-2r}/n^2 = O((1+4/n)^{-r}),$$

so that in this case the suprema $S_n(W_\alpha^r, C)$ and $S_n(W_\alpha^r, L)$ are equal up to a term of order $(1+4/n)^{-r}$. But if r grows more rapidly than $n(\log n)$ as $n \to \infty$, then the factor $\cos^2(\pi/2n)$ on the right-hand side of (1.17) cannot be dropped. For example, if $r = 2n(\log n)$, then

$$(1+4/n)^{-r} = O(n^{-8}), \qquad (1+1/n)^{-2r} = n^{-4} + O(n^{-8}\log n),$$

and hence in this case

$$S_n(W_\alpha^r, C) = n^{-r}\left\{\frac{4}{\pi} + \frac{1}{\pi}n^{-4} + O(n^{-8}\log n)\right\},$$

$$S_n(W_\alpha^r, L) = n^{-r}\left\{\frac{4}{\pi} + \frac{1}{\pi}n^{-4} - \frac{\pi}{4}n^{-6} + O(n^{-8}\log n)\right\},$$

by (1.15) and (1.17).

Theorems 1 and 2 will be proved in §3, where results relating to more general classes of functions are also obtained. The proofs are based on lemmas contained in §2.

In all the estimates with O-symbols in §§2 and 3 the constants are absolute. This will be mentioned only in the statements and not in the proofs.

2. Lemmas on trigonometric polynomials

In this section we establish for trigonometric polynomials of the form

$$g(x) := \cos nx + \sum_{s=1}^{3} a_s \cos((n+s)x + \alpha_s) \tag{2.1}$$

some properties connected with estimates of their norms in the metric of L with some closely related questions. The results obtained are interesting when the coefficients a_s are sufficiently small.

In this section we assume that the coefficients a_s satisfy the conditions

$$a_1 \geq a_2 \geq a_3 \geq 0 \tag{2.2}$$

and

$$\sum_{s=1}^{3}(n+s)a_s < n/2, \tag{2.3}$$

and the numbers α_s are arbitrary. These conditions will not be mentioned in what follows. We remark that (2.2) and (2.3) yield

$$0 \leq a_1 + a_2 + a_3 < 1/2. \tag{2.4}$$

Let

$$b := a_1^3 + a_1 a_2 + a_3; \qquad b_1 := a_1 b + a_2^2. \tag{2.5}$$

We begin with an estimate of the zeros of the polynomial $g(x)$.

Let

$$t_m := \frac{2m-1}{2n}\pi, \qquad m = 1, \dots, 2n, \tag{2.6}$$

be the zeros of $\cos nx$ on $[0, 2\pi]$. The values $\pm 1/2$ are taken by the function $\cos nx$ at points at a distance $\pi/6n$ from t_m. From this, using the estimate (2.4), we find that $g(x)$ has a zero in the $(\pi/6n)$-neighborhood of each point t_m. There is only one such zero, because the modulus of the derivative of $\cos nx$ in this neighborhood exceeds $n/2$, and hence the derivative of $g(x)$ has constant sign in view of (2.3). Denote this zero by x_m.

Let us estimate the difference $x_m - t_m$. For simplification of the expressions to follow we let

$$y_m := n(x_m - t_m). \tag{2.7}$$

We keep the notation $g(x)$, b, b_1, t_m, x_m, y_m until the end of the section.

By the definition of x_m,

$$\cos nx_m = -\sum_{s=1}^{3} a_s \cos((n+s)x_m + \alpha_s). \tag{2.8}$$

But $\cos nx_m = (-1)^m \sin n(x_m - t_m)$. Therefore, it follows from (2.8) that $|\sin y_m| \le a_1 + a_2 + a_3$, and hence $|y_m| \le \arcsin(a_1 + a_2 + a_3)$. From this, using the fact that a_s is monotone, we find that

$$y_m = O(a_1). \tag{2.9}$$

We refine the estimate (2.9).

Lemma 1. *The estimate*

$$y_m = -a_1 \sin(t_m + \alpha_1) - a_2 \sin(2t_m + \alpha_2) + \tfrac{1}{2} a_1^2 (1 + 1/n) \sin(2t_m + 2\alpha_1) + O(b) \tag{2.10}$$

is valid with an absolute constant in the remainder term.

Proof. We represent the polynomial $g(x)$ by the Taylor formula with Lagrange's remainder in a neighborhood of the point t_m for $x = x_m$:

$$0 = g(t_m) + g'(t_m)\frac{y_m}{n} + \frac{1}{2} g''(t_m + \xi_m)\frac{y_m^2}{n^2},$$

where ξ_m is an intermediate point, i.e., $|\xi_m| \le |x_m - t_m| = |y_m|/n$. From this,

$$y_m = -n\frac{g(t_m)}{g'(t_m)} - \frac{g''(t_m + \xi_m) y_m^2}{2n g'(t_m)}. \tag{2.11}$$

We have

$$\begin{aligned} g(t_m) &= \sum_{s=1}^{3} a_s \cos((n+s)t_m + \alpha_s) \\ &= (-1)^m [a_1 \sin(t_m + \alpha_1) + a_2 \sin(2t_m + \alpha_2)] + O(a_3). \end{aligned} \tag{2.12}$$

Further,

$$
\begin{aligned}
g'(t_m) &= -n\sin nt_m - \sum_{s=1}^{3} a_s(n+s)\sin((n+s)t_m+\alpha_s) \\
&= (-1)^m\left[n+\sum_{s=1}^{3} a_s(n+s)\cos(st_m+\alpha_s)\right]\,;
\end{aligned}
\tag{2.13}
$$

therefore,

$$
\begin{aligned}
(-1)^m n/g'(t_m) &= \left[1+\sum_{s=1}^{3} a_s(1+s/n)\cos(st_m+\alpha_s)\right]^{-1} \\
&= 1-\sum_{s=1}^{3} a_s(1+s/n)\cos(st_m+\alpha_s)+O(a_1^2) \\
&= 1-a_1(1+1/n)\cos(t_m+\alpha_1)+O(a_1^2+a_2)\,.
\end{aligned}
\tag{2.14}
$$

By (2.12) and (2.14),

$$
\begin{aligned}
-ng(t_m)\,g'(t_m) &= -[a_1\sin(t_m+\alpha_1)+a_2\sin(2t_m+\alpha_2)+O(a_3)] \\
&\qquad\times[1-a_1(1+1/n)\cos(t_m+\alpha_1)+O(a_1^2+a_2)] \\
&= -a_1\sin(t_m+\alpha_1)-a_2\sin(2t_m+\alpha_2) \\
&\qquad+\tfrac{1}{2}a_1^2(1+1/n)\sin(2t_m+2\alpha_1)+O(b)\,.
\end{aligned}
\tag{2.15}
$$

We now estimate $g''(t_m+\xi_m)$. Since

$$|\cos n(t_m+\xi_m)| = |\sin n\xi_m| \le |y_m|\,,$$

(2.9) gives

$$|g''(t_m+\xi_m)| \le n^2|\cos n(t_m+\xi_m)| + \sum_{s=1}^{3}(n+s)^2 a_s = O(n^2 a_1)\,.$$

We remark that, by (2.3), (2.13) implies that $|g'(t_m)| > n/2$. Therefore, using the estimate (2.9) once more, we get

$$(g''(t_m+\xi_m)/2ng'(t_m))y_m^2 = O(a_1^3)\,. \tag{2.16}$$

The estimate (2.10) follows from (2.11), (2.15), and (2.16). The lemma is proved.

Lemma 2. *The identities*

$$\sum_{k=1}^{N}\cos(ku+\beta) = \sin\frac{N}{2}u\cos\left(\frac{N+1}{2}u+\beta\right)\Big/\sin\frac{u}{2}\,, \tag{2.17}$$

$$\sum_{m=1}^{2n}\sin^2(t_m+\beta) = n\,, \tag{2.18}$$

$$\sum_{m=1}^{2n}\sin(t_m+\beta)\sin(2t_m+\beta_1) = 0 \tag{2.19}$$

are valid, where the numbers β and β_1 are arbitrary, and u is any number that is not a multiple of 2π.

Proof. The proof of the identity (2.17) is elementary, and is well known (see, for example, [19], 420.4).

The proof of (2.18) makes use of (2.17):

$$\sum_{m=1}^{2n} \sin^2(t_m+\beta) = \frac{1}{2}\sum_{m=1}^{2n}(1-\cos(2t_m+2\beta))$$
$$= n - \frac{1}{2}\sum_{m=1}^{2n}\cos((2\pi/n)m - \pi/n + 2\beta) = n\,.$$

Finally,

$$\sum_{m=1}^{2n}\sin(t_m+\beta)\sin(2t_m+\beta_1) = \frac{1}{2}\sum_{m=1}^{2n}[\cos(t_m-\beta+\beta_1)-\cos(3t_m+\beta+\beta_1)],$$

and again the matter reduces to an application of (2.17). The lemma is proved.

Lemma 3. *Suppose that μ_- and μ_+ are the measures of the sets of points in $[0, 2\pi]$ where $g(x)<0$ and $g(x)>0$, respectively. Then*

$$\mu_- = \pi + O(b)\,, \qquad \mu_+ = \pi + O(b) \tag{2.20}$$

with absolute constants in the remainder terms.

Proof. Since $\mu_- + \mu_+ = 2\pi$, it suffices to prove the estimate for μ_-. We have

$$\mu_- = \sum_{k=1}^{n}(x_{2k}-x_{2k-1})\,.$$

Since $x_m = t_m + (1/n)y_m$, it follows that

$$\begin{aligned}\mu_- &= \sum_{k=1}^{n}\left(\frac{4k-1}{2n}\pi + \frac{1}{n}y_{2k} - \frac{4k-3}{2n}\pi - \frac{1}{n}y_{2k-1}\right)\\ &= \pi + \frac{1}{n}\sum_{k=1}^{n}(y_{2k}-y_{2k-1})\,.\end{aligned} \tag{2.21}$$

Using the estimate (2.10), we find that

$$\begin{aligned}&\sum_{k=1}^{n}(y_{2k}-y_{2k-1})\\ &\quad= \sum_{k=1}^{n}\Big\{ a_1[-\sin(t_{2k}+\alpha_1)+\sin(t_{2k-1}+\alpha_1)]\\ &\qquad + a_2[-\sin(2t_{2k}+\alpha_2)+\sin(2t_{2k-1}+\alpha_2)]\\ &\qquad + \frac{1}{2}a_1^2\left(1+\frac{1}{n}\right)[\sin(2t_{2k}+2\alpha_1)-\sin(2t_{2k-1}+2\alpha_1)] + O(b)\Big\}\\ &\quad= -2a_1\sin\frac{\pi}{2n}\sum_{k=1}^{n}\cos\left(\frac{2\pi}{n}k - \frac{\pi}{n}+\alpha_1\right)\\ &\qquad - 2a_2\sin\frac{\pi}{n}\sum_{k=1}^{n}\cos\left(\frac{4\pi}{n}k-\frac{2\pi}{n}+\alpha_2\right)\\ &\qquad + a_1^2\left(1+\frac{1}{n}\right)\sin\frac{\pi}{n}\sum_{k=1}^{n}\cos\left(\frac{4\pi}{n}-\frac{2\pi}{n}+2\alpha_1\right) + O(nb)\,.\end{aligned} \tag{2.22}$$

Using the identity (2.17) for u equal to $2\pi/n$ or $4\pi/n$, we see that all the sums of the cosines on the right-hand side of (2.22) are equal to zero.

Therefore, the estimate (2.20) for μ_- follows from (2.21) and (2.22). The lemma is proved.

Lemma 4. *Suppose that φ runs through the set of functions satisfying the conditions* (1.4). *Then*

$$I := \sup_{\varphi}\left|\int_0^{2\pi} g(x)\varphi(x)\,dx\right| = \int_0^{2\pi} |g(x)|\,dx + O(b^2) \tag{2.23}$$

with an absolute constant in the remainder term.

Proof. The estimate

$$I \le \int_0^{2\pi} |g(x)|\,dx \tag{2.24}$$

is obvious. On the other hand, the function $g_0(x) := \operatorname{sgn} g(x)$ may not satisfy the second condition in (1.4). But as shown by the estimates (2.20), this condition can be achieved by changing the sign of $g_0(x)$ on a set with measure $\le cb$, where c is some absolute constant. This set can be chosen so that on it

$$|g(x)| \le cb. \tag{2.25}$$

Indeed, since $|g'(x)| < 3n/2$ for all x in view of (2.3), the estimate (2.25) holds on the collection of neighborhoods of length $cb/2n$ of each of the $2n$ zeros of $g(x)$.

But if the function $\varphi_0(x)$ is obtained from $g_0(x)$ by the indicated change of sign, then

$$\int_0^{2\pi} g(x)\varphi_0(x)\,dx = \int_0^{2\pi} |g(x)|\,dx + O(b^2).$$

This and (2.24) yield (2.23). The lemma is proved.

Lemma 5. *The estimate*

$$\int_0^{2\pi} |g(x)|\,dx = 4 + a_1^2 + O(b_1) \tag{2.26}$$

is valid with an absolute constant in the remainder term.

Proof. In addition to the notation introduced above let $x_{2n+1} := 2\pi + x_1$. We have

$$\begin{aligned}\int_0^{2\pi} |g(x)|\,dx &= \sum_{k=1}^{n}\left(\int_{x_{2k}}^{x_{2k+1}} g(x)\,dx - \int_{x_{2k-1}}^{x_{2k}} g(x)\,dx\right)\\ &= 2\left[\frac{1}{n}\sum_{k=1}^{n}\sin nx + \sum_{s=1}^{3}\frac{a_s}{n+s}\sum_{k=1}^{n}\sin((n+s)x+\alpha_s)\right]_{x_{2k}}^{x_{2k+1}}.\end{aligned} \tag{2.27}$$

We transform the sums on the right-hand side of (2.27):

$$\begin{aligned}\sum_{k=1}^{n}\sin nx\Big|_{x_{2k}}^{x_{2k+1}} &= \sum_{k=1}^{n}(\sin nx_{2k+1} - \sin nx_{2k})\\ &= \sum_{k=1}^{n}\left[\sin\left(\frac{4k+1}{2}\pi + y_{2k+1}\right) - \sin\left(\frac{4k-1}{2}\pi + y_{2k}\right)\right] = \sum_{m=1}^{2n}\cos y_m.\end{aligned}$$

Similarly,

$$\sum_{k=1}^{n} \sin((n+s)x+\alpha_s)\Big|_{x_{2k}}^{x_{2k+1}} = \sum_{k=1}^{n}\Big[\sin\left(\frac{4k+1}{2}\pi + st_{2k+1} + \left(1+\frac{s}{n}\right)y_{2k+1}+\alpha_s\right)$$
$$-\sin\left(\frac{4k-1}{2}\pi + st_{2k} + \left(1+\frac{s}{n}\right)y_{2k}+\alpha_s\right)\Big]$$
$$= \sum_{m=1}^{2n} \cos\left(st_m + \left(1+\frac{s}{n}\right)y_m + \alpha_s\right).$$

Substituting these expressions in (2.27), we get

(2.28)
$$\int_0^{2\pi} |g(x)|\,dx = 2\left[\frac{1}{n}\sum_{m=1}^{2n}\cos y_m + \sum_{s=1}^{3}\frac{a_s}{n+s}\sum_{m=1}^{2n}\cos\left(st_m + \left(1+\frac{s}{n}\right)y_m+\alpha_s\right)\right].$$

By (2.9),

$$\sum_{m=1}^{2n}\cos y_m = \sum_{m=1}^{2n}\left(1-\tfrac{1}{2}y_m^2\right) + O\left(\sum_{m=1}^{2n} y_m^4\right) = 2n - \tfrac{1}{2}\sum_{m=1}^{2n} y_m^2 + O(na_1^4). \tag{2.29}$$

If we now use the estimate (2.10) for y_m and then the identities (2.18) and (2.19), we obtain

$$\sum_{m=1}^{2n} y_m^2 = \sum_{m=1}^{2n}[a_1^2\sin^2(t_m+\alpha_1) + 2a_1a_2\sin(t_m+\alpha_1)\sin(2t_m+\alpha_2)$$
$$-a_1^3(1+1/n) - \sin(t_m+\alpha_1)\sin(2t_m+2\alpha_1) + O(b_1)]$$
$$= na_1^2 + O(nb_1).$$

Therefore, by (2.29),

$$\sum_{m=1}^{2n}\cos y_m = 2n - \frac{n}{2}a_1^2 + O(nb_1). \tag{2.30}$$

Further, using the Taylor formula with Lagrange's remainder, which we estimate with the help of (2.9), we find the expression

(2.31)
$$\sum_{m=1}^{2n}\cos\left(t_m + \left(1+\frac{1}{n}\right)y_m + \alpha_1\right) = \sum_{m=1}^{2n}\Big[\cos(t_m+\alpha_1) - \sin(t_m+\alpha_1)\left(1+\frac{1}{n}\right)y_m$$
$$-\frac{1}{2}\cos(t_m+\alpha_1)\cdot\left(1+\frac{1}{n}\right)^2 y_m^2 + O(a_1^3)\Big]$$

for the sum in (2.28) corresponding to $s=1$.

Let us consider separately the sums on the right-hand side of (2.31). By (2.17),

$$\sum_{m=1}^{2n}\cos(t_m+\alpha_1) = \sum_{m=1}^{2n}\cos\left(\frac{\pi}{n}m - \frac{\pi}{2n} + \alpha_1\right) = 0. \tag{2.32}$$

Using the estimate (2.10) and the identities (2.18) and (2.19), we get
(2.33)

$$\sum_{m=1}^{2n} \sin(t_m + \alpha_1) \cdot y_m = -a_1 \sum_{m=1}^{2n} \sin^2(t_m + \alpha_1) - a_2 \sum_{m=1}^{2n} \sin(t_m + \alpha_1) \sin(2t_m + \alpha_2)$$
$$+ \frac{1}{2} a_1^2 \left(1 + \frac{1}{n}\right) \sum_{m=1}^{2n} \sin(t_m + \alpha_1) \sin(2t_m + 2\alpha_1) + O(nb)$$
$$= -na_1 + O(nb)$$

and
(2.34)

$$\sum_{m=1}^{2n} \cos(t_m + \alpha_1) \cdot y_m^2 = \sum_{m=1}^{2n} \cos(t_m + \alpha_1)[a_1^2 \sin^2(t_m + \alpha_1) + O(a_1^3 + a_1 a_2)]$$
$$= \frac{1}{2} a_1^2 \sum_{m=1}^{2n} \sin(2t_m + 2\alpha_1) \sin(t_m + \alpha_1) + O(nb) = O(nb).$$

From (2.31)–(2.34),

$$\sum_{m=1}^{2n} \cos\left(t_m + \left(1 + \frac{1}{n}\right) y_m + \alpha_1\right) = (n+1)a_1 + O(nb). \tag{2.35}$$

The same ideas are used in estimating the sums in (2.28) corresponding to $s = 2$ and $s = 3$. But here the situation is simpler, because it suffices to find the values of these sums with coarser remainder terms. We have

$$\sum_{m=1}^{2n} \cos\left(2t_m + \left(1 + \frac{2}{n}\right) y_m + \alpha_2\right)$$
$$= \sum_{m=1}^{2n} \left[\cos(2t_m + \alpha_2) - \sin(2t_m + \alpha_2) \cdot \left(1 + \frac{2}{n}\right) y_m + O(a_1^2)\right],$$

and since

$$\sum_{m=1}^{2n} \cos(2t_m + \alpha_2) = 0$$

and

$$\sum_{m=1}^{2n} \sin(2t_m + \alpha_2) \cdot y_m = \sum_{m=1}^{2n} [-a_1 \sin(2t_m + \alpha_2) \sin(t_m + \alpha_1) + O(a_1^2 + a_2)],$$

it follows that

$$\sum_{m=1}^{2n} \cos\left(2t_m + \left(1 + \frac{2}{n}\right) y_m + \alpha_2\right) = O(n(a_1^2 + a_2)). \tag{2.36}$$

Finally,

$$\sum_{m=1}^{2n} \cos\left(3t_m + \left(1 + \frac{3}{n}\right) y_m + \alpha_3\right) = \sum_{m=1}^{2n} (\cos(3t_m + \alpha_3) + O(a_1)) = O(na_1). \tag{2.37}$$

Substituting the estimates (2.30) and (2.35)–(2.37) in (2.28), we conclude that

$$\int_0^{2\pi} |g(x)|\,dx = 2[2 - \tfrac{1}{2}a_1^2 + O(b_1) + a_1(a_1 + O(b)) + O(a_2(a_1^2 + a_2)) + O(a_3a_1)]$$
$$= 4 + a_1^2 + O(b_1).$$

The lemma is proved.

Lemma 6. *The estimate*

$$I = 4 + a_1^2 + O(b_1) \tag{2.38}$$

is valid with an absolute constant in the remainder term.

This assertion follows from Lemmas 4 and 5, since $b \le 3a_1$, and hence $b^2 \le 3b_1$.

Lemma 7. *The estimate*

$$J := \max_u \frac{1}{2}\int_0^{2\pi} |g(x) - g(x+u)|\,dx = 4 + a_1^2\cos^2\frac{\pi}{2n} + O(b_1) \tag{2.39}$$

is valid with an absolute constant in the remainder term.

Proof. In the definition of J it suffices to consider $u \in [0, \pi]$ in view of the periodicity of $g(x)$. Each such u can be represented in the form

$$u = (p/n)\pi - 2z/n, \tag{2.40}$$

where p is an odd number with $1 \le p \le n$, and $|z| \le \pi/2$.

For u given by (2.40) we estimate the integral

$$\tfrac{1}{2}\int_0^{2\pi} |g(x) - g(x+u)|\,dx, \tag{2.41}$$

with the remainder term uniform with respect to the indicated p and z. First, for a fixed p, we regard the integral (2.41) as a function of z. We have

(2.42)

$$J(p, z) := \frac{1}{2}\int_0^{2\pi} \left|g(x) - g\left(x + \frac{p}{n}\pi - \frac{2z}{n}\right)\right| dx$$
$$= \frac{1}{2}\int_0^{2\pi} \Bigg|\cos nx + \cos(nx - 2z)$$
$$+ \sum_{s=1}^{3} a_s\left[\cos((n+s)x + \alpha_s) + \cos\left((n+s)x + \frac{sp}{n}\pi - 2\left(1 + \frac{s}{n}\right)z + \alpha_s\right)\right]\Bigg|\,dx$$
$$= \int_0^{2\pi} \Bigg|\cos z\cos(nx - z)$$
$$+ \sum_{s=1}^{3} a_s\cos\left(\frac{sp}{2n}\pi - \left(1 + \frac{s}{n}\right)z\right)\cos\left((n+s)x + \frac{sp}{2n}\pi\left(1 + \frac{s}{n}\right)z + \alpha_s\right)\Bigg|\,dx$$
$$= \int_0^{2\pi} \Bigg|\cos z\cos nx$$
$$+ \sum_{s=1}^{3} a_s\cos\left(\frac{sp}{sn}\pi - \left(1 + \frac{2}{n}\right)\right) z\cos\left((n+s)x + \frac{sp}{2n}\pi + \alpha_s\right)\Bigg|\,dx.$$

It follows from (2.42) that for $n > 1$

$$J \geq 4. \tag{2.43}$$

Indeed, $J \geq J(1, 0)$, and since for $n > 1$ the Fourier series of the function $\operatorname{sgn}(\cos nx)$ does not contain harmonics of order $n+s$, $s = 1, 2, 3$, it follows that

$$J(1, 0) \geq \int_0^{2\pi} \operatorname{sgn}\cos nx \left[\cos nx + \sum_{s=1}^{3} a_s \cos\frac{s}{2n}\pi \cos\left((n+s)x + \frac{s}{2n}\pi + \alpha_s\right)\right] dx$$
$$= \int_0^{2\pi} |\cos nx|\, dx = 4.$$

Of course, by (2.43) we can confine ourselves to those values of z such that

$$J(p, z) \geq 4. \tag{2.44}$$

From (2.42) and (2.4),

$$J(p, z) \leq 4\cos z + 4\sum_{s=1}^{3} a_s < 4\cos z + 2,$$

which together with (2.44) leads to the inequality

$$\cos z > 1/2. \tag{2.45}$$

Thus, the condition $|z| \leq \pi/2$ can be replaced by $|z| < \pi/3$ for values of u of the form (2.40).

Further arguments are invoked to narrow down the interval of values of z under consideration by disregarding those z for which the estimate (2.44) automatically cannot hold.

Using the estimate (2.45), from (2.42) with the help of Lemma 5 we derive that

$$\begin{aligned} J(p, z) &= \cos z\left(4 + \frac{\cos^2(p\pi/2n - (1+1/n)z)}{\cos^2 z}a_1^2 + O(b_1)\right) \\ &= 4\cos z + \frac{\cos^2(p\pi/2n - (1+1/n)z)}{\cos z}a_1^2 + O(b_1). \end{aligned} \tag{2.46}$$

From this, using the estimate (2.45), we find that

$$J(p, z) \leq 4\cos z + 2a_1^2 + O(b_1). \tag{2.47}$$

Since we are interested only in those values of z such that (2.44) holds, it suffices to consider those z for which

$$1 - \cos z = O(a_1^2). \tag{2.48}$$

But

$$1 - \cos z = 2\sin^2(z/2) \geq (2/\pi^2)z^2. \tag{2.49}$$

Therefore, it follows from (2.48) that

$$z = O(a_1). \tag{2.50}$$

The estimates thus obtained enable us to simplify the right-hand side of (2.46). Indeed, by (2.48),

$$1/\cos z = 1 + O(a_1^2).$$

Further, using the Taylor formula with Lagrange's remainder, and estimating the latter with the help of (2.50), we find

$$\cos^2(p\pi/2n - (1+1/n)z) = \cos^2(p\pi/2n) + \sin(p\pi/n)\cdot(1+1/n)z + O(a_1^2).$$

Thus, for the values of z such that the estimate (2.48) holds, it follows from (2.46) that

$$J(p, z) = 4\cos z + \left[\cos^2\frac{p\pi}{2n} + \sin(p\pi/n)\cdot\left(1+\frac{1}{n}\right)z\right]a_1^2 + O(b_1). \tag{2.51}$$

From this,

$$J(p, 0) = 4 + a_1^2\cos^2(p\pi/2n) + O(b_1), \tag{2.52}$$

and for $z \neq 0$ we find by using (2.49) that

$$J(p, z) \le 4 + a_1^2\cos^2(p\pi/2n) - (8/\pi^2)z^2 + \sin(p\pi/n)\cdot(1+1/n)a_1^2 z + O(b_1).$$

Comparing this estimate with (2.52), we see that $J(p, z)$ can exceed (with the specified accuracy) the value $J(p, 0)$ only when

$$-(8/\pi^2)z^2 + \sin(p\pi/n)\cdot(1+1/n)a_1^2 z > 0,$$

which implies that

$$z = O(a_1^2). \tag{2.53}$$

But if z satisfies this condition, then, by (2.51),

$$J(p, z) = 4 + a_1^2\cos^2(p\pi/2n) + O(b_1).$$

Thus, for each $p = 1, \dots, n$ the largest (to within the remainder terms considered) value of $J(p, z)$ is attained at $z = 0$.

But since the second term on the right-hand side of (2.52) has its largest value at $p = 1$, it follows that

$$J = J(1, 0) + O(b_1). \tag{2.54}$$

The estimate (2.39) follows from (2.54) and (2.52), and the lemma is proved.

We note that the lemmas proved here refine and generalize the results obtained in [17] for polynomials of the form (2.1) with $a_2 = a_3 = 0$. An estimate of the form (2.9) was established there and was used to show that

$$I = 4 + O(a_1^2), \qquad J = 4 + O(a_1^2)$$

with absolute constants in the remainder terms.

3. Proof of the main results

Following Stepanets (see [20]), we introduce certain classes of functions. Suppose that $\{\psi(k)\}$ is a sequence of positive numbers tending to zero, $\{\alpha_k\}$ is an arbitrary numerical sequence, and

$$\sum_{k=1}^{\infty}\psi(k)\cos(kt+\alpha_k) \tag{3.1}$$

is the Fourier series of some function $\Psi(t)$. We take this function as the kernel in the convolutions (1.2) and (1.5). The class of functions $f(x)$ representable by the formula (1.2), where φ satisfies the conditions (1.4), is denoted by $\Psi_\alpha(C)$, and the class of functions $f(x)$ representable by the formula (1.5), where v satisfies the conditions (1.6), is denoted by $\Psi_\alpha(L)$. If $\psi(k) = k^{-r}$, then we get the corresponding classes W^r_α. We remark that our notation for the classes Ψ_α differs from the notation in [20].

By analogy with the convention adopted in §1, we write $S_n(\Psi_\alpha, C)$ and $S_n(\Psi_\alpha, L)$ instead of $S_n(\Psi_\alpha(C), C)$ and $S_n(\Psi_\alpha(L), L)$, respectively.

The lemmas proved in §2 enable us to get the following statement, which uses the notation

$$\begin{aligned}\psi_n := \left(\frac{\psi(n+1)}{\psi(n)}\right)^4 + \left(\frac{\psi(n+1)}{\psi(n)}\right)^2 \frac{\psi(n+2)}{\psi(n)} \\ + \frac{\psi(n+1)}{\psi(n)}\frac{\psi(n+3)}{\psi(n)} + \left(\frac{\psi(n+2)}{\psi(n)}\right)^2 + \sum_{k=4}^{\infty}\frac{\psi(n+k)}{\psi(n)}.\end{aligned} \tag{3.2}$$

Theorem 3. *Suppose that for some* n

$$\psi(n+1) \geq \psi(n+2) \geq \psi(n+3) \tag{3.3}$$

and

$$\sum_{s=1}^{3}(n+s)\psi(n+s) < \tfrac{1}{2}n\psi(n). \tag{3.4}$$

Then

$$S_n(\Psi_\alpha, C) = \psi(n)\left\{\frac{4}{\pi} + \frac{1}{\pi}\left(\frac{\psi(n+1)}{\psi(n)}\right)^2 + O(\psi_n)\right\}, \tag{3.5}$$

$$S_n(\Psi_\alpha, L) = \psi(n)\left\{\frac{4}{\pi} + \frac{1}{\pi}\left(\frac{\psi(n+1)}{\psi(n)}\right)^2 \cos^2\frac{\pi}{2n} + O(\psi_n)\right\} \tag{3.6}$$

with absolute constants in the remainder terms.

Proof. Let $f \in \Psi_\alpha(C)$. Using the representation (1.2), we obtain

$$f(x) - s_{n-1}(f, x) = \frac{1}{\pi}\int_0^{2\pi} \varphi(x+t)\sum_{k=n}^{\infty}\psi(k)\cos(kt+\alpha_k)\,dt,$$

where φ is some function satisfying the conditions (1.4). Therefore,

$$\begin{aligned}S_n(\Psi_\alpha, C) &= \sup_\varphi \frac{1}{\pi}\left|\int_0^{2\pi}\varphi(x+t)\sum_{k=n}^{\infty}\psi(k)\cos(kt+\alpha_k)\,dt\right| \\ &= \sup_\varphi\frac{1}{\pi}\left|\int_0^{2\pi}\varphi(x+t)\left[\psi(n)\cos(nt+\alpha_n) + \sum_{s=1}^{3}\psi(n+s)\cos((n+s)t+\alpha_{n+s})\right]dt\right| \\ &\quad + O\left(\sum_{k=n+4}^{\infty}\psi(k)\right) \\ &= \frac{1}{\pi}\psi(n)\sup_\varphi\left|\int_0^{2\pi}\varphi(t)\left[\cos nt + \sum_{s=1}^{3}\frac{\psi(n+s)}{\psi(n)}\cos((n+s)t+\beta_s)\right]dt\right| \\ &\quad + O\left(\sum_{k=n+4}^{\infty}\psi(k)\right),\end{aligned} \tag{3.7}$$

where the suprema are over the functions φ satisfying the conditions (1.4), and $\beta_s = \alpha_{n+s} - ((n+s)/n)\alpha_n$.

If Lemma 6 is applied to the supremum on the right-hand side of (3.7), then we get the estimate (3.5).

Let us prove (3.6). Using the representation (1.5) of functions in the class $\Psi_\alpha(L)$, we find

$$S_n(\Psi_\alpha, L) = \sup_v \frac{1}{n}\left|\int_0^{2\pi} \sum_{k=n}^{\infty} \psi(k)\cos(k(t-x)+\alpha_k)\,dv(t)\right|, \tag{3.8}$$

where the supremum is over the functions v of bounded variation satisfying the conditions (1.6).

Let

$$H(t) := \sum_{k=n}^{\infty} \psi(t)\cos(kt+\alpha_k), \quad h(t) := \sum_{k=n}^{n+3} \frac{\psi(k)}{\psi(n)} \cos\left(kt+\alpha_k - \frac{k}{n}\alpha_n\right).$$

Applying Nikol'skiĭ's theorem (Theorem 3 in §2 of [18]; see also [6]) to the supremum in (3.8), we get

$$S_n(\Psi_\alpha, L) = \max_u \frac{1}{2\pi}\int_0^{2\pi} |H(t) - H(t+u)|\,dt. \tag{3.9}$$

But

$$\int_0^{2\pi} |H(t)-H(t+u)|\,du = \psi(n)\int_0^{2\pi} |h(t)-h(t+u)|\,dt + O\left(\sum_{k=n+4}^{\infty} \psi(k)\right);$$

therefore, using Lemma 7 to estimate the last integral, we get (3.6) from (3.9). The theorem is proved.

We derive Theorems 1 and 2 from Theorem 3. Let

$$\psi(k) = k^{-r}. \tag{3.10}$$

Then the inequalities (3.3) hold for all n. We show that the estimate (3.4) holds for $r \geq 2n+1$.

In our case the estimate (3.4) is equivalent to the inequality

$$\sum_{s=1}^{3} (1+s/n)^{1-r} < 1/2,$$

which it suffices to prove for the smallest r, i.e., it must be established that

$$\sum_{s=1}^{3} (1+s/n)^{-2n} < 1/2. \tag{3.11}$$

It is well known that the sequence $(1+1/n)^n$ is increasing. It is also easy to prove that the sequences $(1+2/n)^n$ and $(1+3/n)^n$ are increasing. Therefore, it suffices to verify the inequalities (3.11) for $n=1$, and this reduces to a simple computation.

Let us estimate ψ_n. We have

$$\begin{aligned}\psi_n &= \left(1+\frac{1}{n}\right)^{-4r} + \left(1+\frac{1}{n}\right)^{-2r}\left(1+\frac{2}{n}\right)^{-r} + \left(1+\frac{1}{n}\right)^{-r}\left(1+\frac{3}{n}\right)^{-r} \\ &\quad + \left(1+\frac{2}{n}\right)^{-2r} + n^r\sum_{k=1}^{\infty}(n+k)^{-r} \\ &< 4\left(1+\frac{4}{n}\right)^{-r} + n^r\left[(n+4)^{-r} + \int_{n+4}^{\infty} u^{-r}\,du\right] \\ &= 5\left(1+\frac{4}{n}\right)^{-r} + \frac{n+4}{r-1}\left(1+\frac{4}{n}\right)^{-r} < 8\left(1+\frac{4}{n}\right)^{-r}.\end{aligned}$$

Using this inequality, we see that the estimates (1.15) and (1.17) follow from (3.5) and (3.6), respectively, and this concludes the proof of Theorems 1 and 2.

It can be proved (we do not dwell on this here) that the remainder terms in the estimates (1.15) and (1.17) are somewhat excessive. The expression $(1+4/n)^{-r}$ in these estimates can be replaced by $(1+2/n)^{-2r}$, which gives an improvement in order if $r/n^2 \to \infty$ as $n \to \infty$. Any further improvement in the order of decrease of the remainder terms in (1.15) and (1.17) is impossible.

We mention one case when under additional conditions on the sequence $\{\psi(k)\}$ it is possible to get a simple estimate of the quantity ψ_n in the remainder terms in the estimates (3.5) and (3.6) and to ensure the condition (3.4).

It will be assumed that for $k \geq n$ the sequence $\{\psi(k+1)/\psi(k)\}$ is monotonically decreasing, and

$$\psi(n+1)/\psi(n) \leq 1/6. \tag{3.12}$$

Then

$$\frac{\psi(n+k)}{\psi(n)} = \frac{\psi(n+1)}{\psi(n)}\frac{\psi(n+2)}{\psi(n+1)}\cdots\frac{\psi(n+k)}{\psi(n+k-1)} \leq \left(\frac{\psi(n+1)}{\psi(n)}\right)^k. \tag{3.13}$$

By this estimate and (3.12),

$$\sum_{k=1}^{\infty}\frac{\psi(n+k)}{\psi(n)} \leq \sum_{k=4}^{\infty}\left(\frac{\psi(n+1)}{\psi(n)}\right)^k \leq \frac{6}{5}\left(\frac{\psi(n+1)}{\psi(n)}\right)^4.$$

Applying the estimate (3.13) to the remainder terms on the right-hand side of (3.2), we get

$$\psi_n < 6(\psi(n+1)/\psi(n))^4. \tag{3.14}$$

Let us verify the condition (3.4). Note that it is equivalent to the inequality

$$\sum_{s=1}^{3}(1+s/n)\psi(n+s)/\psi(n) < 1/2.$$

Using the estimates (3.13) and (3.12), we find

$$\sum_{s=1}^{3}\left(1+\frac{s}{n}\right)\frac{\psi(n+s)}{\psi(n)} \leq \sum_{s=1}^{3}\left(1+\frac{s}{n}\right)\left(\frac{\psi(n+1)}{\psi(n)}\right)^s \leq \sum_{s=1}^{3}\left(1+\frac{s}{n}\right)\frac{1}{6^s} < \frac{1}{2}.$$

Of course, the condition (3.3) also holds in our case.

Thus, the following assertion is a consequence of Theorem 3.

Corollary. *Suppose that for* $k \geq n$ *the sequence* $\{\psi(k+1)/\psi(k)\}$ *is monotonically decreasing, and the condition* (3.12) *holds. Then*

$$(3.15)\quad S_n(\Psi_\alpha, C) = \psi(n)\left\{\frac{4}{\pi} + \frac{1}{\pi}\left(\frac{\psi(n+1)}{\psi(n)}\right)^2 + O\left(\left(\frac{\psi(n+1)}{\psi(n)}\right)^4\right)\right\},$$

$$(3.16)\quad S_n(\Psi_\alpha, L) = \psi(n)\left\{\frac{4}{\pi} + \frac{1}{\pi}\left(\frac{\psi(n+1)}{\psi(n)}\right)^2 \cos^2\frac{\pi}{2n} + O\left(\left(\frac{\psi(n+1)}{\psi(n)}\right)^4\right)\right\}$$

with absolute constants in the remainder terms.

We consider two examples.

Corresponding to the sequence

$$(3.17)\qquad \psi(k) = q^k$$

$(0 < q < 1)$ is the problem of approximating a Poisson integral by Fourier sums, which was considered by Stechkin in [16]. He found the principal terms in the asymptotics of the suprema $S_n(\Psi_\alpha, C)$ and $S_n(\Psi_\alpha, L)$, with remainder terms uniform with respect to all n, α, and $0 < q < 1$. In particular, for $q \leq e^{-1/n}$ he proved the estimate

$$(3.18)\qquad S_n(\Psi_\alpha, C) = q^n\left\{\frac{8}{\pi^2}K(q) + O\left(\frac{q}{n(1-q)}\right)\right\},$$

where $K(q)$ is the complete elliptic integral of the first kind, and an analogous estimate for $S_n(\Psi_\alpha, L)$.

In our case $\psi(k+1)/\psi(k) = q$; therefore, the conditions of the corollary hold with $q \leq 1/6$. Therefore, *if the sequence* $\{\psi(k)\}$ *is given by* (3.17), *where* $0 < q \leq 1/6$, *then, uniformly with respect to all the parameters,*

$$(3.19)\qquad S_n(\Psi_\alpha, C) = q^n\left\{\frac{4}{\pi} + \frac{1}{\pi}q^2 + O(q^4)\right\},$$

$$(3.20)\qquad S_n(\Psi_\alpha, L) = q^n\left\{\frac{4}{\pi} + \frac{1}{\pi}q^2\cos^2\frac{\pi}{2n} + O(q^4)\right\}.$$

We compare the estimates (3.18) and (3.19), assuming that $q \leq 1/6$ and $n \to \infty$. For $q = n^{-1/3}$ the remainder terms in these estimates have the same order. For q decreasing more slowly the estimate (3.18) is sharper, but if $q = o(n^{-1/3})$, the estimate (3.19) gives a sharper result.

We remark further that in [16] estimates of the form (3.18) were proved not only for $S_n(\Psi_\alpha, C)$ and $S_n(\Psi_\alpha, L)$, but also for the corresponding suprema, taken over broader classes of functions when the second conditions in (1.4) and (1.6) are not required. With the help of Lemma 5 it is easy to establish that estimates of the form (3.19) also hold for these suprema.

As a second example of the use of the corollary we consider the sequence $\{\psi(k)\}$ given by

$$(3.21)\qquad \psi(k) = \exp(-Ak^R), \qquad A > 0, \quad R > 1.$$

Then the sequence $\{\psi(k+1)/\psi(k)\}$ is monotonically decreasing. Indeed,

$$\frac{\psi(k+1)}{\psi(k)} : \frac{\psi(k+2)}{\psi(k+1)} = \exp A((k+2)^R - 2(k+1)^R + k^R) > 1,$$

since the sequence $\{k^R\}$ is convex for $R > 1$. Further,

$$\frac{\psi(n+1)}{\psi(n)} = \exp A(-(n+1)^R + n^R) = \exp(-AR(n+\theta)^{R-1}) < \exp(-ARn^{R-1});$$

therefore, the condition (3.12) holds if

$$ARn^{R-1} \geq \log 6\,. \tag{3.22}$$

For example, we can take

$$n \geq (\log 6/A)^{1/(R-1)}\,.$$

Thus, *if the sequence* $\{\psi(k)\}$ *is defined by the conditions* (3.21) *and the estimate* (3.22) *holds, then, uniformly with respect to all parameters,*

$$\begin{aligned} S_n(\Psi_\alpha\,, C) = \exp(-An^R)\cdot \Big\{ \frac{4}{\pi} + \frac{1}{\pi} \exp 2A(-(n+1)^R + n^R) \\ + O(\exp 4A(-(n+1)^R + n^R))\Big\} \end{aligned} \tag{3.23}$$

$$\begin{aligned} S_n(\Psi_\alpha\,, L) = \exp(-An^R)\cdot \Big\{ \frac{4}{\pi} + \frac{1}{\pi} \exp 2A(-(n+1)^R + n^R)\cdot \cos^2 \frac{\pi}{2n} \\ + O(\exp 4A(-(n+1)^R + n^R)) \Big\}\,. \end{aligned} \tag{3.24}$$

For sequences $\{\psi(k)\}$ given by conditions (3.21), the suprema $S_n(\Psi_\alpha\,, C)$ and $S_n(\Psi_\alpha\,, L)$ were considered by Stepanets, who found the principal terms of the asymptotics of these quantities ([20], Theorems 3.9.1 and 3.13.31). The remainder terms in these estimates were improved by the author [17]. In the case when the condition (3.22) holds, the results obtained in [17] give the estimate

$$S_n(\Psi_\alpha\,, C) = \exp(-An^R)\cdot \Big\{\frac{4}{\pi} + O(\exp 2A(-(n+1)^R + n^R))\Big\}$$

and an analogous estimate for $S_n(\Psi_\alpha\,, L)$ with absolute constants in the remainder terms.

Bibliography

1. Henri Lebesgue, *Sur la représentation trigonométrique approchée des fonctions satisfaisant à une condition de Lipschitz*, Bull. Soc. Math. France **38** (1910), 184–210.
2. A. Kolmogoroff [A. N. Kolmogorov], *Zur Grössenordnung des Restgliedes Fourierscher Reihen differenzierbarer Funktionen*, Ann. of Math. (2) **36** (1935), 521–526.
3. Bela Sz.-Nagy, *Über gewisse Extremalfragen bei transformierten trigonometrischen Entwicklungen. I: Periodischer Fall*, Ber. Verh. Sächs. Akad. Wiss. Leipzig Math.-Phys. Kl. **90** (1938), 103–134.
4. S. B. Stechkin, *On best approximation of certain classes of periodic functions by trigonometric polynomials*, Izv. Akad. Nauk SSSR Ser. Mat. **20** (1956), 643–648. (Russian)
5. N. I. Akhiezer, *Lectures on the theory of approximation*, 2nd rev. ed., "Nauka", Moscow, 1965; English transl. of 1st ed., Ungar, New York, 1956.
6. S. B. Stechkin and S. A. Telyakovskiĭ, *On approximation of differentiable functions by trigonometric polynomials in the metric of* L, Trudy Mat. Inst. Steklov. **88** (1967), 20–29; English transl. in Proc. Steklov Inst. Math. **88** (1967).
7. V. T. Pinkevich, *On the order of the remainder term of the Fourier series of a function differentiable in the sense of Weyl*, Izv. Akad. Nauk SSSR Ser. Mat. **4** (1940), 521–528. (Russian; French summary)
8. S. Nikolsky [S. M. Nikol'skiĭ], *Sur l'évaluation asymptotique du reste dans l'approximation au moyen des sommes de Fourier*, C. R. (Dokl.) Acad. Sci. URSS **32** (1941), 386–389.
9. ———, *Approximation of periodic functions by trigonometric polynomials*, Trudy Mat. Inst. Steklov. **15** (1945). (Russian; English summary)
10. A. V. Efimov, *Approximation of continuous periodic functions by Fourier sums*, Izv. Akad. Nauk SSSR Ser. Mat. **24** (1960), 243–296. (Russian)

11. S. A. Telyakovskiĭ, *On the norms of trigonometric polynomials, and approximation of differentiable functions by linear averages of their Fourier series.* I, Trudy Mat. Inst. Steklov. **62** (1961), 61–97; English transl. in Amer. Math. Soc. Transl. (2) **28** (1963).

12. I. G. Sokolov, *The remainder term of the Fourier series of a differentiable function*, Dokl. Akad. Nauk SSSR **103** (1955), 23–26. (Russian)

13. S. G. Selivanova, *Approximation by Fourier sums of functions having derivatives satisfying a Lipschitz condition*, Dokl. Akad. Nauk SSSR **105** (1955), 909–912. (Russian)

14. G. I. Natanson, *Approximation by Fourier sums of functions having different structural properties on different parts of the domain*, Vestnik Leningrad. Univ. **1961**, no. 19 (Ser. Mat. Mekh. Astr. vyp. 4), 20–35. (Russian)

15. S. A. Telyakovskiĭ, *Approximation of differentiable functions by partial sums of their Fourier series*, Mat. Zametki **4** (1968), 291–300; English transl. in Math Notes **4** (1968).

16. S. B. Stechkin, *An estimate of the remainder of the Fourier series of a differentiable function*, Trudy Mat. Inst. Steklov. **145** (1980), 126–151; English transl. in Proc. Steklov Inst. Math. **1981**, no. 1 (145).

17. S. A. Telyakovskiĭ, *On approximation of functions of high smoothness by Fourier sums*, Ukrain. Mat. Zh. **41** (1989), 510–518; English transl. in Ukrainian Math. J. **41** (1989).

18. S. M. Nikol′skiĭ, *Approximation of functions by trigonometric polynomials in the mean*, Izv. Akad. Nauk SSSR Ser. Mat. **10** (1946), 207–256; English transl. in Amer. Math. Soc. Transl. (2) **102** (1973).

19. Herbert Bristol Dwight, *Tables of integrals and other mathematical data*, 4th ed., Macmillan, New York, 1961.

20. S. I. Stepanets, *Classification and approximation of periodic functions*, "Naukova Dumka", Kiev, 1987. (Russian)

Translated by H. H. McFADEN

Труды Матем. Инст. Стеклова
198 (1992)

Proceedings of the Steklov
Institute of Mathematics
1994, Issue 1

ON SOME THIN SETS IN THE THEORY OF FUNCTIONS AND TOPOLOGY

UDC 517.5

N. N. KHOLSHCHEVNIKOVA

1. Introduction

In this paper we consider some well-known special subsets of the real line **R** that have been studied in set theory and topology, and the use of them in solving problems arising in the theory of trigonometric series. We prove that it is consistent to assume that every set concentrated near an F_σ-set A in **R** is contained in A with the possible exception of a countable number of points. On the basis of this we show that the question of whether a set concentrated near an N-set is itself an N-set is undecidable in the framework of the usual ZFC (Zermelo-Fraenkel) system of axioms of set theory.

We recall some definitions and results. (A contemporary survey of special subsets of **R** is given by Miller in [1] (pp. 201–233), and the basic concepts are presented in Kuratowski's book [2], §40.)

In 1914 Luzin used the continuum hypothesis (CH) to construct an uncountable set of real numbers intersecting each nowhere dense set in a countable set. Such uncountable sets are now called *Luzin sets*.

A set $X \subset \mathbf{R}$ is called a *set of strong measure* 0 if for every sequence $\{\varepsilon_n\}$ of positive numbers there exists a sequence $\{I_n\}_{n=1}^{\infty}$ of intervals such that $X \subset \bigcup_{n=1}^{\infty} I_n$, and the length of I_n is less than ε_n for $n = 1, 2, \dots$.

Such sets are also said to have the *C-property* (see [2], §40, VII, Theorem 7). In 1919 E. Borel gave the definition of a set of strong measure 0 and conjectured that every set of strong measure 0 is countable.

A set X is said to be *concentrated near a set* A if the difference $X \setminus G$ is countable for every open set $G \supset A$.

This definition was given by Besicovitch in 1934. Mainly, sets concentrated near a countable set have been studied.

The relations between these sets are as follows: Luzin set $\Rightarrow$ set concentrated near a countable set $\Rightarrow$ set of strong measure 0.

If CH holds, then Borel's conjecture is false, because a Luzin set gives an example of an uncountable set of strong measure 0. However, the question of the existence of a Luzin set is undecidable in ZFC. Under the assumption of Martin's axiom (an additional set-theoretic assumption consistent with ZFC) plus the negation of CH $(\mathrm{MA} + \neg\mathrm{CH})$ there do not exist Luzin sets in **R**, because in this case every set of cardinality less than the continuum $\mathfrak{c}$ has first category. But under $\mathrm{MA}+\neg\mathrm{CH}$ every set of cardinality less than $\mathfrak{c}$ has strong measure 0 (see [3]). We remark that under $\mathrm{MA}+\neg\mathrm{CH}$ every set concentrated near a countable set is countable. It is known that

1991 *Mathematics Subject Classification*. Primary 28A05, 03E15, 43A46, 54H05; Secondary 03E35, 03E50.

0081-5438/94 $1.00 + $.25 per page

the existence of a Luzin set is consistent with the negation of CH. In 1976 Laver [3] constructed a model of set theory in which every set of strong measure 0 is countable. Thus, Borel's conjecture is independent.

2. On concentrated sets

We now consider some simple properties of concentrated sets. The following propositions are obvious.

Proposition 1. *If* $A_1 \subset A_2 \subset \mathbf{R}$ *and the set* X *is concentrated near* A_1, *then* X *is also concentrated near* A_2.

Proposition 2. *If a set* X *is concentrated near a* G_δ*-set* $A \subset \mathbf{R}$, *then* $X \setminus A$ *is countable.*

We prove

Proposition 3. *If CH holds, then for every set* $A \subset \mathbf{R}$ *with* $A \notin G_\delta$ *there exists a set* X *concentrated near* A *such that* $X \setminus A$ *is uncountable.*

Proof. We well-order the set of all open sets $\mathscr{U}_\alpha$ $(\alpha < \omega_1)$ containing A, where ω_1 is the first uncountable ordinal. Then

$$A = \bigcap_{\alpha<\omega_1} \mathscr{U}_\alpha = \bigcap_{\alpha<\omega_1} \left(\bigcap_{\beta<\alpha} \mathscr{U}_\beta \right).$$

Choose an $x_\alpha \in (\bigcap_{\beta<\alpha} \mathscr{U}_\beta) \setminus \bigcap_{\beta<\alpha+1} \mathscr{U}_\beta$ if the corresponding difference is nonempty.

The set X consisting of all such x_α is concentrated near A. Indeed, if $G \supset A$, then $G = \mathscr{U}_\beta$ for some $\beta < \omega_1$. Consequently, x_α belongs to U_β for $\alpha > \beta$, i.e., only countably many x_γ $(\gamma \le \beta \le \omega_1)$ can be outside G. If $A \notin G_\delta$, then the number of nonempty differences $(\bigcap_{\beta<\alpha} \mathscr{U}_\beta) \setminus \bigcap_{\beta<\alpha+1} \mathscr{U}_\beta$ is uncountable, and the set $X = X \setminus A$ is uncountable. The proposition is proved.

As follows from a result of Laver in [3], Proposition 3 can be false under the negation of CH. It is interesting that the following proposition can be proved without additional assumptions.

Proposition 4. *If* $A \subset \mathbf{R}$ *is an analytic set that is not a Borel set, then there exists a set* X *concentrated near* A *such that the cardinality of* $X \setminus A$ *is* $\aleph_1$ (*the first uncountable cardinal*).

Proof. It is known that under the given conditions the complement of A can be partitioned into $\aleph_1$ nonempty Borel sets (constituents), i.e., $\complement A = \bigcup_{\alpha<\omega_1} B_\alpha$, where the B_α are Borel sets and $B_\alpha \cap B_{\alpha'} = \varnothing$ (for $\alpha \neq \alpha'$). Choose a point x_α in each B_α. The set $X = \{x_\alpha\}_{\alpha<\omega_1}$ is concentrated near A. Indeed, suppose that $G \supset A$, $F = \mathbf{R} \setminus G \subset \complement A$. By virtue of the Luzin-Sierpiński principle of restriction of indices (see [2], §39.VIII, Theorem 5, or [9], p. 283), there is a $\beta < \omega_1$ such that $F \subset \bigcup_{\alpha<\beta} B_\alpha$, i.e., F can contain at most countably many points of X. The proposition is proved.

This argument was given by Luzin [4] in constructing a set that is of first category on every perfect set; the construction gives an example of a set of universal measure 0 and cardinality $\aleph_1$ ([1], p. 213).

A set is said to be *totally imperfect* if it does not contain a nonempty perfect subset. The following results are easy to see.

Proposition 5. *If* $A \subset \mathbf{R}$ *and* X *is concentrated near* A, *then* $X \setminus A$ *is totally imperfect.*

Proposition 6. *If the complement of a set* $A \subset \mathbf{R}$ *is totally imperfect, then every set is concentrated near* A.

To formulate the next proposition we give two definitions.

A set $A \subset \mathbf{R}$ has *universal measure* 0 if for every nonatomic nonnegative Borel measure μ there exists a Borel set B such that $A \subset B$ and $\mu(B) = 0$.

It is known that every set of strong measure 0 has universal measure 0.

A set $A \subset \mathbf{R}$ is said to be *universally measurable* if for every nonnegative Borel measure μ it is a Borel set modulo sets of μ-measure 0.

Borel sets, analytic sets, and coanalytic sets are universally measurable.

Proposition 7. *If a set* A *is universally measurable and* X *is concentrated near* A, *then the set* $X \setminus A$ *has universal measure* 0. *Every set of universal measure* 0 *is concentrated near some set of Lebesgue measure* 0.

Proof. Let μ be a nonatomic nonnegative Borel measure, and let $\varepsilon > 0$. There exist a closed set F and an open set G such that $F \subset A \subset G$ and $\mu(G \setminus F) < \varepsilon$.

Since $X \setminus G$ is countable, it follows that

$$X \setminus A \subset (G \setminus F) \cup (X \setminus G), \qquad \text{where } \mu((G \setminus F) \cup (X \setminus G)) < \varepsilon.$$

Hence, $X \setminus A$ is of universal measure 0.

Suppose that X has universal measure 0. There exist open sets G_n with $X \subset G_n$ and with Lebesgue measure $m(G_n) < 1/n$ $(n = 1, 2, \ldots)$. Let $A = \bigcap_{i=1}^{\infty} G_n$; then $m(A) = 0$. We show that X is concentrated near $A \setminus X$. Indeed, let $G \supset A \setminus X$, $F = \mathbf{R} \setminus G$. Then $X \setminus G = X \cap F$. But F is a closed set, and X is totally imperfect, since it has universal measure 0 (see, for example, [3]); therefore, $X \setminus G$ is countable.

The proposition is proved.

For what follows we need a set-theoretic assertion that is a consequence of MA. Martin's axiom is a nontrivial generalization of CH, and, as already mentioned, is consistent with the negation of CH.

Corollary to MA ([5], p. 496). *Let* $\mathscr{A}$ *and* $\mathscr{B}$ *be families of cardinality less than* $\mathfrak{c}$ *that consist of subsets of the natural sequence* ω, *with* $B \setminus \bigcup \mathscr{A}'$ *infinite for any* $B \in \mathscr{B}$ *and any finite* $\mathscr{A}' \subseteq \mathscr{A}$. *Then there is an infinite set* $L \subseteq \omega$ *such that* $A \cap L$ *is finite for each* $A \in \mathscr{A}$, *but infinite for each* $B \in \mathscr{B}$.

Theorem 1 (MA + $\neg$CH).([1]) *Suppose that* X *is a set concentrated near an* F_σ*-set* $A \subset \mathbf{R}$. *Then* $X \setminus A$ *is countable.*

Proof. Assume that $X \setminus A$ is uncountable. We take a subset $X_1 \subset X \setminus A$ of cardinality $\aleph_1$. Well-ordering the elements of X_1, we get $X_1 = \{x_\alpha\}_{\alpha < \omega_1}$. Let the set A be represented in the form $A = \bigcup_{k=1}^{\infty} F_k$, where the F_k are compact subsets of $\mathbf{R}$. Let $\mathscr{U} = \{U_n : n \in \omega\}$ be a base of open sets in $\mathbf{R}$. We take the new base $\mathscr{V} = \{V_n : n \in \omega\}$, where each of the sets V_n is a union of finitely many elements of the base $\mathscr{U}$. Now let

$$B_n = \{m \in \omega : F_n \subset V_m\}, \qquad A_\alpha = \{n \in \omega : x_\alpha \in V_n\},$$
$$\mathscr{B} = \{B_n\}_{n \in \omega}, \qquad \mathscr{A} = \{A_\alpha\}_{\alpha < \omega_1}.$$

([1])This means that the theorem is proved under the assumption that Martin's axiom holds along with the negation of CH.

Let us show that each set $B_n \setminus \bigcup_{i=1}^{k} A_{\alpha_i}$ is infinite $(n, k \in \omega,\ \alpha_1, \dots, \alpha_k \in \omega_1)$. We cover F_n by open sets in the base $\mathscr{U}$ that do not contain the points $x_{\alpha_1}, \dots, x_{\alpha_k}$. Since F_n is compact, there is a finite subcovering, which is an element V_{i_1} of the base $\mathscr{V}$ that contains F_n and does not contain $x_{\alpha_1}, \dots, x_{\alpha_k}$. At the next step it is possible to take a covering of F_n by open sets in $\mathscr{U}$ that not only does not contain the points $x_{\alpha_1}, \dots, x_{\alpha_k}$, but also does not contain any point in $V_{i_1} \setminus F_n$, and so on. Therefore, the number of different elements of $\mathscr{V}$ containing F_n and not containing $x_{\alpha_1}, \dots, x_{\alpha_k}$ is infinite, and the set $B_n \setminus \bigcup_{i=1}^{k} A_{\alpha_i}$ is infinite. Consequently, the families $\mathscr{B}$ and $\mathscr{A}$ satisfy the conditions of the corollary to MA. In view of this, there exists a set $L \subseteq \omega$ such that $L \cap B_n$ is infinite for all $n \in \omega$, whereas $L \cap A_\alpha$ is finite for all $\alpha < \omega_1$. Let

$$G_n = \bigcup \{V_m : m \in L,\ m > n\}, \qquad Y_n = \mathbf{R} \setminus G_n \qquad (n = 1, 2, \dots).$$

For each k the set $B_k \cap L$ is infinite; therefore, $F_k \subset G_n$, and hence $F_k \cap Y_n = \varnothing$ $(k, n \in \omega)$. But since $L \cap A_\alpha$ is finite, for each $\alpha < \omega_1$ there exists an $n \in \omega$ such that $x_\alpha \in Y_n$. Consequently, $X_1 \subset \bigcup_{n=1}^{\infty} Y_n$, $A \cap \bigcup_{n=1}^{\infty} Y_n = \varnothing$. Then one of the sets Y_n contains an uncountable set of points in X_1. The corresponding open set G_n contains A, but $X_1 \setminus G_n = X_1 \cap Y_n$ is uncountable. Thus, the set X_1, hence also X, is not concentrated near A. This yields the theorem.

We remark that the central part of the proof of Theorem 1 is analogous to the proof of the theorem that the union of fewer than $\mathfrak{c}$ sets of first category is a set of first category (MA) (see [5], p. 497).

Corollary to Theorem 1 $(\mathrm{MA} + \neg\mathrm{CH})$. *Every set concentrated near a countable set is countable.*

This follows also from Theorem 12 in [5] (p. 497).

Conjecture. It is consistent to assume that the set $X \setminus A$ is countable for every set X concentrated near a Borel set A.

3. Applications to U-, R-, and N-sets

We turn now to special sets considered in the theory of trigonometric series.

All sets below will be assumed to be in $[0, 2\pi]$. The definitions and assertions given earlier carry over in a natural way to this case.

We recall some definitions.

A set E is called a *set of uniqueness* (briefly, a *U-set*) for trigonometric series if the convergence of a series of the form

$$\frac{a_0}{2} + \sum_{n=1}^{\infty} a_n \cos nx + b_n \sin nx \tag{1}$$

to zero for all $x \in [0, 2\pi] \setminus E$ implies that $a_0 = a_n = b_n = 0$ $(n = 1, 2, \dots)$. Otherwise, E is called an *M-set.*

A set E is called an *R-set* if there exists a trigonometric series of the form (1) that converges on E and has coefficients that do not tend to zero.

A set E is called an *N-set* if there exists a series of the form (1) that is absolutely convergent on E and such that the sum of the moduli of its coefficients diverges.

It is known (see [6], Chapter XIV, §23) that countable sets are U-, R-, and N-sets; R- and N-sets have measure 0 and are of first category. Every set of positive measure is an M-set.

Since an M-set contains an uncountable Borel subset, every totally imperfect set is a U-set. The sets considered earlier (Luzin sets, sets of strong measure 0, sets of universal measure 0) are totally imperfect; therefore, they are U-sets. Every set of cardinality less than $\mathfrak{c}$ is a U-set.

The question of whether a set of cardinality less than $\mathfrak{c}$ is an R- or an N-set is undecidable in ZFC. Under the assumption of CH the answer to this question is trivially positive, but under the negation of CH the answer is ambiguous (see [7]). We remark that a negative answer to this question is obtained if we consider (under the negation of CH) the axiomatics or a model of set theory in which there exists a Luzin set on the line (and hence also in $[0, 2\pi]$). A subset of it of cardinality $\aleph_1 < \mathfrak{c}$ that is also a Luzin set is neither an R-set nor an N-set, because it is not a first-category set.

Kechris and Louveau ([8], p. 300) recently obtained a result which implies that every Borel U-set has first category. In view of this, every Borel set containing a Luzin set is an M-set.

It is known (see [6], Chapter XIII, §8) that the union of an N-set and a countable set is an N-set. For U- and R-sets it is unknown whether they remain U- and R-sets upon addition of a countable set of points. But if E is a U-set of type F_σ, then, as follows from a theorem of Bari ([6], Chapter XIV, §6, Theorem 3), it remains a U-set upon addition of a countable set of closed U-sets. Next, we consider questions relating to sets concentrated near U-sets of type F_σ and N-sets.

Proposition 8. *The union of a U-set of type F_σ and a totally imperfect set is a U-set.*

Proof. Let A be a U-set of type F_σ, and X a totally imperfect set. Assume that $A \cup X$ is an M-set, i.e., there is a nonzero series of the form (1) that converges to 0 on $[0, 2\pi] \setminus (A \cup X)$. We remark that this series is a zero-series, i.e., it converges to 0 almost everywhere (indeed, the series converges to 0 on a set of outer measure 2π, and since the set of points where a trigonometric series converges to 0 is a Borel set, there is convergence to 0 on a set of full measure). Let E be the kernel of the zero-series: the set where the series does not converge to 0. It is known that E is an uncountable Borel set. We have

$$E = E \cap (A \cup X) = (E \cap A) \cup (E \cap (X \setminus A)). \tag{2}$$

The set $E \cap (X \setminus A) = E \setminus (E \cap A)$ is a Borel set, and hence is countable, because X is totally imperfect, and an uncountable Borel set contains a perfect subset. Then it follows from (2) that E is a subset of a U-set of type F_σ, and thus is a U-set. This contradicts the fact that E is the kernel of a zero-series. Hence, $A \cup X$ is a U-set. The proposition is proved.

Corollary. *Every set concentrated near a U-set of type F_σ is a U-set* (*see Proposition* 5).

The following result is proved in [8] (p. 233).

Theorem (Kechris, Louveau, Debs, Saint Raymond). *Let E be closed, and let E_n $(n = 1, 2, \dots)$ be closed U-sets. If every closed subset of the set $E \setminus \bigcup_{n=1}^{\infty} E_n$ is a U-set, then E is a U-set.*

We have

Theorem 2. *If E is a closed M-set, and X is concentrated near a U-set A of type F_σ, then $E \setminus X$ is an M-set.*

Proof. By the preceding theorem, the set $E \setminus A$ contains a closed M-set F. Let $G = [0, 2\pi] \setminus F$. Then $A \subset G$, and hence $X \setminus G$ is countable. This implies that the set $F \setminus X$ also contains a closed M-set, but $E \setminus X \supset F \setminus X$, hence $E \setminus X$ is an M-set. The theorem is proved.

In Theorem 2 it is impossible to replace X by a totally imperfect set, because every set can be represented as the union of two totally imperfect U-sets.

We now show that the question of whether a set concentrated near an N-set is an N-set is undecidable in ZFC.

Indeed, under the assumption of CH there exists a Luzin set in $[0, 2\pi]$ that is concentrated near the set of rational numbers, which is an N-set, but is not itself an N-set.

Theorem 3 (MA + $\neg$CH). *A set concentrated near an N-set is an N-set.*

Proof. Suppose that E is an N-set. Since the set of points of absolute convergence of a trigonometric series has type F_σ, it follows that $E \subset E_1 \in F_\sigma$, where E_1 is an N-set. If X is concentrated near E, then X is also concentrated near E_1, by Proposition 1. It follows from Theorem 1 that $X \setminus E_1$ is countable. Then $E_1 \cup X$ is an N-set, being the union of an N-set and a countable set. Hence, X is also an N-set. The theorem is proved.

Bibliography

1. K. Kunen and J. E. Vaughan (editors), *Handbook of set-theoretic topology*, North-Holland, Amsterdam, 1984.
2. K. Kuratowski, *Topology*. Vol. I, 5th ed., PWN, Warsaw, and Academic Press, New York, 1966.
3. Richard Laver, *On the consistency of Borel's conjecture*, Acta Math. **137** (1976), 151–169.
4. N. N. Luzin, *Sur un ensemble non dénombrable qui est de la première categorie sur tout ensemble parfait*, Atti Accad. Naz. Lincei Rend. Cl. Sci. Fis. Mat. Nat. (6) **7** (1928), 214–215.
5. J. Barwise (editor), *Handbook of mathematical logic*, North-Holland, Amsterdam, 1977.
6. N. K. Bari, *Trigonometric series*, Fizmatgiz, Moscow, 1961; English transl., Vols. I, II, Macmillan, New York, and Pergamon Press, Oxford, 1964.
7. N. N. Kholshchevnikova, *On uncountable R- and N-sets*, Mat. Zametki **38** (1985), 271–277; English transl. in Math. Notes **38** (1985).
8. Alexander S. Kechris and Alain Louveau, *Descriptive set theory and the structure of sets of uniqueness*, London Math. Soc. Lecture Note Ser., vol. 128, Cambridge Univ. Press, Cambridge, 1987.
9. V. G. Kanoveĭ, *The projective hierarchy of N. N. Luzin: the current state of the theory*, Supplement to the Russian transl. of [5], Part B: *Set theory*, "Nauka", Moscow, 1982, pp. 273–364. (Russian)

Translated by H. H. McFADEN

Труды Матем. Инст. Стеклова
198 (1992)

Proceedings of the Steklov
Institute of Mathematics
1994, Issue 1

WIDTHS AND AN INEQUALITY OF JACKSON TYPE FOR ABSTRACT FUNCTIONS

UDC 517.51

I. G. TSAR'KOV

1. Introduction

Let (B) be the class of all real Banach spaces, and let $X \in (B)$. For an arbitrary set $M \subset X$, $M \neq \varnothing$, let $\operatorname{diam} M$ denote the quantity $\sup_{x,y\in M} \|x-y\|$.

Let $X, Y \in (B)$, let $M \subset X$ be an arbitrary nonempty convex set, and let $f\colon M \to Y$. For any $k \in \mathbf{N}$, $x \in M$, and $h \in X$ with $(x+kh) \in M$ the *kth difference of f at x with step h* is defined to be the expression

$$\Delta_h^k(f; x) = \sum_{i=0}^{k} (-1)^{k-i} \binom{k}{i} f(x+ih).$$

The *kth-order modulus of continuity* is defined to be the function $\omega_k(f, u)$ given on $\mathbf{R}_+$ by

$$\omega_k(f; u) = \sup\{\|\Delta_h^k(f; x)\|_Y \mid h \in X,\ \|h\| \le u,\ [x, x+kh] \subset M\}.$$

For an arbitrary nonempty compact set $M \subset X$ denote by $C(M) = C(M; Y)$ the space of all continuous mappings $f\colon M \to Y$, with norm $\|f\| = \max_{x\in M} \|f(x)\|_Y$. In the case when $M \subset X$ is a finite-dimensional body, denote by $C^r(M; Y)$ $(r \in \mathbf{Z}_+)$ the space of all mappings $f\colon M \to Y$ such that the rth Fréchet derivative $f^{(r)}$ is continuous on M. Denote by Φ^k $(k \in \mathbf{N})$ the set of all monotone continuous functions $\omega\colon [0, \infty) \to \mathbf{R}_+$ such that $\omega(0) = 0$ and $\omega(nt) \le n^k \omega(t)$ for any $t \in [0, \infty)$ and $n \in \mathbf{N}$. For arbitrary $X, Y \in (B)$ and any nonempty finite-dimensional convex body $M \subset X$, $k \in \mathbf{N}$, $r \in \mathbf{Z}_+$, $\omega \in \Phi^k$, denote by $W^r H_k^\omega[M; Y]$ the set of all mappings $f \in C^r(M; Y)$ such that $\omega_k(f^{(r)}, t) \le \omega(t)$ for any $t \in [0; \frac{1}{k}\operatorname{diam} M] \cap [0, 1]$. In the case when $r = 0$ we write $H_k^\omega[M; Y]$ instead of $W^0 H_k^\omega[M; Y]$.

For arbitrary $X \in (B)$ and $n \in \mathbf{N}$ let X^n stand for $X \times \cdots \times X$ (n factors). A mapping $N\colon X^n \to Y \in (B)$ is said to be *n-linear* if it is linear in each argument when the others are fixed and if it is bounded, i.e., there is a $C = C(N) > 0$ such that for any $x_i \in X$ $(i = 1, \dots, n)$

$$\|N(x_1, \dots, x_n)\|_Y \le C\|x_1\| \cdots \|x_n\|.$$

Denote by $\mathscr{P}_n = \mathscr{P}_n(X; Y) \subset C(M; Y)$ the space of all polynomials of degree $\le n$, i.e., the set of all mappings of the form

$$p_n(x) = \sum_{i=0}^{n} b_i(x, \dots, x) \qquad (x \in X),$$

where $b_i\colon X^i \to Y$ is an i-linear mapping.

1991 *Mathematics Subject Classification.* Primary 41A46, 41A17, 41A65.

For arbitrary $X, Y \in (B)$, a nonempty compact set $M \subset X$, and $f \in C(M; Y)$ let

$$E_n(f) = E_n(f; M) = \inf_{p \in \mathscr{P}_n} \|f - p\|_{C(M; Y)}$$

and

$$E_n(W^r H_k^\omega[M; Y]) = \sup\{E_n(f) \mid f \in W^r H_k^\omega[M; Y]\}.$$

For an arbitrary set $K \subset X \in (B)$, $K \neq \varnothing$, and a subspace $L \subset X$ let

$$E(K, L)_X = \sup_{x \in K} \inf_{y \in L} \|x - y\|_X,$$

$$d_n(K, X) = \inf_{L, \dim L = n} E(K, L) \qquad (n \in Z_+).$$

Furthermore, let

$$B_p^m = \left\{(x_1, \dots, x_m) \in \mathbf{R}^m \,\middle|\, \sum_{i=1}^m |x_i|^p \leq 1\right\} \qquad (1 \leq p < \infty),$$

$$B_\infty^m = \left\{(x_1, \dots, x_m) \in \mathbf{R}^m \,\middle|\, \max_{i=1, \dots, n} |x_i| \leq 1\right\}.$$

In this paper we estimate the best approximation of classes of smooth mappings on various sets (including m-dimensional balls B_p^m, $1 \leq p \leq \infty$) by polynomials of degree $\leq n$ as n and m increase.

This problem was investigated by Newman and Shapiro [1], who proved that there exist absolute constants $C_1, C_2 > 0$ such that for any $m, n \in \mathbf{N}$ with $m \leq n$ and any $f \in C(B; \mathbf{R})$ the inequality

$$E_n(f; B) \leq C_1 \omega_1(f; \theta)$$

of Jackson type holds for the subset $B = B_2^m$ $(B = B_\infty^m)$ of the m-dimensional Euclidean space $\mathbf{R}^m$, where $\theta = m/n$ $(\theta = m^{3/2}/n)$. They also showed that this inequality is sharp with respect to the order of growth of m and n on the class H_1^ω, i.e., the estimate

$$E_n(H_1^\omega[B; \mathbf{R}]) \geq C_2 \omega(\theta)$$

holds for any $m, n \in \mathbf{N}$, $m \leq n$, and $\omega \in \Phi^1$. An analogous result was subsequently announced by Fedorov [2] for the case of real functions with integrable pth powers on the set B_2^m $(1 \leq p \leq \infty)$. In [3] the author proved that for any $k \in \mathbf{N}$ with $k \geq 2$ there is a $C = C(k) > 0$ such that for any $m, n \in \mathbf{N}$ with $n \geq Cm$, for any $Y \in (B)$, and for the unit ball $B \subset \mathbf{R}^m$ there exists a continuous linear operator $A_n \colon C(B; Y) \to \mathscr{P}_n$ such that

$$\|f - A_n f\|_{C(B; Y)} \leq C \omega_k(f; m/n)$$

for an arbitrary function $f \in C(B; Y)$. It was then shown that for any $k \in \mathbf{N}$ and $r \in \mathbf{Z}_+$ there is a $C > 0$ such that

$$E_n(W^r H_k^\omega[B; Y]) \geq C(m/n)^r \omega(m/n)$$

for any $m, n \in \mathbf{N}$, $m \leq n$, $\omega \in \Phi^k$, and $Y \in (B)$.

In this article we prove that there exists a $C_1 > 0$ such that for any $p \in [2, \infty]$, any $m, n \in \mathbf{N}$ with $n \geq C_1 m$, and any $Y \subset (B)$ there exists a continuous linear operator $A_n \colon C(B_p^m; Y) \to \mathscr{P}_n(\mathbf{R}^m; Y)$ for which

$$\|f - A_n f\|_{C(B_p^m; Y)} \leq C_1 \omega_1(f m^{3/2 - 1/p}/n)$$

for any $f \in C(B_p^m; Y)$. We remark that the constant C_1 is the same for $Y = \mathbf{R}$ and for an arbitrary $Y \in (B)$. It will then be shown that for any $p \in [2, \infty]$ there exists a $C > 0$ such that

$$E_n(H_1^\omega[B_p^m; Y]) \geq C\omega(m^{3/2-1/p}/n)$$

for any $m, n \in \mathbf{N}$, $n \geq m^{3/2-1/p}$, $Y \in (B)$, and $\omega \in \Phi^1$. In the case when $\dim Y < \infty$ we get estimates for the Kolmogorov width

$$d_n(W^r H_k^\omega[B_p^m; Y]; C(B_p^m; Y)) \qquad (B_p^m \subset \mathbf{R}^m)$$

that are sharp with respect to the order of growth of n, m, and $l = \dim Y$.

2. An extension theorem

For arbitrary $x \in X \in (B)$ and $r > 0$ we consider the ball $B(x, r) = \{y \in X \mid \|x - y\| \leq r\}$ and the sphere $S(x, r) = \{y \in X \mid \|x - y\| = r\}$. For an arbitrary nonempty set $M \subset X$ and $\delta > 0$ we denote by $O_\delta(M)$ the δ-neighborhood of it, and by ∂M its boundary. Denote by $UC(M) = UC(M; Y)$ the space of uniformly continuous mappings $f\colon M \to Y$, with $\|f\| = \sup_{x \in M} \|f(x)\|_Y$.

For any $X, Y \in (B)$, $d, r > 0$, $k \in \mathbf{N}$, $k \geq 2$, and any closed body $M \subset X$ with $B(0, r) \subset M \subset B(0, d)$ we construct a mapping $F_f\colon O_{r/(k-1)}(M) \to Y$ extending $f\colon M \to Y$, by setting

$$F_f(x) = F_f^k(x) = \begin{cases} f(x) & \text{if } x \in M, \\ -\sum_{i=0}^{k-1}(-1)^{k-i}\binom{k}{i} f(x + (k-i)(y - x)) & \\ \qquad \text{if } x \in O_{r/(k-1)}(M) \backslash M \text{ and } \{y\} = [0, x] \cap \partial M. \end{cases}$$

It is not hard to verify that $F_{\lambda f + \mu g} = \lambda F_f + \mu F_g$ for any mappings $f, g\colon M \to Y$ and any $\lambda, \mu \in \mathbf{R}$.

Theorem 1. *For any $k \in \mathbf{N}$ with $k \geq 2$ there exists a $C > 0$ such that for any $X, Y \in (B)$, any $d, r > 0$, any points $t \in M \subset X$, and $h \in X$ with $\|h\| \leq r/(2k^2 d)$, and a mapping $f\colon M \to Y$*

$$\|\Delta_h^k(F_f^k; t)\| \leq C\omega_k(f; (d/r)\|h\|); \tag{1}$$

moreover, for any $u, z \in X$ with $[u, u + kz] \subset O_{r/(k-1)}(M)$

$$\|\Delta_z^k(F_f^k; u)\| \leq C\omega_k(f, d/kr).$$

Proof. It will be assumed without loss of generality that $r = 1$. We first prove the inequality (1). Consider the case when t and h are collinear, $t + (k-1)h \in M$, and $\|h\| \leq 1/(k-1)$. If $t + kh \in M$, then $\|\Delta_h^k(F_f; t)\| \leq \omega_k(f; \|h\|)$.

Assume that $x = t + kh \notin M$. Then

$$F_f(x) = (-1)^{k+1} \sum_{i=0}^{k-1} (-1)^{k-i} \binom{k}{i} f(y + iz),$$

where z is the vector from the point x' where $[0, x]$ intersects ∂M to the point x, and $y = x - kz$. We introduce the notation

$$h_i = \frac{1}{k}\left\{\left(1 - \frac{i}{k}\right)x + \frac{i}{k}y - t\right\}, \qquad \tau_j = \frac{j}{k^2}(y - x) \qquad (i, j = 0, \dots, k).$$

Since $h = (x - t)/k$, we have the identity

$$\begin{aligned}
&\sum_{i=0}^{k}(-1)^{k-i}\binom{k}{i}\Delta_{h_i}^k(F_f; t) \\
&\quad = \sum_{i=0}^{k}(-1)^{k-i}\binom{k}{i}\sum_{j=0}^{k}(-1)^{k-j}\binom{k}{j}F_f\left(t + \frac{j}{k}\left\{\left(1 - \frac{i}{k}\right)x + \frac{i}{k}y - t\right\}\right) \\
&\quad = \sum_{j=0}^{k}(-1)^{k-j}\binom{k}{j}\sum_{i=0}^{k}(-1)^{k-i}\binom{k}{i}F_f\left(t + jh + \frac{ij}{k^2}(y - x)\right) \\
&\quad = \sum_{j=0}^{k}(-1)^{k-j}\binom{k}{j}\Delta_{\tau_j}^k(F_f; t + jh).
\end{aligned}$$

Therefore,

$$\begin{aligned}
(-1)^k\Delta_{h_0}^k(F_f; t) = &-\sum_{i=1}^{k}(-1)^{k-i}\binom{k}{i}\Delta_{h_i}^k(Ff; t) \\
&+\sum_{i=0}^{k}(-1)^{k-j}\binom{k}{j}\Delta_{\tau_j}^k(F_f; t + jh).
\end{aligned} \tag{2}$$

From the construction of F_f,

$$\begin{aligned}
\Delta_{\tau_k}^k(F_f; t + kh) &= \sum_{i=0}^{k}(-1)^{k-i}\binom{k}{i}F_f\left(x + \frac{i}{k}(y - x)\right) \\
&= \sum_{i=1}^{k}(-1)^{k-i}\binom{k}{i}f(y + (k - i)z) + (-1)^k F_f(x) = 0.
\end{aligned} \tag{3}$$

Since $x' \in [t + (k - 1)h; t + kh]$, it follows that $\|z\| \le \|h\|$ and

$$\begin{aligned}
\|h_i\| &\le \left(1 - \frac{i}{k}\right)\left\|\frac{x - t}{k}\right\| + \frac{i}{k}\left\|\frac{y - t}{k}\right\| = \left(1 - \frac{i}{k}\right)\|h\| + \frac{i}{k}\|h - z\| \\
&= \left(1 - \frac{i}{k}\right)\|h\| + \frac{i}{k}(\|h\| - \|z\|) \le \|h\|, \\
\|\tau_j\| &= \frac{1}{k}\|y - x\| = \|z\| \le \|h\| \qquad (i, j = 0, \dots, k).
\end{aligned}$$

Since $h_0 = h$ and $t + jh_i \in M$ $(j = 0, \dots, k - 1,\ i = 0, \dots, k)$, (2) and (3) give $\|\Delta_h^k(F_f; t)\| \le 2^{k+1}\omega_k(f; \|h\|)$.

We now consider the general case. Take arbitrary $t \in M$ and $h \in X$ with $\|h\| \le 1/(2k^2 d)$. If $x = t + kh \in M$, then $F_f(t + ih) = f(t + ih)$ $(i = 0, \dots, k)$, and hence (1) holds. Assume that $x \notin M$. We introduce the notation

$$\begin{aligned}
h_i = k\|h\|\frac{d}{\|x\|}(t + ih), \qquad &\tau_j = h\left(1 + \frac{kd\|h\|(j - k)}{\|x\|}\right), \\
\omega_i = t + ih - kh_i \qquad &(i, j = 0, \dots, k).
\end{aligned}$$

It is not hard to establish the relations

$$\begin{aligned}
\omega_0 + kh_0 = t, \qquad &\omega_i + jh_i = \omega_0 + jh_0 + i\tau_j, \\
\omega_0 + jh_0 + k\tau_j = \omega_k + jh_k \qquad &(i, j = 0, \dots, k).
\end{aligned}$$

We have the identity

$$\begin{aligned}\sum_{i=0}^{k}(-1)^{k-i}\binom{k}{i}\Delta_{h_i}^k(F_f;\omega_i) &= \sum_{i=0}^{k}(-1)^{k-i}\binom{k}{i}\sum_{j=0}^{k}(-1)^{k-j}\binom{k}{j}F_f(\omega_i+jh_i)\\ &= \sum_{j=0}^{k}(-1)^{k-j}\binom{k}{j}\sum_{i=0}^{k}(-1)^{k-i}\binom{k}{i}F_f(\omega_0+jh_0+i\tau_j)\\ &= \sum_{j=0}^{k}(-1)^{k-j}\binom{k}{j}\Delta_{\tau_j}^k(F_f;\omega_0+jh_0).\end{aligned}$$

Consequently,

$$\begin{aligned}\Delta_{\tau_k}^k(F_f;t) &= \Delta_{\tau_k}^k(F_f;\omega_0+kh_0)\\ &= \sum_{i=0}^{k}(-1)^{k-i}\binom{k}{i}\Delta_{h_i}^k(F_f;\omega_i)-\sum_{j=0}^{k-1}(-1)^{k-j}\binom{k}{j}\Delta_{\tau_j}^k(F_f;\omega_0+jh_0).\end{aligned}\tag{4}$$

Since t, $h/\|h\|\in M$, it follows that

$$t\left(\frac{1}{k\|h\|+1}\right)+\left(\frac{k\|h\|}{k\|h\|+1}\right)\frac{h}{\|h\|}=\frac{x}{k\|h\|+1}=q\in M,$$

$$\|x\|\le\|t\|+k\|h\|\le d+k\|h\|,\qquad d\ge1,$$

$$\|t\|\ge\|x\|-k\|h\|\ge1-1/kd\ge1/2.$$

And since $\omega_k+(k-1)h_k=x(1-kd\|h\|/\|x\|)$ and

$$\|\omega_k+(k-1)h_k\|=\|x\|\left(1-\frac{kd\|h\|}{\|x\|}\right)\le\|x\|\frac{1}{k\|h\|+1}=\|q\|,$$

it follows that $\omega_k+(k-1)h_k\in M$. Moreover,

$$\begin{aligned}&(\omega_k+jh_k)\in[0,\omega_k+(k-1)h_k]\subset M\qquad(j=0,\dots,k-1),\\ &(\omega_0+jh_0)\in[0,t]\subset M\qquad(j=0,\dots,k),\\ &(\omega_i+jh_i)=\omega_0+jh_0+i\tau_j\in[\omega_0+jh_0;\omega_k+jh_k]\qquad(j=0,\dots,k-1).\end{aligned}\tag{5}$$

Therefore,

$$\|\Delta_{\tau_j}^k(F_f;\omega_0+jh_0)\|\le\omega_k(f;\|\tau_j\|)\qquad(j=0,\dots,k-1).$$

In view of (5) and the fact that ω_i and h_i are collinear,

$$\Delta_{h_i}^k(F_f;\omega_i)\le2^{k+1}\omega_k(f;\|h_i\|)\qquad(i=0,\dots,k).$$

And since $\|\tau_j\|\le\|h\|(1+kd(k-j)\|h\|)\le2\|h\|$,

$$\begin{aligned}\|h_i\| &= k\|h\|d\frac{\|t+ih\|}{\|x\|}\le k\|h\|d\frac{\|t\|+k\|h\|}{\|t\|-k\|h\|}\\ &\le dk\|h\|\cdot\frac{1+k\|h\|/\|t\|}{1-k\|h\|/\|t\|}\le dk\|h\|\frac{1+2k\|h\|}{1-2k\|h\|}\le4k\|h\|d,\end{aligned}$$

it follows from (4) that

$$\begin{aligned}\|\Delta_{\tau_k}^k(F_f;t)\| &\le 2^k\omega_k(f;2\|h\|)+2^k\cdot2^{k+1}\omega_k(f;4kd\|h\|)\\ &\le(2^{2k+1}(4k)^k+2^{2k})\omega_k(f;\|h\|d),\end{aligned}$$

i.e., (1) holds.

Analogously, for any points $u, h \in X$ with

$$[u, u+kh] \subset O_{1/(k-1)}(M), \qquad h_i = (u+ih)/k, \qquad \tau_j = hj/k,$$

we have the identity

$$\sum_{i=0}^{k}(-1)^{k-i}\binom{k}{i}\Delta_{h_i}^k(F_f; 0) = \sum_{i=0}^{k}(-1)^{k-j}\binom{k}{j}\Delta_{\tau_j}^k\left(F_f; \frac{uj}{k}\right).$$

But since jh_i and $(uj/k + i\tau_j)$ belong to M $(i = 0, \dots, k,\ j = 0, \dots, k-1)$, by repeating the above arguments we conclude that there is a $C > 0$ depending only on k such that $\|\Delta_h^k(F_f; u)\| \le C\omega_k(f; d/k)$. The theorem is proved.

Lemma 1. *Let H be a Hilbert space, $K \subset H$ a nonempty closed convex set, and $Y \in (B)$. Then there exists a continuous linear operator $A\colon UC(K; Y) \to UC(H; Y)$ such that for any $f \in UC(K; Y)$ and any $t \in [0, \infty)$*

$$Af|_K = f \quad \textit{and} \quad \omega_1(Af; t) \le \omega_1(f; t).$$

Proof. For each $x \in H$ we consider $Px = \{y \in K \mid \|x-y\| = \inf_{z\in K}\|x-y\|\}$. It is known that P is a single-valued mapping and that $\|Pz_1 - Pz_2\| \le \|z_1 - z_2\|$ for any $z_1, z_2 \in H$ (see [4]). Let $AF(x) = \tilde{f}(x) = f(P(x))$ $(x \in H)$. We show that $\tilde{f}$ is the desired mapping. Indeed, for any $x, h \in H$

$$\begin{aligned}\|\Delta_h^1\tilde{f}(x)\|_Y &= \|f(P(x+h)) - f(P(x))\|_Y \\ &\le \omega_1(f; \|P(x+h) - P(x)\|) = \omega_1(f; \|h\|).\end{aligned}$$

This proves the lemma.

Lemma 2. *Let $X, Y \in (B)$. Then there exists a continuous linear operator*

$$A\colon UC(B(0,1); Y) \to UC(B(0,2); Y)$$

such that $Af|_{B(0,1)} = f$ and $Af|_{S(0,2)} \equiv f(0)$ for any $f \in UC(B(0,1); Y)$, and for any $t > 0$

$$\omega_1(Af; t) \le 6\omega_1(f; t).$$

Proof. Let $Af(x) = f(2x/\|x\| - x)$ for each $x \in X$, $1 \le \|x\| \le 2$. For any $x, y \in X$, $\|x\| \ge 1$, $\|y\| \ge 1$ we have the estimate

$$\begin{aligned}\left\|\frac{x}{\|x\|} - \frac{y}{\|y\|}\right\| &= \left\|\frac{x}{\|x\|} - \frac{x}{\|y\|} + \frac{1}{\|y\|}(x-y)\right\| \\ &\le \frac{\|x\|}{\|x\|\,\|y\|}|\,\|x\| - \|y\|\,| + \frac{1}{\|y\|}\|x-y\| \le \frac{2}{\|y\|}\|x-y\| \le 2\|x-y\|;\end{aligned}$$

therefore,

$$\begin{aligned}\|Af(x) - Af(y)\|_Y &\le \omega_1\left(f; \left\|\frac{2x}{\|x\|} - x - \frac{2y}{\|y\|} + y\right\|\right) \\ &\le \omega_1(f; 5\|x-y\|) \le 5\omega_1(f; \|x-y\|).\end{aligned}$$

We consider the case when $x \in B(0,1)$ and $y \in B(0,2)\backslash B(0,1)$. Then there exists an $x' \in [x, y] \cap S(0,1)$, and

$$\begin{aligned}\|Af(x) - Af(y)\|_Y &\le \|Af(x) - Af(x')\| + \|Af(x') - Af(y)\| \\ &\le \omega_1(f; \|x-x'\|) + \omega_1(Af; \|x'-y\|) \\ &\le \omega_1(f; \|x-x'\|) + 5\omega_1(f; \|x'-y\|) \le 6\omega_1(f; \|x-y\|).\end{aligned}$$

The case when $x, y \in B(0,1)$ is trivial, and the lemma is proved.

3. An inequality of Jackson type

Theorem 2. *There exists a* $C_1 > 0$ *such that for any* $m, n \in \mathbf{N}$, $n \geq C_1 m$, *any nonempty compact convex set* $K \subset \mathbf{R}^m$ *with* $d = \operatorname{diam} K > 0$, *and any space* $Y \in (B)$ *there exists a continuous linear operator* $A_n\colon C(K; Y) \to \mathscr{P}_n(\mathbf{R}^m; Y)$ *such that for any mapping* $f \in C(K; Y)$

$$\|f - A_n f\| \leq C_1 \omega_1(f; md/n).$$

Proof. Without loss of generality it will be assumed that $\{0\} \in K$. Let $\varphi(x) = T_1 f(x) = f(x/d)$, $x \in K$. We use Lemma 1 and construct an operator $T_2\colon C(K; Y) \to C(H; Y)$ such that $\|T_2\varphi(x) - T_2\varphi(y)\| \leq \omega_1(\varphi; \|x - y\|)$ for any $x, y \in H$. It was proved in [3] that for any $m, n \in \mathbf{N}$ with $n \geq Cm$ there exists a continuous linear operator $T_3\colon C(B(0; 1); Y) \to \mathscr{P}_n(B(0; 1);\ H)$ such that for any mapping $g \in C(B(0; 1); Y)$

$$\|T_3 g - g\|_{C(B(0;1))} \leq C\omega_1(g; m/n),$$

where $C > 0$ is an absolute constant. Consequently, the operator $A = T_0^{-1} \circ T_3 \circ T_2 \circ T_1$ satisfies the inequality

$$\begin{aligned}\|Af - f\|_{C(K)} &= \|T_1^{-1}(T_3 \circ T_2 \circ T_1 f - T_1 f)\|_{C(K)} \\ &\leq \|T_1^{-1}\|\, \|T_3 \circ T_2 \circ T_1 f - T_2 \circ T_1 f\|_{C(B(0;1))} \leq C\omega_1(T_2 \circ T_1 f, m/n).\end{aligned}$$

By Lemma 1, this yields

$$\|Af - f\|_{C(K)} \leq C\omega_1(T_1 f; m/n) \leq C\omega_1(f; md/n).$$

The theorem is proved.

Corollary 1. *There exists a* $C_1 > 0$ *such that for any* $p \in [2, \infty]$, $m, n \in \mathbf{N}$, $n \geq Cm$, *and any space* $Y \in (B)$ *there exists a continuous linear operator* $A_n\colon C(B_p^m; Y) \to \mathscr{P}_n(\mathbf{R}^m; Y)$ *such that for any mapping* $f \in C(B_p^m; Y)$

$$\|f - A_n f\| \leq C_1 \omega_1(f; m^{3/2 - 1/p}/n).$$

Corollary 2. *There exists a* $C_1 > 0$ *such that for any* $m, n \in \mathbf{N}$, *any nonempty compact set* $K \subset \mathbf{R}^m$, *any* $Y \in (B)$, *and any* $f \in C(K; Y)$

$$E_n(f, K) \leq C_1 \omega_1(f, \theta), \qquad \text{where } \theta = \min\{d, md/n\},\ d = \operatorname{diam} K.$$

Proof. For a proof it suffices to note that $\|f(x) - f(x_0)\|_Y \leq \omega_1(f; d)$ for any particular point $x_0 \in K$ and every other $x \in K$.

Theorem 3. *Suppose that* $k \in \mathbf{N}$, $k \geq 2$, *and* $d, r > 0$ $(d \geq 2r)$. *Then there exists a* $C = C(k, d/r) > 0$ *such that for any* $m, n \in \mathbf{N}$ *with* $n \geq Cm$, *any body* $M \subset \mathbf{R}^m$ *with diameter* d *and inscribed ball of radius* r *there is a continuous linear operator* $A_n\colon C(M; Y) \to \mathscr{P}_n(\mathbf{R}^m; Y)$ *such that for any mapping* $f \in C(M; Y)$

$$\|A_n f - f\| \leq C\omega_k(f; d^2 m/rn).$$

Proof. It will be assumed without loss of generality that $d = 1$. By Theorem 1, there exist a $C_1 = C_1(k) > 0$ and a continuous linear operator $A\colon C(M; Y) \to C(O_{r/(k-1)}(M))$ such that, for any function $f \in C(M; Y)$, we have $Af|_M \equiv f$ and

$$\|\Delta_h^k(Af; t)\| \leq C_1 \omega_k(f, (d/r)\|h\|)$$

for all $t \in M$ and $h \in \mathbf{R}^m$ with $\|h\| \leq r/2k^2$, and

$$\|\Delta_z^k(Af; u)\| \leq C_1 \omega_k(f; 1/(kr))$$

for any $u, z \in \mathbf{R}^m$ with $[u, u + kz] \subset O_{r/(k-1)}(M)$.

Using Theorem 1 successively $[\ln((2k+4)/r)\ln(k/(k-1))+1]$ times, we construct a linear operator $T_1\colon C(M;Y)\to C(B(0;2k+4);Y)$ and a number $C_2 = C_2(k, d/r)$ such that, for any $f\in C(M;Y)$, $f_1 = T_1 f|_M \equiv f$ and the inequality $\|\Delta_h^k(f_1;t)\|\le C_2\omega_k(f;\|h\|)$ holds for any $t\in M$ and $h\in\mathbf{R}^m$, $\|h\|\le r/2k^2$; moreover, the inequality $\|\Delta_h^k(f_1;u)\|\le C_2\omega_k(f;1/k)$ holds for any $u, h\in H$ with $[u, u+kh]\subset B(0;2k+4)$. We construct a continuous linear operator

$$T_2\colon C(B(0;2k+4);Y)\to C(B(0;((2k+4)/(k-1))k))$$

by setting $T_2(\varphi)(x) = \varphi((k/(k-1))x)$ for any $\varphi\in C(B(0;2k+4);Y)$ and $x\in B(0;2k+4)$.

In [3] (see Lemma 6) it was proved that there exist a continuous linear operator $T_3\colon C(B(0;2k+4);Y)\to\mathscr{P}_n$ and a constant $\alpha>1$ depending on k such that for any $\varphi\in C(B(0;2k+4);Y)$

$$\|\varphi - T_3\varphi\|_{C(B(0;1))}\le\alpha^m\omega_k(\varphi;(2k+4)/k)\le\alpha^m(2k+4)^k\omega_1(\varphi;1/k).$$

Let

$$N=\max\{[n/2(m+k)],1\},\qquad V=B(0,1/4k^2).$$

For any $f\in C(B(0;2k+4);Y)$ and $t\in(k/(k-1))M$ and for the set $B=B(0;1)$ we consider

$$\begin{aligned} T_4(f)(t) = \frac{1}{\gamma_{N,m}}\Bigg\{ \int_{B\setminus V} T_3(f)(t)J_{N,m}^k(x)\,dx \\ +(-1)^k\sum_{i=1}^{k}(-1)^{k-i}\binom{k}{i}\Bigg\{\int_V T_3(f)(ix+t)J_{N,m}^k(x)\,dx \\ -\int_B T_3(f)(ix)J_{N,m}^k\left(x-\frac{t}{i}\right)dx \\ -\int_B f(ix)J_{N,m}^k\left(x-\frac{t}{i}\right)dx\Bigg\}\Bigg\}, \end{aligned}$$

where

$$J_{N,m}=J_{N,m}^k=\left[\frac{\sin(n/2)\arccos(1-\|x\|^2/2)}{\sin(1/2)\arccos(1-\|x\|^2/2)}\right]^{2(m+k)},$$

$$\gamma_{N,m}=\int_B J_{N,m}(x)\,dx.$$

In [3] (Lemma 5) it was proved that for any $k\in\mathbf{N}$ and $\alpha>1$ there is a $C>0$ such that for any $m, n\in\mathbf{N}$ with $n\ge C$ and for

$$P_j=\frac{1}{\gamma_{n,m}}\int_B\|x\|^j J_{n,m}(x)\,dx,\qquad R=\frac{1}{\gamma_{n,m}}\int_{B(0;(k+1)/k)\setminus V}J_{n,m}(x)\,dx$$

we have the inequalities

$$|P_j|\le C/n^j,\qquad |R|\le(C/n^k)\alpha^{-m}\qquad(j=0,\dots,k).$$

Moreover, it was shown in [3] that $T_4(f)$ is a polynomial of degree at most n on $B(0;(k-1)/k)$. For an arbitrary function $f\in C(M;Y)$ let $\varphi=T_2\circ T_1(f)$. Then for any point $t \in M$

$$
\begin{aligned}
\|\varphi(t) - T_4(\varphi)(t)\| \leq &\left\| \frac{1}{\gamma_{N,m}} \int_{B\setminus V} \{\varphi(t) - T_3(\varphi)(t)\} J^k_{N,m}(x)\,dx \right\| \\
&+ \left\| \frac{1}{\gamma_{N,m}} \left\{ \int_V \varphi(t) J^k_{N,m}(x)\,dx \right.\right. \\
&\qquad + (-1)^k \sum_{i=1}^{k} (-1)^{k-i} \binom{k}{i} \left\{ - \int_V T_3(\varphi)(ix+t) J^k_{N,m}(x)\,dx \right. \\
&\qquad \left.\left.\left. + \int_B T_3(\varphi)(ix) J^k_{N,m}\left(x - \frac{t}{i}\right) dx + \int_B \varphi(ix) J^k_{N,m}\left(x - \frac{t}{i}\right) dx \right\}\right\}\right\| \\
\leq &(2k+4)^k \alpha^m \omega_k(\varphi;\, 1/(kr))|R| \\
&+ \left\| \frac{(-1)^k}{\gamma_{N,m}} \left\{ \int_V \Delta^k_x(\varphi;\, t) J^k_{N,m}(x)\,dx \right.\right. \\
&\qquad \left.\left. + \sum_{i=1}^{k} (-1)^{k-i} \binom{k}{i} \left\{ \int_{B(-t/i;\,1)\setminus V} (\varphi - T_3(\varphi))(ix+t) J^k_{N,m}(x)\,dx \right\}\right\}\right\| \\
\leq &\frac{C(2k+4)^k}{N^k} \omega_k\left(\varphi;\, \frac{1}{rk}\right) + \left\| \frac{1}{\gamma_{N,m}} \int_V C_2 \omega_2\left(f;\, \frac{k}{r(k-1)}\|x\|\right) J^k_{N,m}(x)\,dx \right\| \\
&+ \sum_{i=1}^{k} \binom{k}{i} \left\| \frac{1}{\gamma_{N,m}} \int_{B(0;\,(k+1)/k)\setminus V} (\varphi - T_3(\varphi))(ix+t) J^k_{N,m}(x)\,dx \right\| \\
\leq &\frac{C(2k+4)^k}{N^k} \omega_k\left(f_1;\, \frac{1}{r(k-1)}\right) \\
&+ \left\| \frac{1}{\gamma_{N,m}} \int_V (N\|x\|+1)^k \left(\frac{k}{k-1}+1\right)^k C_2 \omega_k\left(f;\, \frac{1}{rN}\right) J^k_{N,m}(x)\,dx \right\| \\
&+ 2^k (2k+4)^k a^m \omega_k(\varphi;\, 1/(rk))|R| \\
\leq &CC_2 \frac{(2k+4)^k}{N^k} \omega_k\left(f;\, \frac{1}{kr}\right) \\
&+ CC_2 \left(\frac{k}{k-1}+1\right)^k \omega_k\left(f;\, \frac{1}{Nr}\right) \sum_{i=0}^{k} \binom{k}{i} N^{k-i} \frac{1}{N^{k-i}} \\
&+ 2^k (2k+4)^k C N^{-k} \omega_k\left(f_1;\, \frac{1}{(k-1)r}\right) \leq C_3 \omega_k\left(f;\, \frac{1}{Nr}\right),
\end{aligned}
$$

where $C_3 = C_2 C(2k+4)^k + C_2 C(k/(k-1)+1)^k 2^k + 2^k C_2 C(2k+4)^k$.

The inequalities

$$
\|f - Ff\|_{C(M)} = \|T_1 f - Ff\|_{C(M)} \leq \|T_2^{-1}\|\,\|\varphi - T_4\varphi\| \leq \|T_2^{-1}\| C_2 \omega_k(f;\, 1/(Nr))
$$

hold for an arbitrary mapping $f \in C(M;\, Y)$, $\varphi = T_2 \circ T_1(f)$, and the continuous linear operator $F = T_2^{-1} \circ T_4 \circ T_2 \circ T_1$.

If $n/2(m+k) \leq 2$, then

$$
\omega_k\left(f;\, \frac{1}{Nr}\right) \leq \omega_k\left(f;\, \frac{1}{r}\right) \leq \left(\frac{n}{m}+1\right)^k \omega_k\left(f;\, \frac{m}{rn}\right) \leq (4k+5)^k \omega_k\left(f;\, \frac{m}{rn}\right).
$$

If $n/2(m+k) > 2$, then $m/n < \frac{1}{4}(1+k/m)^{-1}$, and

$$\omega_k\left(f;\frac{1}{rN}\right) \le \omega_k\left(f;\frac{1}{r(n/2(m+k)-1)}\right) \le \left(\frac{n/m}{n/2(m+k)-1}+1\right)^k \omega_k\left(f;\frac{m}{rn}\right)$$
$$\le \left(\frac{1}{\frac{1}{2}(1+k/m)^{-1}-m/n}+1\right)^k \omega_k\left(f;\frac{m}{rn}\right)$$
$$\le \left(4\left(1+\frac{k}{m}\right)+1\right)^k \omega_k\left(f;\frac{m}{rn}\right) \le (5+4k)^k\omega_k\left(f;\frac{m}{rn}\right).$$

The assertion of the theorem follows from the arbitrariness of f.

4. Widths of the classes $W^rH_k^\omega$

In this section we consider the quantity

$$d_n(W^rH_k^\omega[B_p^m;Y]) = d_n(W^rH_k^\omega[B_p^m;Y];C(B_p^m;Y))$$

for finite-dimensional $Y \in (B)$.

Theorem 4. *Suppose that $k \in \mathbf{N}$, $p \in [2,\infty)$, and $r \in \mathbf{Z}_+$. Then there exist $C, C_1, C_2 > 0$ such that*

$$C_1\theta^r\omega(\theta) \le d_n(W^rH_k^\omega[B_p^m;Y]) \le C_2\theta^r\omega(\theta) \tag{6}$$

for any $l, m, n \in \mathbf{N}$, $n \ge C^m l m^{m(1/2-1/p)}$, $Y \in (B)$, $\dim Y = l$, and $\omega \in \Phi^k$, where $\theta = m^{3/2-1/p}/\sqrt[m]{n/l}$.

Proof. Let $T_i = T_i(\varepsilon, p, m) \subset B_p^m$ be a minimal ε-net for the set B_p^m in $\mathbf{R}^m$. Then (see [5]) there exist $\alpha_1 = \alpha_1(p) > 0$ and $\alpha_2 = \alpha_2(p) > 0$ such that

$$(\alpha_1)^m(1/\varepsilon)^m m^{m(1/2-1/p)} \le |T_i| \le (\alpha_2)^m(1/\varepsilon)^m m^{m(1/2-1/p)}.$$

We prove the upper estimate in (6). Take a minimal ε-net $T_i = \{x_i\}_{i=1}^s$, $\varepsilon \in (0,1]$. The balls $B(x_i,\varepsilon)$ $(i=1,\dots,s)$ cover the whole of B_p^m, and for any $i \in \mathbf{N}$ the set $M_i = B_p^m \cap B(x_i;2\varepsilon)$ contains a ball of radius γ_p^ε (γ_p depends only on p). Therefore, by Theorem 3, for any mapping $f \in C(B_p^m;Y)$ there is a polynomial $p_N^i \in \mathscr{P}_N(\mathbf{R}^m;Y)$ ($N = Cm$, and C depends only on r and k) such that $\|f-p_N^i\|_{C(B(x_i;2\varepsilon))} \le A_1\omega_k(\operatorname{diam} M_i) \le A_2\omega_k(f;\varepsilon)$, where $A_1, A_2 > 0$ depend only on k and r. It is not hard to see that $(A_3)^m l \le \dim \mathscr{P}_N(\mathbf{R}^m;Y) \le (A_4)^m l$, where $A_3, A_4 > 0$ are absolute constants. Let $\Psi_i \in C(\mathbf{R}^m;\mathbf{R})$ be a partition of unity, i.e., for each $i \in \mathbf{N}$ we have $\Psi_i \ge 0$ on B_p^m, $\Psi_i \equiv 1$ on $B(x_i,\varepsilon)$, $\Psi_i \equiv 0$ outside $B(x_i,2\varepsilon)$, and $\sum_{i=1}^s \Psi_i \equiv 1$ on B_p^m. We take an arbitrary point $x \in B_p^m$ and consider the set $I = \{i \in \mathbf{N},\ i \le s \mid \Psi_i(x) \ne 0\}$. Then $\|x-x_i\| \le 2\varepsilon$ for any index $i \in I$. Therefore,

$$\left\|f(x) - \sum_{i=1}^s p_N^i\Psi_i(x)\right\| = \left\|\sum_{i=1}^s f(x)\Psi_i(x) - \sum_{i=1}^s p_N^i\Psi_i(x)\right\|$$
$$\le \sum_{i=1}^s \Psi_i(x)\|f(x)-p_N^i\|_Y = \sum_{i\in I}\Psi_i(x)\|f(x)-p_N^i(x)\|_Y$$
$$\le A_2\omega_k(f;\varepsilon)\sum_{i\in I}\Psi_i(x) = A_2\omega_k(f;\varepsilon).$$

This implies that $\|f - \sum_{i=1}^{s} p_N^i \Psi_i\|_{C(B_p^m; Y)} \le A_2 \omega(f; \varepsilon)$. We consider the space

$$L = \left\{ f\colon B_p^m \to Y \;\middle|\; f = \sum_{i=1}^{s} \Psi_i p_i ,\ p_i \in \mathscr{P}_N(\mathbf{R}^m; Y),\ i = 1, \dots, s \right\}.$$

Then

$$\dim L \le [(\alpha_2 A_4)^m (1/\varepsilon)^m m^{m(1/2-1/p)} l + 1] = n .$$

Consequently, for all $n \ge (\alpha_2 A_4)^m l m^{m(1/2-1/p)}$

$$d_n(W^r H_k^\omega[B_p^m; Y]) \le A_1 \omega_k(\varepsilon) \le C_1 \omega_k \left(m^{3/2-1/p} / \sqrt[m]{n/l} \right),$$

where C_1 depends only on k, r, and p.

We get a lower estimate. In [3] (Lemma 4) it was proved that for any $r \in \mathbf{Z}_+$ and $k \in \mathbf{N}$ there exist $N_1, N_2, N_3 > 0$ such that for all $\omega \in \Phi^k$ and $\alpha \in (0, 1)$ there is a function $\varphi_\alpha\colon B(0; \alpha) \to \mathbf{R}$ such that $\varphi_\alpha \equiv 0$ outside $B(0; \alpha)$, and for any $t \in [0; 1/k]$

$$\omega_k(\varphi_\alpha^{(r)}; t) \le N_1 \omega(t), \qquad \|\varphi_\alpha^{(r)}\| \le N_2 \omega_k(\varphi_\alpha^{(r)}; \alpha), \qquad \|\varphi_\alpha\| \ge N_3 \alpha^r \omega(\alpha).$$

We let $\alpha = \varepsilon/3$ and consider the space

$$\mathscr{L} = \left\{ \sum_{i=1}^{s} y_i \varphi_\alpha(x - x_i),\ y_i \in Y,\ i = 1, \dots, s \right\}.$$

Then $\dim \mathscr{L} \ge l \cdot s \ge l \cdot (\alpha_1)^m m^{m(1/2-1/p)} (1/\varepsilon)^m$. A theorem of Tikhomirov (see [6]) asserts that for any Banach space Z and any subspaces $L_1, L_2 \subset Z$ the condition $\dim L_1 < \dim L_2$ implies the existence of an element $x \in L_2$ such that $\|x\| = 1 = \inf_{y \in L_1} \|x - y\|$. Consequently, for $n = [l(\alpha_1)^m m^{m(1/2-1/p)} (1/\varepsilon)^m - 1]$ and for any subspace L with $\dim L \le n$ there exists a function $\varphi \in \mathscr{L}$ such that $\|\varphi\| = 1 = \inf_{f \in L} \|\varphi - f\|$. Since $\|\varphi\| = \max_{i=1,\dots,s} \|y_i\| \varphi_\alpha(x - x_i)$, and since the supports of the functions $\varphi_\alpha(x - x_i)$ are disjoint, it follows that $1 = \|\varphi\| = \|\varphi_\alpha\| \max_{i=1,\dots,s} \|y_i\|$. Moreover, for all $t < \alpha/k$

$$\omega_k(\varphi^{(r)}; t) \le \omega_k(\varphi_\alpha^{(r)}; t),$$

and for all $t \ge \alpha/k$

$$\omega_k(\varphi^{(r)}, t) \le k \|\varphi_\alpha^{(r)}\| \le k N_2 \omega(\alpha),$$

$$E(\overline{\varphi}, L) = \left(D \max_{i=1,\dots,s} \|y_i\| \right)^{-1} = \frac{\|\varphi_\alpha\|}{D} \ge \frac{N_3}{D} \alpha^r \omega(\alpha),$$

where $D = \max\{N_1, k N_2\}$ and $\overline{\varphi} = \varphi \cdot (D \max_{i=1,\dots,s} \|y_i\|)^{-1} \in W^r H_k^\omega[B_p^m; Y]$. Therefore, $E(W^r H_k^\omega[B_p^m; Y]) \ge E(\overline{\varphi}; L)$, and the arbitrariness of L gives

$$\begin{aligned} d_n(W^r H_k^\omega[B_p^m; Y]) &\ge \frac{N_3}{D} \left(\frac{\varepsilon}{3}\right)^r \omega\left(\frac{\varepsilon}{3}\right) \\ &\ge C_2 \left(m^{3/2-1/p} \Big/ \sqrt[m]{\frac{n}{l}} \right)^r \omega\left(m^{3/2-1/p} \Big/ \sqrt[m]{\frac{n}{l}} \right), \end{aligned}$$

where C_2 depends only on k, r, and p.

The theorem is proved.

We remark that the assertion of Theorem 4 remains true if instead of the Kolmogorov width we consider the linear width of the same class.

Theorem 5. *Suppose that* $p \in [2, \infty]$. *Then there exist* $C_1, C_2 > 0$ *such that*

$$C_1\omega\left(\frac{m^{3/2-1/p}}{n}\right) \le E_n(H_1^\omega[B_p^m; Y]) \le C_2\omega\left(\frac{m^{3/2-1/p}}{n}\right)$$

for all $n, m \in \mathbf{N}$, $n \ge m^{3/2-1/p}$, $Y \in (B)$, *and* $\omega \in \Phi^1$.

Proof. The upper estimate follows from Corollary 2.

We prove the lower estimate. It was proved in [3] (Lemma 4) that there exist $C_1, C_2, C_3 > 0$ such that for all $\omega \in \Phi^1$ and $\alpha \in (0, 1)$ there is a function $\varphi_\alpha : B(0; \alpha) \to \mathbf{R}$ with $\varphi_\alpha \equiv 0$ on $B(0, \alpha)$ and with

$$\omega_1(\varphi_\alpha, t) \le C_1\omega(t), \qquad C_2\omega(\alpha) \le \|\varphi_\alpha\| \le C_2\omega_1(\varphi_\alpha; \alpha)$$

for all $t \in [0, 1]$. There exists (see [5]) an $\alpha_1 = \alpha_1(p) > 0$ such that for any $m \in \mathbf{N}$ and $\alpha > 0$ the largest number of balls of radius 2α belonging to B_p^m is at least $(\alpha_1)^m(\alpha)^{-m}m^{m(1/2-1/p)}$. Let $\{x_i\}_{i=1}^s$ be the centers of these balls. The number $m^{3/2-1/p}/(4e^2\alpha_1 N)$ is taken as α. For each point $x \in B_p^m$ we set $R_i(x) = \varphi_\alpha(x - x_i)$. For any arbitrary $y \in Y$ with $\|y\| = 1$ let

$$\mathscr{L} = \left\{\sum_{i=1}^s \beta_i R_i(x) \mid \beta_i \in \mathbf{R}\right\}.$$

Then

$$\dim \mathscr{L} \ge (\alpha_1)^m(\alpha)^{-m}m^{m(1/2-1/p)} \ge (4e^2N/m)^m.$$

We have

$$\dim \mathscr{P}_N(\mathbf{R}^m; \mathbf{R}) = \binom{N+m}{m} \le 2\frac{(N+m)^{N+m}}{m^m N^N}\sqrt{\frac{N+m}{mN}}$$
$$\le 2\sqrt{2}\frac{N^m(1+m/N)^{N+m}}{m^m} \le 2\sqrt{2}\frac{N^m e^{2m}}{m^m} < \left(4e^2\frac{N}{m}\right)^m \le \dim \mathscr{L}.$$

By a theorem of Tikhomirov (see [6]), there is a function $f \in \mathscr{L}$, $\|f\|_{C(B_p^m; \mathbf{R})} = 1$, such that $E(f, \mathscr{P}_N) = E(f, \mathscr{P}_N(\mathbf{R}^m; \mathbf{R})) = 1$. Just as in the proof of Theorem 4, it can be shown that there is an absolute constant $D > 0$ such that

$$\tilde{f} = D^{-1}f \in H_1^\omega[B_p^m; \mathbf{R}] \quad \text{and} \quad E(\tilde{f}, \mathscr{P}_N) \ge C_3 D^{-1}\omega(\alpha).$$

The relations $y \cdot \tilde{f} \in H_1^\omega[B_p^m; Y]$ and

$$E(y \cdot \tilde{f}, \mathscr{P}_N(\mathbf{R}^m; Y)) \ge \inf_{p \in \mathscr{P}_N(\mathbf{R}^m; Y)} \|y^*(y \cdot \tilde{f} - p)\|$$
$$\ge \inf_{p \in \mathscr{P}_N(\mathbf{R}^m; Y)} \|\tilde{f} - y^*(p)\| = E(\tilde{f}, \mathscr{P}_N) \ge C_3 D^{-1}\omega(\alpha)$$

hold for arbitrary $y \in Y$ with $\|y\| = 1$ and any linear functional y^*, $y^*(y) = \|y^*\| = 1$. This implies the assertion of the theorem.

Bibliography

1. D. J. Newman and H. S. Shapiro, *Jackson's theorem in higher dimensions*, On Approximation Theory (Proc. Conf., Oberwolfach, 1963; P. L. Butzer and J. Korevaar, editors), Birkhäuser, Basel, 1964, pp. 208–219.
2. V. M. Fedorov, *Approximation of functions on a sphere and ball in a Euclidean space of high dimension*, All-Union School Theory of Functions, Abstracts of Reports, Erevan, 1987, p. 98. (Russian)
3. I. G. Tsar'kov, *An inequality of Jackson type for abstract functions*, Mat. Sb. **180** (1989), 676–699; English transl. in Math. USSR Sb. **67** (1990).

4. Pierre-Jean Laurent, *Approximation et optimisation*, Hermann, Paris, 1972.

5. A. N. Kolmogorov and V. M. Tikhomirov, *ε-entropy and ε-capacity of sets in function spaces*, Uspekhi Mat. Nauk **14** (1959), no. 2 (86), 3–86; English transl. in Amer. Math. Soc. Transl. (2) **17** (1961).

6. V. M. Tikhomirov, *Some questions in approximation theory*, Izdat. Moskov. Gos. Univ., Moscow, 1976. (Russian)

Translated by H. H. McFADEN

Труды Матем. Инст. Стеклова
198 (1992)

Proceedings of the Steklov
Institute of Mathematics
1994, Issue 1

JACKSON'S INEQUALITY IN $L_p(0, 2\pi)$ $(1 \le p < 2)$ WITH SHARP CONSTANT

UDC 517.5

N. I. CHERNYKH

Let $L_p(0, 2\pi)$ $(1 \le p < \infty)$ be the space of real 2π-periodic functions $f(x)$ such that $|f(x)|^p$ is integrable over the period, with the usual norm

$$\|f\|_p = \left(\int_{-\pi}^{\pi} |f(x)|^p\, dx\right)^{1/p};$$

let $t_{n-1}(x)$ be 2π-periodic trigonometric polynomials of order $n-1$,

$$E_n(f)_p = \inf_{t_{n-1}} \|f - t_{n-1}\|_p, \qquad \omega(\delta, f)_p = \sup_{|t|<\delta} \|\Delta_t f\|_p,$$

where $\Delta_t f(x)$ equals $f(x+t) - f(x)$ or $f(x+t/2) - f(x-t/2)$. For $p = \infty$ by $L_\infty(0, 2\pi)$ we mean the space $C_{2\pi}$ of continuous 2π-periodic functions with the norm $\|f\|_\infty = \max |f(x)|$.

The (first) Jackson inequality is an inequality of the form

$$E_n(f)_p \le M\omega(c/n, f)_p, \tag{1}$$

where M and c are constants. Jackson actually proved that, for $p = \infty$, for any $c > 0$ there exists an $M(c)$ such that inequality (1) is valid for all $f(x) \in C_{2\pi}$ and all $n \in \mathbf{N}$ (the extension of this result to the case $p < \infty$ is due to Quade). It follows from (1) that for any $\delta > 0$ and any $n \in \mathbf{N}$ there exists a $K(\delta)$ such that the inequality

$$E_n(f)_p \le K(\delta)\omega(\delta, f)_p \tag{2}$$

holds on the whole space $L_p(0, 2\pi)$. Let $K_n(\delta)_p$ be the smallest among the all possible constants $K(\delta)$ in (2) (for fixed n, δ, p); $K_n(\delta)_p$ is a nonincreasing function of the parameter $\delta > 0$. It is known that $K_n(\delta)_\infty \ge 1 - 1/2n$ for $p = \infty$, and that $K_n(\delta)_p \ge 2^{|1/p-1/2|-1/2}$ for $1 \le p < \infty$. This follows from the lower estimates for the constant $M(c)$ in (1) at $c = \pi$ of Korneĭchuk [1], for $p = \infty$, and of Berdyshev [2], for $1 \le p < \infty$, since the functions constructed by them for those estimates are $(2\pi/n)$-periodic. Thus, for all $\delta > 0$, $K_n(\delta)_p$ is separated away from zero.

Let us call

$$K_{n,p} = \inf_{\delta>0} K_n(\delta)_p \tag{3}$$

the best constant in the Jackson inequality. By the very definition of $K_{n,p}$, for no $c > 0$ is it possible to make M smaller than $K_{n,p}$ in the inequality (1) without violating it for some $f \in L_p(0, 2\pi)$.

In this paper we justify an upper estimate of the smallest constant $K_n(2\pi/n)_p = M(2\pi/n)$ in the inequality (1) for $c = 2\pi$ and $1 \le p < 2$, close to the above-mentioned lower estimate of Berdyshev (see Theorem 2). A direct (needing no further justification) corollary of this is the following theorem.

1991 *Mathematics Subject Classification*. Primary 42A10.

Theorem 1. *For any* $p \in [1, 2)$ *and* $n \in \mathbf{N}$ *the best constant* $K_{n,p}$ *in* (3) *and the smallest constants* $K_n(\delta)_p$ *in* (2) *for* $\delta \geq 2\pi/n$ *satisfy the following equality*:

$$(4) \qquad K_{n,p} = K_n(\delta)_p = 2^{-1/p'} \qquad (1/p' + 1/p = 1).$$

In the case $p > 2$, on the basis of S. B. Stechkin's well-known estimate $K_n(\pi/n) < 3/2$ (uniformly in p) and our lower estimates of $K_n(\delta)_p$, for now it can only be said that $2^{-1/p} \leq K_{n,p} < 3/2$. The cases $p = 2$ and $p = \infty$ are exceptional. N. P. Korneĭchuk proved that $K_n(\pi/n)_\infty \leq 1$, which together with his lower estimate of $K_n(\pi/n)_\infty$ implies the estimates

$$1 - 1/2n \leq K_{n,\infty} \leq K_n(\delta)_\infty \leq K_n(\pi/n)_\infty \leq 1 \qquad (\delta \geq \pi/n).$$

In the case $p = 2$ the author proved (see [3] and [4]) that for $\delta' < \pi/n$ and $\delta > \pi/n$

$$K_n(\delta')_2 > K_n(\delta)_2 = K_{n,2} = 1/\sqrt{2}.$$

Korneĭchuk gave two proofs of the estimate $K_n(\pi/n)_\infty \leq 1$, one of which is based on intermediate approximation of the class $H^\omega(C_{2\pi})$ by the class $K\,\mathrm{Lip}\,1(C_{2\pi})$, the other on his profound comparison theorems (see [1]). In the proof of the estimate $K_n(\pi/n)_2 \leq 1/\sqrt{2}$ the author [3] used the fact, trivial in $L_2(0, 2\pi) = L_2$, that

$$\|\Delta_t f\|_2 \geq \|\Delta_t(f - t_{n-1}(f)\|_2 \quad \text{for any } t$$

for the polynomial of best approximation $t_{n-1}(x; f)_2 = t_{n-1}(f)$, and also the specific fact that in the expansion of the even function $\|\Delta_t(f - t_{n-1}(f)\|_2^2 = u(t)$ the coefficients at the harmonics $\cos kt$ are nonpositive for $k \geq n$ and are equal to zero for $k = 1, 2, \ldots, n-1$.

It turned out to be possible to extend (with modification) the method, applied in L_2, to L_p for $p < 2$ thanks to a theorem of Konyagin [5], according to which the expansions of the functions $\|\Delta_t f\|_p^p$ $(p < 2,\ f \in L_p(0, 2\pi))$ in Fourier series have the same form as those of the functions $\|\Delta_t g\|_2^2$ $(g \in L_2(0, 2\pi))$: *for any function* $f \in L_p(0, 2\pi)$ *for* $0 < p < 2$

$$(5) \qquad \|\Delta_t f\|_p^p = \sum_{k=1}^{\infty} A_k^2 (1 - \cos kt),$$

where $A_k = A_k(f) \geq 0$ $(k \in \mathbf{N})$.

Konyagin based his proof of this theorem on this elegant idea: represent $|\Delta y|^p = |y - y_1|^p$ $(y, y_1 \in \mathbf{R})$ as the "sum" of the squares of the increments of a collection of functions $g_\xi(y)$, thus reducing the matter to L_2. This turned out to be possible for $0 < p < 2$, and as a result he found the following representation for $\|\Delta_t f\|_p^p$:

$$(6) \qquad \|\Delta_t f\|_p^p = \gamma_p \int_{-\infty}^{\infty} \int_{-\pi}^{\pi} \left\{(\xi - f(x+t))_+^{(p-1)/2} - (\xi - f(x))_+^{(p-1)/2}\right\}^2 dx\, d\xi$$

$(f \in L_p(0, 2\pi),\ 0 < p < 2,\ y_+^\alpha = y^\alpha$ for $y > 0$, $y_+^\alpha = 0$ for $y < 0$, and γ_p is a normalizing factor). The equality (5) follows easily from this.

In view of the significance of formula (5), it appears useful to indicate here one more expansion of $\|\Delta_t f\|_p^p$, simpler than (6), it seems to me, as the sum of norms of increments in L_2:

$$(7) \qquad \|\Delta_t f\|_p^p = \delta_p \int_{-\infty}^{\infty} \int_{-\pi}^{\pi} |\Delta_t e^{i\xi f(x)}|^2\, dx \frac{d\xi}{|\xi|^{p+1}} \qquad (f \in L_p(0, 2\pi),\ 0 < p < 2),$$

where

$$(8) \qquad \delta_p = \left(\int_{-\infty}^{\infty} 4 \sin^2 \frac{u}{2} \frac{du}{|u|^{p+1}} \right)^{-1}$$

(in (7) and (8) the interval of integration $(-\infty, \infty)$ may be replaced by $(0, \infty)$).

Indeed, since for any real y, y_1

$$|e^{i\xi y} - e^{i\xi y_1}|^2 = |e^{i\xi(y-y_1)} - 1|^2 = 4 \sin^2 \frac{\xi |y - y_1|}{2},$$

the function $|\xi|^{-(p+1)}|e^{i\xi y} - e^{i\xi y_1}|$ for $0 < p < 2$ has integrable singularities at the point $\xi = 0$ and at infinity, and the change $\xi|y-y_1| = u$ of the variable of integration gives the equality

$$\int_{-\infty}^{\infty} |e^{i\xi y} - e^{i\xi y_1}|^2 \frac{d\xi}{|\xi|^{p+1}} = |y - y_1|^p \delta_p^{-1} \qquad (\delta_p^{-1} > 0).$$

Setting here $y = f(x+t)$, $y_1 = f(x)$, and integrating with respect to x, after changing the order of integration we obtain (7).

Below we prove several lemmas, from which we subsequently derive, in Theorem 2, the Jackson inequality for $1 \le p < 2$ with sharp constant. We shall use the notation

$$x^{\langle \alpha \rangle} = |x|^\alpha \operatorname{sgn} x \qquad (x, \alpha \in \mathbf{R})$$

in order to abbreviate the writing.

Lemma 1. *For any real numbers a, b the numerical inequality*

$$(9) \qquad |a^{\langle p-1 \rangle} - b^{\langle p-1 \rangle}|^{p'} \le 2^{p'-2}(a-b)(a^{\langle p-1 \rangle} - b^{\langle p-1 \rangle})$$

holds for $1 < p < 2$, where $p' = p/(p-1) > 2$. Equality is attained only for $a = \pm b$.

Proof. It is clear that the right-hand side of (9) is nonnegative. Without loss of generality we may assume that $|a| \ge |b|$. Dividing both sides of (9) by $|a|^p = a \cdot a^{\langle p-1 \rangle}$ and taking into account that for $a = 0$ and for $a = \pm b$ (9) is trivial, we rewrite the assertion of the lemma in the equivalent form

$$(10) \qquad \psi_p(x) \overset{\text{def}}{=} (1 - x^{\langle p-1 \rangle})^{p'-1} - 2^{p'-2}(1-x) < 0 \qquad (-1 < x < 1),$$

where $x = b/a$. Since $x^{\langle \alpha \rangle}$ is an odd function, it follows that

$$dx^{\langle \alpha \rangle}/dx = \alpha |x|^{\alpha - 1} \qquad (x \ne 0 \text{ for } \alpha < 1).$$

Therefore, since $(p-1)(p'-1) = 1$,

$$\psi_p'(x) = -(1 - x^{\langle p-1 \rangle})^{p'-2}|x|^{p-2} + 2^{p'-2},$$

$$\begin{aligned} \psi_p''(x) &= (2-p)(1 - x^{\langle p-1 \rangle})^{p'-3}|x|^{2(p-2)} - (1 - x^{\langle p-1 \rangle})^{p'-2}(p-2)x^{\langle p-3 \rangle} \\ &= (2-p)(1 - x^{\langle p-1 \rangle})^{p'-3}x^{\langle p-3 \rangle}, \end{aligned}$$

whence

$$\psi_p''(x) < 0 \quad \text{for } x < 0, \qquad \psi_p''(x) > 0 \quad \text{for } x > 0,$$
$$\psi_p'(x) \le \psi_p'(-1) = 0 \quad \text{for } -1 \le x < 0$$

and for some $x_p \in (0, 1)$

$$\psi_p'(x) < 0 \quad \text{for } 0 < x < x_p, \qquad \psi_p'(x) > 0 \quad \text{for } x_p < x < 1.$$

Consequently,

$$\psi_p(x) < \max(\psi_p(-1), \psi_p(1)) = 0 \qquad (|x| < 1).$$

Inequality (10) and Lemma 1 are proved.

Let $t^*_{n-1}(x, f)$ be a trigonometric polynomial of order $n-1$ of best approximation for a function $f \in L_p(0, 2\pi)$ in $L_p(0, 2\pi)$, and let

$$f_n(x) = f(x) - t^*_{n-1}(x, f). \tag{11}$$

Lemma 2. *For any function* $f \in L_p(0, 2\pi)$

$$\|\Delta_t f\|_p^p \geq 2^{p-2} \int_{-\pi}^{\pi} \Delta_t f_u(x) \Delta_t f_n^{\langle p-1\rangle}(x)\, dx \tag{12}$$

for $1 < p < 2$ *and any* t.

Proof. By Nikol′skiĭ's well-known criterion the function $v(x) = \Delta_t f_n^{\langle p-1\rangle}(x)$ is orthogonal to any trigonometric polynomial $t_{n-1}(x)$, including $\Delta_t t^*_{n-1}(x, f)$. Therefore

$$J_n \overset{\text{def}}{=} \int_{-\pi}^{\pi} \Delta_t f_n(x) \Delta_t f_n^{\langle p-1\rangle}(x)\, dx = \int_{-\pi}^{\pi} \Delta_t f(x) \Delta_t f_n^{\langle p-1\rangle}(x)\, dx,$$

$$J_n \leq \|\Delta_t f(x)\|_p \cdot \|\Delta_t f_n^{\langle p-1\rangle}(x)\|_{p'}.$$

Further, by Lemma 1,

$$|\Delta_t f_n^{\langle p-1\rangle}(x)|^{p'} \leq 2^{p'-2} \Delta_t f_n(x) \Delta_t f_n^{\langle p-1\rangle}(x)$$

for any x and t. Consequently,

$$J_n \leq \|\Delta_t f(x)\|_p (2^{p'-2} J_n)^{1/p'},$$

whence for $J_n \neq 0$

$$J_n^{1/p} \leq 2^{1-2/p'} \|\Delta_t f(x)\|_p = 2^{2/p-1} \|\Delta_t f(x)\|_p,$$
$$\|\Delta_t f(x)\|_p^p \geq 2^{p-2} J_n.$$

For $J_n = 0$ the inequality (12) is trivial. Lemma 2 is proved.

Lemma 3. *The function*

$$\varphi_{n,\theta}(t) = \frac{1}{2} + \sum_{k=1}^{\infty} d_k(\theta) \cos kt, \tag{13}$$

where $0 < \theta \leq n$, $n \in \mathbf{N}$, *and*

$$d_k(\theta) = (1 - \theta^2 k^2/n^2)^{-1} \cos^2(\theta k\pi/2n), \qquad d_k(\theta) = 0 \quad \text{for } k = n/\theta, \tag{14}$$

is zero for $\theta\pi/n < |t| \leq \pi$, *and for* $|t| \leq \theta\pi/n$ *it is nonnegative:*

$$\varphi_{n,\theta}(t) = (\pi n/4\theta)|\sin(nt/\theta)| \qquad (|t| \leq \theta\pi/n). \tag{15}$$

For $\theta = 1$ this lemma was given in [3]. In the present general form the lemma is also proved by direct calculation of the Fourier coefficients of the function (15) extended by zero to the remaining part of the interval $(-\pi, \pi)$:

$$\frac{2}{\pi} \int_0^{\theta\pi/n} \frac{\pi n}{4\theta} \sin\frac{nt}{\theta} \cos kt\, dt = \frac{n}{4\theta} \left\{ -\frac{\cos\left(\frac{n}{\theta} + k\right)t}{\frac{n}{\theta} + k} - \frac{\cos\left(\frac{n}{\theta} - k\right)t}{\frac{n}{\theta} - k} \right\} \Bigg|_0^{\theta\pi/n}$$
$$= \frac{n}{4\theta} \frac{2n}{\theta} \frac{2\cos^2\frac{k\theta\pi}{2n}}{\left(\frac{n}{\theta}\right)^2 - k^2} = d_k(\theta) \qquad (k \neq n/\theta),$$

and for $n/\theta = k$

$$\int_0^{\theta\pi/n} \sin\frac{nt}{\theta}\cos kt\,dt = \frac{1}{2}\int_0^{k\pi}\sin 2kt\,dt = 0.$$

Denote by $\tau_{n-2}(t)$ the Fejér polynomial with the largest coefficient c_1 among the nonnegative trigonometric polynomials of order $n-2$ of the form

$$t_{n-2}(t) = \frac{1}{2} + \sum_{k=1}^{n-2} c_k \cos kt \qquad (n > 2);$$

that is,

$$\tau_{n-2}(t) = \frac{1}{n}\sin^2\frac{\pi}{n}\frac{1+\cos nt}{(\cos t - \cos\frac{\pi}{n})^2} = \frac{1}{2} + \sum_{k=1}^{n-2} a_k \cos kt, \tag{16}$$

where

$$a_k = \left(1 - \frac{k}{n}\right)\cos\frac{k\pi}{n} + \frac{1}{n}\cot\frac{\pi}{n}\sin\frac{k\pi}{n} \tag{17}$$

(see, for example, [6], §3 and the references listed there). Consider the polynomial

$$\begin{aligned}\tau_{n-1}^*(t) = \tau_{n-2}(t)\frac{1+\cos t}{1+\cos\frac{\pi}{n}} &= \frac{2}{n}\sin^2\frac{\pi}{2n}\frac{(1+\cos nt)(1+\cos t)}{(\cos t - \cos\frac{\pi}{n})^2} \\ &= \frac{1}{2} + \sum_{k=1}^{n-1} a_k^* \cos kt.\end{aligned} \tag{18}$$

It is easy to verify that

$$a_k^* = \left(1 - \frac{k}{n}\right)\cos\frac{k\pi}{n} + \frac{1}{n}\cos\frac{\pi}{n}\cdot\sin\frac{k\pi}{n}. \tag{19}$$

Indeed, by (17), $a_n = a_{n-1} = 0$, $a_0 = 1$, whence we get

$$\begin{aligned}(1+\cos t)\tau_{n-2}(t) &= \left(\frac{1}{2} + \sum_{k=1}^{n-2} a_k\cos kt\right)(1+\cos t) \\ &= \frac{1}{2} + \frac{a_1}{2} + \frac{1}{2}\sum_{k=1}^{n-1}(a_{k-1} + 2a_k + a_{k+1})\cos kt \\ &= \frac{1}{2}\left(1+\cos\frac{\pi}{n}\right) + \sum_{k=1}^{n-1}\left\{\left(\left(1-\frac{k}{n}\right)\cos\frac{k\pi}{n} + \frac{1}{n}\cot\frac{\pi}{n}\sin\frac{k\pi}{n}\right)\right. \\ &\qquad\qquad \left.\times\left(1+\cos\frac{\pi}{n}\right) + \frac{1}{n}\sin\frac{k\pi}{n}\cdot\sin\frac{\pi}{n}\right\}\cos kt \\ &= \left(1+\cos\frac{\pi}{n}\right)\left\{\frac{1}{2} + \sum_{k=1}^{n-1} a_k^*\cos kt\right\}.\end{aligned}$$

Lemma 4. *The Fourier coefficients $d_k(\theta)$ (14) of the function (13) for $\theta \geq 2$ and the coefficients a_k^* (19) of the polynomial $\tau_{n-1}^*(t)$ (18) are related by the inequalities*

$$d_k(\theta) \leq a_k^* \qquad (k = 1, 2, \ldots, n-1). \tag{20}$$

Proof. Since

$$\sin(k\pi/n) \geq \sin(\pi/n)\cdot(1-k/n)(1-1/n)^{-1} > \sin(\pi/n)\cdot(1-k/n)$$

for $1 \le k < n$, it follows that $a_k^* > 0$ $(k \le n-1)$, and it suffices to verify the inequality (20) only for $1 \le k < n/\theta$, where $d_k(\theta) > 0$. Further, for $\theta < \theta'$ and $k < n/\theta$ we have $d_k(\theta') < d_k(\theta)$. Therefore it is enough to prove that

$$d_k(2) \le a_k^* \qquad (1 \le k \le n/2).$$

The last inequality is equivalent to

$$\cos\frac{k\pi}{n} \le \left(\left(1-\frac{k}{n}\right) + \frac{1}{n\sin(\pi/n)}\tan\frac{k\pi}{n}\right)\left(1-\frac{4k^2}{n^2}\right) \qquad (1 \le k < n/2),$$

which follows from the crude estimates

$$\tan\frac{k\pi}{n} \ge \frac{k\pi}{n} \qquad (1 \le k < n/2), \qquad \sin\frac{\pi}{n} < \frac{\pi}{n},$$
$$\cos\frac{k\pi}{n} \le 1 - \frac{4k^2}{n^2} \qquad (1 \le k \le n/2).$$

The lemma is proved.

The estimates of $E_n(f)_p$ in the following theorem follow from (15) and Lemmas 2–4. The sharpness of these estimates was proved long ago by Berdyshev [2].

Theorem 2. *For any function $f(x) \in L_p(0, 2\pi)$, $1 \le p < 2$, $n \ge 2$, and $2 \le \theta \le n$, the following inequalities hold:*

$$E_n^p(f)_p \le \frac{1}{2^{p/p'}} \int_0^{\theta\pi/n} \frac{n}{2\theta} \sin\frac{nt}{\theta} \|\Delta_t f\|_p^p \, dt, \tag{21}$$

$$E_n^p(f)_p \le \frac{1}{2^{p/p'}} \int_0^{2\pi/n} \frac{n}{4} \sin\frac{nt}{2} \omega^p(t, f)_p \, dt, \tag{22}$$

$$E_n(f)_p \le \frac{t}{2^{1/p'}} \omega\left(\frac{2\pi}{n}, f\right)_p \qquad (1/p + 1/p' = 1). \tag{23}$$

If $f(x)$ is not equivalent to a constant, then the last of these inequalities is strict. The constant in front of the integrals and the constant in front of the modulus of continuity on the right-hand sides of these inequalities cannot be lowered without violating the inequalities for some functions in $L_p(0, 2\pi)$.

Proof. Let $1 < p < 2$ and let $f_n(x)$ be defined by $f(x)$ via formula (11). Then, as observed above, $f_n(x)$ is orthogonal to any trigonometric polynomial $t_{n-1}(x)$ of order $n-1$, and together with it the function

$$\int_{-\pi}^{\pi} (f_n(x) f_n^{\langle p-1\rangle}(x+t) + f_n(x+t) f_n^{\langle p-1\rangle}(x))\, dx$$

will be orthogonal to any $t_{n-1}(t)$.

Therefore, in particular for the polynomial $\tau_{n-1}^*(t)$ (18), we have

$$2E_n^p(f)_p = 2\|f_n\|_p^p = \frac{2}{\pi}\int_0^{\pi}\int_{-\pi}^{\pi} \Delta_t f_n(x) \cdot \Delta_t f_n^{\langle p-1\rangle}(x)\, dx\, \tau_{n-1}^*(t)\, dt.$$

By Lemma 2, in view of the nonnegativity of $\tau_{n-1}^*(t)$,

$$E_n^p(f)_p \le 2^{1-p} \cdot \frac{2}{\pi}\int_0^{\pi} \|\Delta_t f(x)\|_p^p \, \tau_{n-1}^*(t)\, dt.$$

Applying now Konyagin's formula (5) and the inequality (20) from Lemma 4, we have

$$E_n^p(f)_p \leq 2^{1-p}\left\{\sum_{k=1}^{\infty} A_k^2(f) - \sum_{k=1}^{n-1} A_k^2(f)a_k^*\right\}$$
$$\leq 2^{1-p}\left\{\sum_{k=1}^{\infty} A_k^2(f) - \sum_{k=1}^{n-1} A_k^2(f)d_k(\theta)\right\} \qquad (\theta \geq 2).$$

Finally, by Lemma 3 $d_k(\theta) \leq 0$ for $k > n/\theta$ and, the more so, for $k \geq n$, which implies the estimate

$$E_n^p(f)_p \leq 2^{1-p}\sum_{k=1}^{\infty} A_k^2(f)(1-d_k(\theta)) = 2^{1-p}\cdot\frac{2}{\pi}\int_0^{\pi} \|\Delta_t f\|_p^p\, \varphi_{n,\theta}(t)\,dt$$
$$= 2^{1-p}\int_0^{\pi\theta/n} \frac{n}{2\theta}\sin\frac{nt}{\theta}\|\Delta_t f\|_p^p\,dt,$$

equivalent to the estimate (21) for $p > 1$.

To prove (21) for $p = 1$, we note that any function $f(x)$ in $L_1(0, 2\pi)$ can be approximated in $L_1(0, 2\pi)$ with any accuracy ε by continuous functions $f_\varepsilon(x)$, moreover with $\|\Delta_t f_\varepsilon\|_1 \to \|\Delta_t f\|_1$ and $E_n(f_\varepsilon)_1 \to E_n(f)_1$ $(\varepsilon \to 0)$. Therefore it is enough to prove the inequality (21) with $p = 1$ for continuous functions. But in that case we can pass to the limit as $p \to 1$ in the already proved inequality (21) with $1 < p < 2$. Thus, (21) is valid for any function $f(x) \in L_p(0, 2\pi)$ for any $1 \leq p < 2$. Below we assume that $1 \leq p < 2$.

It is possible to replace $\|\Delta_t f\|_p^p$ in (21) by $\omega^p(t, f)_p$, and since $\omega^p(t, f)_p$ is monotone the best of the resulting estimates will be the one with smallest θ, for which (20) was proved, i.e., the estimate (22). If $f(x) \not\sim$ const, then on a part of the interval $(0, 2\pi/n)$ the inequality $\omega(t) \leq \omega(2\pi/n)$ is strict, which together with (22) guarantees (23) with strict inequality. But if $f \sim$ const, then (23) turns into a trivial equality.

In [2] Berdyshev formulated and proved a theorem to the effect that the constant K in the inequality

$$E_n(f)_p \leq K\omega(\pi/n, f)_p \qquad (1 \leq p < 2), \tag{24}$$

valid on the whole of $L_p(0, 2\pi)$, cannot be smaller than $2^{-1/p'}$. In fact, he proved more: the functions $f_{n\varepsilon} \in L_p$ he constructed, for which

$$E_n(f_{n\varepsilon})_p > 2^{-1/p'}(1-\varepsilon)\omega(\pi/n, f_{n\varepsilon})_p \qquad (0 < \varepsilon < 1),$$

are $(2\pi/n)$-periodic, and so, if $\theta \geq 1$, then $\omega(\pi\theta/n, f_{n\varepsilon})_p \equiv \omega(\pi/n, f_{n\varepsilon})_p$ and

$$E_n(f_{n\varepsilon})_p > 2^{-1/p'}(1-\varepsilon)\omega(\pi\theta/n, f_{n\varepsilon})_p.$$

Using this fact, we conclude that in the estimates

$$E_n(f)_p \leq 2^{-1/p'}\omega(\theta\pi/n, f)_p \qquad (\theta \geq 2) \tag{25}$$

resulting from (21) and (22) the constant $2^{-1/p'}$ cannot be lowered, and the more so for the constants in (21)–(23). The theorem is completely proved.

Remark 1. By sharper and more complicated estimates it is possible to show that the inequalities (20) in Lemma 4 are satisfied for $\theta \geq \theta^* = (1/2 - 2/\pi^2)^{-1/2} \approx 1.834$, in view of which the inequality (21) in Theorem 2 remains valid for $\theta \geq \theta^*$. Therefore, the value $2\pi/n$ of the argument of $\omega(\delta, f)_p$ on the right in (22) and (23) is clearly not optimal. The question of the smallest $\theta = \theta_{\min}(p)$ in (25) for $1 \le p < 2$ remains open. In the case $p = 2$ it was proved in [4] that $\theta_{\min}(2) = 1$.

Remark 2. For best approximation by constants, integration over $(0, \pi)$ of both sides of the inequality (12) of Lemma 2 gives the unimprovable estimates

$$E_1^p(f)_p \le 2^{-p/p'} \int_0^\pi \frac{1}{\pi} \|\Delta_t f\|_p^p \, dt, \tag{26}$$

$$E_1(f)_p \le 2^{-1/p'} \omega(\pi, f)_p \qquad (1 \le p < 2). \tag{27}$$

Analogously, integrating for $p > 2$ and $n = 1$ both sides of the inequality

$$\int_{-\pi}^{\pi} \Delta_t f_n(x) \cdot \Delta_t f_n^{\langle p-1 \rangle}(x) \, dx \le \|\Delta_t f\|_p \|\Delta_t f_n^{\langle p-1 \rangle}\|_{p'}, \tag{28}$$

we get

$$2E_1^p(f)_p \le \left(\int_0^\pi \frac{1}{\pi} \|\Delta_t f\|_p^p \, dt \right)^{1/p} \left(\int_0^\pi \frac{1}{\pi} \|\Delta_t f_1^{\langle p-1 \rangle}\|_{p'}^{p'} \, dt \right)^{1/p'}. \tag{29}$$

To estimate the second integral in (29) we first prove one more lemma.

Lemma 5. *Let $0 < p < 2$. Then for any function $f(x) \in L_p(0, 2\pi)$*

$$\frac{1}{\pi} \int_0^\pi \|\Delta_t f\|_p^p \, dt \le 2E_1^p(f)_p \le 2\|f\|_p^p. \tag{30}$$

Proof. We use the representation of

$$\|\Delta_t f\|_p^p = \sum_{k=1}^\infty A_k^2(f)(1 - \cos kt)$$

in the form (7). From this it follows that

$$A_k^2(f) = 4\pi\delta_p \int_{-\infty}^\infty (|\rho_k(\xi, f)|^2 + |\rho_{-k}(\xi, f)|^2) \frac{d\xi}{|\xi|^{p+1}},$$

where

$$\rho_k(\xi, f) = \frac{1}{2\pi} \int_{-\pi}^{\pi} (e^{i\xi f(x)} - 1) e^{-ikx} \, dx.$$

Therefore,

$$\begin{aligned} \frac{1}{\pi} \int_0^\pi \|\Delta_t f\|_p^p \, dt &= \sum_{k=1}^\infty A_k^2(f) = 2\delta_p \int_{-\infty}^\infty 2\pi \sum_{\substack{k=-\infty \\ k \ne 0}}^\infty |\rho_k^2(\xi, f)| \frac{d\xi}{|\xi|^{p+1}} \\ &\le 2\delta_p \int_{-\infty}^\infty \|e^{i\xi f(x)} - 1\|_2^2 \frac{d\xi}{|\xi|^{p+1}} = 2\int_{-\pi}^{\pi} dx \, \delta_p \int_{-\infty}^\infty |e^{i\xi f(x)} - 1|^2 \frac{d\xi}{|\xi|^{p+1}} \\ &= 2\int_{-\pi}^{\pi} |f(x)|^p \, dx, \end{aligned}$$

and

$$\frac{1}{\pi} \int_0^\pi \|\Delta_t f\|_p^p \le 2\|f\|_p^p.$$

Replacing $f(x)$ by $f(x) - \text{const}$, we obtain (30).

We continue with the estimate (29). Since $p > 2$ and $p' < 2$ in it, applying Lemma 5 with p' instead of p to the function $f_1^{\langle p-1\rangle}(x) \in L_{p'}(0, 2\pi)$, we obtain

$$2E_1^p(f)_p \leq \left(\frac{1}{\pi}\int_0^\pi \|\Delta_t f\|_p^p\,dt\right)^{1/p} (2\|f_1^{\langle p-1\rangle}\|_{p'}^{p'})^{1/p'}.$$

Since $\|f_1^{\langle p-1\rangle}\|_{p'}^{p'} = \|f_1\|_p^p = E_1^p(f)_p$, from this we finally deduce that *for $p > 2$ the following sharp estimates are valid*:

$$E_1^p(f)_p \leq \frac{1}{2}\int_0^\pi \frac{1}{\pi}\|\Delta_t f\|_p^p\,dt, \tag{31}$$

$$E_1(f)_p \leq \frac{1}{2^{1/p}}\omega(\pi, f)_p. \tag{32}$$

That the last estimate is sharp was proved by Berdyshev in [2]. Estimates for $E_1(f)_p$ by a different method were also obtained in [7].

Bibliography

1. N. P. Korneĭchuk, *Extremal problems in approximation theory*, "Nauka", Moscow, 1976. (Russian)
2. V. I. Berdyshev, *On Jackson's theorem in L_p*, Trudy Mat. Inst. Steklov. **88** (1967), 3–16; English transl. in Proc. Steklov Inst. Math. **88** (1967).
3. N. I. Chernykh, *On Jackson's inequality in L_p*, Trudy Mat. Inst. Steklov. **88** (1967), 71–74; English transl. in Proc. Steklov Inst. Math. **88** (1967).
4. V. V. Arestov and N. I. Chernykh, *On the L_2-approximation of periodic functions by trigonometric polynomials*, Approximation and Function Spaces (Proc. Internat. Conf., Gdańsk, 1979; Z. Ciesielski, editor), PWN, Warsaw, and North-Holland, Amsterdam, 1981, pp. 25–43.
5. S. V. Konyagin, *On the moduli of continuity of functions*, All-Union School on Theory of Functions (100th Anniversary of the Birth of N. N. Luzin), Abstracts of Reports, Kemerovo, 1983. (Russian).
6. S. B. Stechkin, *Some extremal properties of postive trigonometric polynomials*, Mat. Zametki **7** (1970), 411–422; English transl. in Math. Notes **7** (1970).
7. S. A. Pichugov, *The Jung constant of the space L_p*, Mat. Zametki **43** (1988), 604–614; English transl. in Math. Notes **43** (1988).

Translated by S. IVANOV

Труды Матем. Инст. Стеклова
198 (1992)

Proceedings of the Steklov
Institute of Mathematics
1994, Issue 1

LOWER ESTIMATES OF THE WIDTHS OF SOME CLASSES OF PERIODIC FUNCTIONS

UDC 517.5

V. T. SHEVALDIN

INTRODUCTION

Let $L_p = L_p[0; 2\pi]$ $(1 \le p \le \infty)$ be the space of 2π-periodic functions whose pth power is absolutely integrable on $[0; 2\pi]$ for $1 \le p < \infty$, and essentially bounded for $p = \infty$, let $C_{2\pi}$ be the space of continuous 2π-periodic functions, and let $C[a, b]$ be the space of continuous functions on the segment $[a, b]$. As usual, let $W_p^{r,\alpha}$ $(r > 0, \alpha \in \mathbf{R}, 1 \le p \le \infty)$ be the class of 2π-periodic functions f of the form

$$f(x) = C + \frac{1}{\pi}\int_0^{2\pi} K_{r,\alpha}(x-t)\varphi(t)\,dt = C + \pi^{-1}K_{r,\alpha} * \varphi,$$

where C is an arbitrary constant, $K_{r,\alpha}(t) = \sum_{k=1}^{\infty} k^{-r}\cos(kt - \alpha\pi/2)$, and φ is such that $\|\varphi\|_p \le 1$ and $\varphi \perp C$ (i.e., $\int_0^{2\pi}\varphi(t)\,dt = 0$). For $\alpha = r \in \mathbf{N}$ the class $W_p^{r,\alpha}$ is the class of 2π-periodic functions f whose $(r-1)$st derivative is absolutely continuous and $\|f^{(r)}\|_p \le 1$; it is usually denoted by W_p^r. For $\alpha + 1 = r \in \mathbf{N}$ the class $W_p^{r,\alpha}$ is the class of the trigonometric conjugates of the functions of the class W_p^r and is denoted by $\widetilde{W_p^r}$.

Let $\omega(f, \delta)$ be the modulus of continuity of a function $f \in C_{2\pi}$ and let H^ω be the class of functions $f \in C_{2\pi}$ such that $\omega(f, \delta) \le \omega(\delta)$ for any $\delta \ge 0$, where $\omega = \omega(\delta)$ is a given modulus of continuity.

Also, let W^rH^ω $(r \in \mathbf{N})$ be the class of functions $f \in C_{2\pi}$ whose rth derivative is continuous and $\omega(f^{(r)}, \delta) \le \omega(\delta)$. We denote by $\widetilde{W^rH^\omega}$ the class of the trigonometric conjugates of the functions of the class W^rH^ω. It will be convenient to us to use another definition of the class $\widetilde{W^rH^\omega}$, namely:

$$\widetilde{W^rH^\omega} = \{f \in C_{2\pi} : f = C + \pi^{-1}K_{r,r-1} * \varphi,\ \varphi \in H^\omega,\ \varphi \perp C\}.$$

The Kolmogorov width $d_m(F)_X$ of a set F in a Banach space X is the quantity

$$d_m(F)_X = \inf_{F_m\,:\,\dim F_m \le m}\ \sup_{x \in F}\ \inf_{y \in F_m} \|x - y\|_X. \tag{1}$$

A subspace F_m realizing the infimum in (1) is called an extremal subspace for the set F.

This paper is dedicated to obtaining lower estimates of the widths of the classes of periodic functions $W_\infty^{r,\alpha}$ $(r \in \mathbf{N}, \alpha \in \mathbf{R})$ and $\widetilde{W^rH^\omega}$ $(r \in \mathbf{N},\ \omega$ a convex modulus of continuity) in the uniform metric. Many papers have been dedicated to the solution of the problem of calculating the widths of these classes and of their generalizations.

1991 *Mathematics Subject Classification*. Primary 41A46, 42A10.

We give a brief historical review.

Denote by $E_{m-1}(W_p^{r,\alpha})_q$ $(1 \le q \le \infty)$ the magnitude of the best approximation of the class $W_p^{r,\alpha}$ in the metric of L_q by the space of trigonometric polynomials of order $m-1$, the calculation of which permits one to obtain an upper estimate for $d_{2m-1}(W_p^{r,\alpha})_q$. The first results on the calculation of $E_{m-1}(W_p^{r,\alpha})_q$ were obtained in 1937 by J. Favard [1], [2] $(p = q = \infty, \alpha = r \in \mathbf{N})$, and by N. I. Akhiezer and M. G. Kreĭn [3] $(p = q = \infty, \alpha + 1 = r \in \mathbf{N})$. Subsequently this problem was taken up by B. Sz.-Nagy, S. M. Nikol'skiĭ, V. K. Dzyadyk, S. B. Stechkin, and Sun Yunshen. The most general result was obtained by Dzyadyk [4], [5], who calculated $E_{m-1}(W_p^{r,\alpha})_q$ for any $\alpha \in \mathbf{R}$, $r > 0$ for $p = q = \infty$ and $p = q = 1$. Furthermore, Babenko [6] calculated this quantity for $p = \infty$, $q = 1$, $\alpha + 1 = r \in \mathbf{N}$.

In 1969 Tikhomirov [7] showed that the widths $d_{2m-1}(W_\infty^r)_\infty$ and $d_{2m}(W_\infty^r)_\infty$ are equal to $E_{m-1}(W_\infty^r)_\infty$ and, consequently, that the space T_{m-1} of trigonometric polynomials of order at most $m-1$ is an extremal subspace. This was followed by a series of papers dedicated to the calculation of the widths of the classes W_p^r $(1 \le p \le \infty)$ in various metrics, both for the case of periodic functions and for the nonperiodic case, and also of the widths of certain more general classes of functions. The strongest result on the calculation of $d_m(W_p^r)_q$ for $p \ge q$ was announced in 1985 by Buslaev and Tikhomirov in [8]. We note that in many cases, along with the space of trigonometric polynomials, for the widths of even order of the classes W_p^r the space of periodic polynomial splines of degree $r-1$ with uniform nodes is also extremal. A complete bibliography on the indicated results can be found in Korneĭchuk's monographs [9], [10] (for some generalizations, see also [11]–[17]).

For $\alpha \ne r$ few results are known on the exact calculation of the widths of the class $W_p^{r,\alpha}$ in the metric of L_q. An upper estimate of the width is given by the quantity $E_{m-1}(W_p^{r,\alpha})_q$, which, as we already indicated, was calculated by Dzyadyk [4], [5] for $p = q = \infty$ and $p = q = 1$ for any $\alpha \in \mathbf{R}$ and $r > 0$. In 1983 Babenko [18] proved a lower estimate for $d_{2m-1}(W_p^{r,\alpha})_q$, for $p = q = 1$ and $p = q = \infty$, for any $\alpha \in \mathbf{R}$ and $r = 1$; this estimate coincided with the upper estimate obtained by Dzyadyk and, in particular, with the Akhiezer-Kreĭn estimate for $\alpha = 0$. For any $m \in \mathbf{N}$ and any $r \in \mathbf{N}$, $r \ge r_0(m)$ ($r_0(m)$ a certain function of m), the author [19] obtained lower estimates for $d_{2m-1}(\widetilde{W_\infty^r})_\infty$ and $d_{2m}(\widetilde{W_\infty^r})_\infty$, also coinciding with the quantity $E_{m-1}(\widetilde{W_\infty^r})_\infty$ calculated by Akhiezer and Kreĭn.

We now list known results for function classes defined by a modulus of continuity $\omega = \omega(\delta)$, assumed convex throughout. The quantity $E_{m-1}(W^r H^\omega)_q$ of best approximation of the class $W^r H^\omega$ by the space of trigonometric polynomials of order $m-1$ (i.e., an upper estimate for $d_{2m-1}(W^r H^\omega)_q$) was found by Korneĭchuk for any $r \in \mathbf{N}$, $q = 1, \infty$. A lower estimate for $d_{2m-1}(W^r H^\omega)_\infty$ was obtained by Yu. N. Subbotin and G. G. Lorentz, and for $d_{2m}(W^r H^\omega)_\infty$ by V. I. Ruban. References on these results and also certain estimates in other metrics can be found in Korneĭchuk's monographs [9], [10] (for some generalizations see [20]–[22]).

A result on the magnitude of the approximation $E_{m-1}(\widetilde{W^r H^\omega})_q$ $(q = 1, \infty;\ \omega = \omega(\delta)$ a convex modulus of continuity) by the space of trigonometric polynomials of order $m-1$ was announced by Feschiev and Khemeamin in 1983 at the International Conference on the Theory of Functions. However, their subsequent publications [23], [24] rely on unproven assertions.

Now we turn to a brief exposition of the results of this paper. It is a continuation of the author's investigations begun in [19] (see also [13]). Here, for any $\alpha \in \mathbf{R}$, $m \in \mathbf{N}$, $r \in \mathbf{N}$, $m \le C(r+1)$ (C an absolute constant), we obtain lower estimates for the widths $d_{2m-1}(W_\infty^{r,\alpha})_\infty$ and $d_{2m}(W_\infty^{r,\alpha})_\infty$ that are sharp, since they coincide with $E_{m-1}(W^{r,\alpha}_{\infty})_{\infty}$. For $\alpha = r-1$ our estimates for the widths are obtained under weaker restrictions on r and m than in [19]. Finally, in the last section, for $m \leq C(r+1)$ (C an absolute constant) we obtain lower estimates for $d_{2m-1}(W^rH^{\omega})_{\infty}$ and $d_{2m}(W^rH^{\omega})_{\infty}$ (ω a convex modulus of continuity).

We believe that the restrictions of the type $m \leq C(r+1)$ in our estimation of the widths are related to the method of proof and are not definitive.

The essense of the method consists in the following. In §§2 and 3 we investigate the solvability of the problem of interpolation of periodic sequences by splines corresponding to the kernel $K_{r,\alpha}$. These results are obtained for any $\alpha \in \mathbf{R}$, $r \in \mathbf{N}$ and $m \in \mathbf{N}$.

In the course of our work we have to study the properties of a number of auxiliary functions that are eigenvalues of certain matrices (§1 deals with that). Further, with the aid of the results of §§2 and 3 we prove Markov-Bernstein type inequalities for the spaces of corresponding periodic splines, and with the aid of Tikhomirov's theorem on the width of a sphere we prove lower estimates for the widths. We note that our method has been used earlier for other function classes by Taĭkov [25], Subbotin [26] and others.

1. Properties of auxiliary functions

Let $r \in \mathbf{N}$, $m \in \mathbf{N}$, $h = \pi/m$, $\varphi \in \mathbf{R}$. We define the functions

$$(2) \qquad a_j(t) = a_{j,r}(t) = \sum_{\nu=0}^{\infty} \frac{(-1)^{\nu} e^{i((2m\nu+j)t-\varphi)}}{(2m\nu+j)^{r+1}} - \sum_{\nu=1}^{\infty} \frac{(-1)^{\nu} e^{-i((2m\nu-j)t-\varphi)}}{(2m\nu-j)^{r+1}}$$

$$(j = 1, \dots, 2m-1).$$

Our main goal in this section is to demonstrate that at some point $\tau \in [-h/2, h/2)$ (how τ is chosen will be indicated below) for any φ all the functions $a_{j,r}(\tau)$ ($j = 1, \dots, 2m-1$) are different from zero. First we study the function $a_m(t)$. This function is real and has the form

$$(3) \qquad a_m(t) = a_{m,r}(t) = \frac{2}{m^{r+1}} \sum_{\nu=0}^{\infty} \frac{(-1)^{\nu} \cos((2\nu+1)mt - \varphi)}{(2\nu+1)^{r+1}}.$$

For $\varphi = \pi r/2$ we obtain

$$a_m\left(t + \frac{h}{2}\right) = -\frac{\pi}{2m}\varphi_{m,r}(t),$$

and for $\varphi = \pi(r-1)/2$

$$a_m\left(t + \frac{h}{2}\right) = -\frac{\pi}{2m}\tilde{\varphi}_{m,r}(t),$$

where $\varphi_{m,r}$ is an ideal Euler spline (for its definition and properties see, for example, [9] or [10]), and $\tilde{\varphi}_{m,r}$ is the function trigonometrically conjugate to $\varphi_{m,r}$. Thus, clearly, in its properties the function a_m must be close to the functions $\varphi_{m,r}$ and $\tilde{\varphi}_{m,r}$. In the sequel, whenever we need to take into account the dependence of the function $a_{m,r}(t)$ on the parameter φ, we shall write $a_{m,r}(\varphi, t)$.

Lemma 1. *Let $r \in \mathbf{N}$ and $\varphi \in \mathbf{R}$. The function $a_{m,r}(t)$ has the following properties:*

1) *It has exactly $2m$ equidistant zeros on the half-open interval $[0, 2\pi)$, and it changes sign passing through each zero.*

2) *It attains its extrema* $\pm\|a_{m,r}\|_C$ *on the half-open interval* $[0, 2\pi)$ *at exactly* $2m$ *equidistant points, and the points of maximum and minimum alternate.*

Proof. It follows from (3) that

$$a_{m,r}(t+h) = -a_{m,r}(t). \tag{4}$$

Thus, to prove the lemma it suffices to establish that on $[-h/2, h/2)$ the function $a_{m,r}$ has precisely one zero and that this zero is simple, and that its derivative $a'_{m,r}$ also has the same property. The following equalities are easily verified:

$$a''_{m,r+2}(t) = -a_{m,r}(t), \tag{5}$$

$$a_{m,r}(\varphi+\pi, t) = -a_{m,r}(\varphi, t), \tag{6}$$

$$a_{m,r}\left(\frac{h}{2}\right) = \frac{2\sin\varphi}{m^{r+1}}\sum_{\nu=0}^{\infty}(2\nu+1)^{-r-1}, \qquad a_{m,r}\left(-\frac{h}{2}\right) = -a_{m,r}\left(\frac{h}{2}\right). \tag{7}$$

We introduce the function

$$\overline{a}_{m,r}(t) = \frac{2}{m^{r+1}}\sum_{\nu=0}^{\infty}\frac{(-1)^{\nu}e^{i((2\nu+1)mt-\varphi)}}{(2\nu+1)^{r+1}}. \tag{8}$$

Multiplying the νth term in the sum by y^{ν} $(0<y<1)$, differentiating, and passing to the limit as $y\to 1$, we obtain

$$\overline{a}''_{m,1}(t) = -\frac{\cos\varphi}{\cos mt} + i\frac{\sin\varphi}{\cos mt}, \qquad t\in\left(-\frac{h}{2}, \frac{h}{2}\right), \tag{9}$$

for $r=1$, and

$$\overline{a}'_{m,0}(t) = \frac{\sin\varphi}{\cos mt} + i\frac{\cos\varphi}{\cos mt}, \qquad t\in\left(-\frac{h}{2}, \frac{h}{2}\right), \tag{10}$$

for $r=0$. In view of (6) it suffices to prove Lemma 1 for $0\le\varphi<\pi$.

We consider two cases.

1. r is odd. First let $0<\varphi<\pi/2$. Since

$$a_{m,r}(t) = \operatorname{Re}\overline{a}_{m,r}(t),$$

from (7)–(9) we have that $a''_{m,1}(t)<0$ for $t\in(-h/2, h/2)$, $a_{m,1}(-h/2)<0$, and $a_{m,1}(h/2)>0$. Therefore the function $a_{m,1}$ is convex on $(-h/2, h/2)$, has precisely one zero in this interval (we denote it by t_0), and when passing through this zero $a_{m,1}$ changes its sign from minus to plus. With (5) and (7) taken into account we obtain from this that the function $a_{m,3}$ has the following properties: $a_{m,3}(-h/2)<0$; $a_{m,3}(h/2)>0$; $a''_{m,3}(t)>0$ for $-h/2<t<t_0$; $a''_{m,3}(t)<0$ for $t_0<t<h/2$. Thus, on the interval $(-h/2, h/2)$ the function $a_{m,3}$ is concave for $t<t_0$ and convex for $t>t_0$; therefore, it can have at most three zeros on this interval. But since it is negative at the left endpoint of the interval and positive at the right endpoint, we see that $a_{m,3}$ has precisely one zero on this interval and changes its sign from minus to plus on passing through it. (The argument becomes perfectly clear if a figure is drawn.)

The property 1) for $a_{m,r}$ (r odd), $0<\varphi<\pi/2$, is proved similarly by induction, with (5) taken into account.

Let $\pi/2<\varphi<\pi$. In this case we have $a''_{m,1}(t)>0$ $(t\in(-h/2, h/2))$, $a_{m,1}(-h/2)<0$, $a_{m,1}(h/2)>0$. Therefore $a_{m,1}$ has exactly one zero and a simple one at that (we denote it by t_0), and on passing through the zero $a_{m,1}$ changes its sign from minus to plus. As for the functions $a_{m,3}$, $a_{m,5}$ etc., we apply the same arguments as in the case $0<\varphi<\pi/2$.

The cases $\varphi=0$ and $\varphi=\pi/2$ lead to the functions $\varphi_{m,r}$ and $\tilde{\varphi}_{m,r}$, and the property 1) for $a_{m,r}$ is easily deduced either from their properties or directly, as in the previous cases.

2. r is even. It follows from (10) that $a'_{m,0}(t)=\operatorname{Re}\overline{a}_{m,0}(t)>0$ for $0<\varphi<\pi$ and $t\in(-h/2,h/2)$. Thus, the function $a_{m,0}$ is increasing on the interval $(-h/2,h/2)$, has different signs at the endpoints $(a_{m,0}(-h/2)<0,\ a_{m,0}(h/2)>0)$ and, therefore, becomes zero precisely once. Let t_0 be this zero. Then $a_{m,0}(t)<0$ for $t\in(-h/2,t_0)$ and $a_{m,0}(t)>0$ for $t\in(t_0,h/2)$. Consider the function $a_{m,2}$. Taking (5) and (7) into account, we get $a_{m,2}(-h/2)<0$, $a_{m,2}(h/2)>0$, $a''_{m,2}(t)>0$ for $t\in(-h/2,t_0)$, and $a''_{m,2}(t)<0$ for $t\in(t_0,h/2)$. From here, as in case 1), we obtain that $a_{m,2}$ has exactly one zero in $(-h/2,h/2)$, and that it changes its sign from minus to plus when passing through it. By induction, using (5), we conclude that all the functions $a_{m,r}$ (r even) have that property.

The case $\varphi=0$ leads us to the function $\varphi_{m,r}$. Property 1) for $a_{m,r}$ is completely proven. To prove property 2), note that for a fixed φ

$$a'_{m,r}(\varphi,t)=-a_{m,r-1}(\varphi+\pi/2,t)$$

and so, in view of property 1), the function $a'_{m,r}(\varphi,t)$ for a fixed φ, as a function of t, has precisely $2m$ equidistant zeros on the period, and changes its sign on passing through each one of them.

Lemma 1 is proved.

The next lemma concerns the location on the half-open interval $[-h/2,h/2)$ of the extremum of the function $a_{m,r}(t)=a_{m,r}(\varphi,t)$, i.e., of the point at which $a_{m,r}$ takes the value $\pm\|a_{m,r}\|_C$ depending on the parameter φ. Denote that point by τ; by Lemma 1, τ is unique.

Lemma 2. *If $\varphi=0$, then $\tau=0$; if $\varphi=\pi/2$ then $\tau=-h/2$; if $\pi/2<\varphi<\pi$ then $-h/2<\tau<0$; if $0<\varphi<\pi/2$, then $0<\tau<h/2$.*

Proof. Denote

$$A(t)=a'_{m,r}(t)=-\frac{2}{m^r}\sum_{\nu=0}^{\infty}\frac{(-1)^\nu\sin((2\nu+1)mt-\varphi)}{(2\nu+1)^r}.$$

For $0<\varphi<\pi/2$ we obtain $A(0)>0$ and $A(h/2)<0$, and for $\pi/2<\varphi<\pi$ we obtain $A(-h/2)<0$ and $A(0)>0$. In view of Lemma 1 these inequalities imply that $0<\tau<h/2$ for $0<\varphi<\pi/2$ and $-h/2<\tau<0$ for $\pi/2<\varphi<\pi$. The cases $\varphi=0$ and $\varphi=\pi/2$ are trivial. Lemma 2 is proved.

Remark. Let $-\pi/2\le\varphi<\pi/2$ and let $t=\tau$ be the extremum of $a_{m,r}(\varphi,t)$ on $[-h/2,h/2)$. Then for $r\ge 2$ from the representation of $A(t)$ we get

$$|\sin(m\tau-\varphi)|\le A_r=\sum_{k=1}^{\infty}(2k+1)^{-r}.$$

It is known ([35], §5.1.4.1) that $A_r<0.000156$ for $r\ge 8$.

For convenience we denote

$$b_{j,r}(t)=\operatorname{Re}a_{j,r}(t),\qquad j=1,\dots,2m-1. \tag{11}$$

Lemma 3. *Let* $r \in \mathbf{N}$ *and* $j = 1, \dots, 2m-1$. *Then*

$$\begin{array}{lll} \text{for } 0<\varphi<\pi/2 & b_{j,r}(t)>0, & t\in[0,h/2), \\ \text{for } \pi/2<\varphi<\pi & b_{j,r}(t)<0, & t\in(-h/2,0], \\ \text{for } \varphi=0 & b_{j,r}(t)>0, & t\in(-h/2,h/2) \\ \text{for } \varphi=\pi/2 & b_{j,r}(t)<0, & t\in[-h/2,0), \\ & b_{j,r}(t)>0, & t\in(0,h/2]. \end{array}$$

Proof. Denote

$$\alpha_{j,r} = \sum_{\nu=0}^{\infty} \frac{(-1)^\nu}{(2m\nu+j)^{r+1}}, \quad \beta_{j,r} = \sum_{\nu=1}^{\infty} \frac{(-1)^{\nu+1}}{(2m\nu-j)^{r+1}},$$

$$\gamma_{j,r} = \sum_{\nu=0}^{\infty} \frac{1}{(2m\nu+j)^{r+1}}, \quad \delta_{j,r} = \sum_{\nu=1}^{\infty} \frac{1}{(2m\nu-j)^{r+1}}.$$

We have

$$\begin{gathered} \alpha_{j,r}>0, \quad \beta_{j,r}>0, \quad \gamma_{j,r}>0, \quad \delta_{j,r}>0 \qquad (j=1,\dots,2m-1), \\ \alpha_{j,r}>\beta_{j,r}, \quad \gamma_{j,r}>\delta_{j,r} \qquad (j=1,\dots,m-1), \\ \alpha_{j,r}<\beta_{j,r}, \quad \gamma_{j,r}<\delta_{j,r} \qquad (j=m+1,\dots,2m-1). \end{gathered} \tag{12}$$

From (2) and (11) for $t=0$ we obtain

$$b_{j,r}(0) = (\alpha_{j,r}+\beta_{j,r})\cos\varphi. \tag{13}$$

Substituting $t = \pm h/2$ in (2), we obtain

$$\begin{aligned} b_{j,r}\left(\frac{h}{2}\right) &= (\gamma_{j,r}-\delta_{j,r})\cos\frac{jh}{2}\cos\varphi + (\gamma_{j,r}+\delta_{j,r})\sin\frac{jh}{2}\sin\varphi, \\ b_{j,r}\left(-\frac{h}{2}\right) &= (\gamma_{j,r}-\delta_{j,r})\cos\frac{jh}{2}\cos\varphi - (\gamma_{j,r}+\delta_{j,r})\sin\frac{jh}{2}\sin\varphi. \end{aligned} \tag{14}$$

The following equality is easily verified:

$$b''_{j,r+2}(t) = -b_{j,r}(t) \qquad (j=1,\dots,2m-1,\ r\in\mathbf{N}). \tag{15}$$

For $j=m$ Lemma 3 follows from Lemmas 1 and 2. Therefore in what follows we assume that $j \neq m$. We divide the proof into two cases.

1. r is odd. Just as in the proof of (9) and (10), from (2) and (11) we get

$$b''_{j,1}(t) = -\frac{\cos\varphi}{\cos mt}\cos(m-j)t, \qquad t\in\left(-\frac{h}{2},\frac{h}{2}\right), \tag{16}$$

$$b'_{j,0}(t) = \frac{\sin\varphi}{\cos mt}\cos(m-j)t, \qquad t\in\left(-\frac{h}{2},\frac{h}{2}\right). \tag{17}$$

Let $0<\varphi<\pi/2$. From (12) we have

$$(\gamma_{j,r}-\delta_{j,r})\cos(jh/2)>0 \qquad (j\neq m).$$

From this and (12)–(14) we obtain $b_{j,r}(0)>0$ and $b_{j,r}(h/2)>0$ $(r\in\mathbf{N})$. Moreover, from [16] we get that $b''_{j,1}(t)<0$ for $-h/2<t<h/2$. Hence, $b_{j,1}$ is strictly positive on the interval $(0, h/2)$ (it is convex and takes positive values at the endpoints). Further, in view of (15) $b''_{j,3} = -b_{j,1} < 0$, and since $b_{j,3}(0)>0$ and $b_{j,3}(h/2)>0$, it follows that $b_{j,3}(t)>0$ for $0\le t\le h/2$. By induction we obtain the proof of Lemma 3 for $0<\varphi<\pi/2$ for all odd r.

Let $\pi/2 < \varphi < \pi$. We have $b_{j,r}(0) < 0$, $b_{j,r}(-h/2) < 0$, and $b''_{j,1}(t) > 0$ for $-h/2 < t < h/2$. Hence, the function $b_{j,1}$ is concave on $(-h/2, h/2)$ and is negative at $-h/2$ and at 0. It follows that $b_{j,1}(t) < 0$ for $-h/2 \le t \le 0$. Using (15) we get $b_{j,r}(t) < 0$ for $-h/2 \le t \le 0$ for r odd and $\pi/2 < \varphi < \pi$.

For $\varphi = 0$ we have $b_{j,r}(h/2) > 0$, $b_{j,r}(-h/2) > 0$, and $b''_{j,r}(t) < 0$ for $-h/2 < t < h/2$. Applying (15) we obtain Lemma 3 for $\varphi = 0$ and any odd r.

For $\varphi = \pi/2$ we obtain $b_{j,r}(-h/2) < 0$, $b_{j,r}(h/2) > 0$, and $b_{j,r}(0) = 0$. Hence, by (15) and (16), $b_{j,r}(t) > 0$ for $0 < t < h/2$ and $b_{j,r}(t) < 0$ for $-h/2 < t < 0$.

2. r is even. We first investigate the function $b_{j,0}(t)$. For $0 < \varphi < \pi/2$ from (13) and (17) we have $b_{j,0}(0) > 0$, $b'_{j,0}(t) > 0$ $(0 < t < h/2)$ and, thus, $b_{j,0}(t) > 0$ for $0 \le t < h/2$. Analogously, if $\pi/2 < \varphi < \pi$, then $b_{j,0}(t) < 0$ for $-h/2 < t \le 0$, since $b_{j,0}(0) < 0$ and $b'_{j,0}(t) > 0$ for $-h/2 < t < 0$. For $\varphi = 0$ we have $b_{j,0}(t) \equiv b_{j,0}(0) > 0$ $(-h/2 < t < h/2)$. For $\varphi = \pi/2$ we have $b'_{j,0}(t) > 0$ $(-h/2 < t < h/2)$ and $b_{j,0}(0) = 0$, i.e., the function $b_{j,0}$ increases on the interval $-h/2 < t < h/2$ and takes the value 0 at zero. We use the information about $b_{j,0}$ in the study of the properties of the functions $b_{j,r}$ (r even). The rest of the proof is completely analogous to case 1. Here the main roles are played by (15) and the signs of the numbers $b_{j,0}(0)$, $b_{j,r}(-h/2)$ and $b_{j,r}(h/2)$. Lemma 3 is proved.

Recall that τ is the unique point in $[-h/2, h/2)$ such that $|a_{m,r}(\tau)| = \|a_{m,r}\|_C$. Calculating the derivative of $a_{m,r}$ and using Lemma 1, we can also define the point τ as the unique zero of the function

$$\sum_{\nu=0}^{\infty} \frac{(-1)^\nu \sin((2\nu+1)mt - \varphi)}{(2\nu+1)^r}$$

in $[-h/2, h/2)$.

Lemma 4. *Let* $r \in \mathbf{N}$ *and* $j = 1, \dots, 2m-1$. *Then* $a_{j,r}(\tau) \ne 0$.

The proof follows from the equality (11) and Lemmas 2 and 3.

We introduce one more sequence of functions

$$\tilde{b}_{j,2l-1}(t) = \sum_{\nu=0}^{\infty} \frac{\cos(2m\nu + j)t}{(2m\nu+j)^{2l-1}} + \sum_{\nu=1}^{\infty} \frac{\cos(2m\nu - j)t}{(2m\nu-j)^{2l-1}}$$
$$(j = 1, \dots, 2m-1;\ l \in \mathbf{N}).$$

Lemma 5. *The functions* $\tilde{b}_{j,2l-1}$ *are positive on the interval* $(0, h/2)$.

Proof. For $l = 1$ it is not hard to verify that

$$\tilde{b}'_{j,1}(t) = -\frac{\cos(m-j)t}{\sin mt}.$$

Hence $\tilde{b}'_{j,1} < 0$ for $0 < t < h/2$, and, consequently, the function $\tilde{b}_{j,1}$ decreases on the interval $(0, h/2)$. Of interest to us is the sign of $\tilde{b}_{j,2l-1}$ at $t = h/2$:

$$\tilde{b}_{j,2l-1}\left(\frac{h}{2}\right) = \left(\sum_{\nu=0}^{\infty} \frac{(-1)^\nu}{(2m\nu+j)^{2l-1}} + \sum_{\nu=1}^{\infty} \frac{(-1)^\nu}{(2m\nu-j)^{2l-1}}\right) \cos\frac{jh}{2}.$$

From this we have $\tilde{b}_{j,2l-1}(h/2) > 0$ for $j \ne m$ and $\tilde{b}_{m,2l-1}(h/2) = 0$. Thus, we have proved Lemma 5 for $l = 1$. For the other l we need to use the identity

$$\tilde{b}''_{j,2l+1} = -\tilde{b}_{j,2l-1}$$

and the arguments applied in the proof of Lemma 3.

2. Interpolation splines corresponding to the kernel $K_{r,\alpha}$

Let $m, r \in \mathbf{N}$, $\alpha \in \mathbf{R}$, $\varphi = \alpha\pi/2$ and $h = \pi/m$, and let $\tau \in [-h/2, h/2)$ be such that $|a_{m,r}(\tau)| = \|a_{m,r}\|_C$.

Definition. We call a function $S \in C_{2\pi}$ of the form

$$S = C + \pi^{-1} K_{r,\alpha} * \Psi, \qquad \Psi \perp C,$$

where $\Psi(t) = Z_j$, $(j-1)h \le t < jh$, $j \in \mathbf{Z}$, $Z_{j+2m} = Z_j$, $Z_1, Z_2, \dots, Z_{2m}$ and C are arbitrary constants, a *periodic spline corresponding to the kernel* $K_{r,\alpha}$ (briefly, an *SK-spline*).

For $\alpha = r$ an SK-spline is a periodic polynomial spline of degree r with equidistant nodes jh, $j \in \mathbf{Z}$ (see, for example, [9], [27]). Splines with such, and also other, kernels have been studied in many papers (see [11], [13], [19], [20], [27]–[29]).

Theorem 1. *For any sequence $y = \{y_\nu\}_{\nu=-\infty}^{\infty}$ satisfying the condition $y_{\nu+2m} = y_\nu$ there exists a unique SK-spline S such that*

$$S(r + (\nu + 1/2)h) = y_\nu, \qquad \nu \in \mathbf{Z}.$$

Proof. For $\alpha = r \in \mathbf{N}$ Theorem 1 was proved in [28] and [29]. From the definition of a spline we have

$$\begin{aligned}
S(x) &= C + \frac{1}{\pi}\int_0^{2\pi} K_{r,\alpha}(x-t)\Psi(t)\,dt = C + \frac{1}{\pi}\sum_{j=1}^{2m} Z_j \int_0^h K_{r,\alpha}(x - t - (j-1)h)\,dt \\
&= C + \frac{1}{\pi}\sum_{j=1}^{2m} Z_j \sum_{k=1}^{\infty} \frac{1}{k^r}\int_0^h \cos\left(k(x-t-(j-1)h) - \frac{\alpha\pi}{2}\right) dt \\
&= C + \frac{1}{\pi}\sum_{j=1}^{2m} Z_j \sum_{k=1}^{\infty} \frac{1}{k^{r+1}}\left(-\sin\left(k(x-t-(j-1)h) - \frac{\alpha\pi}{2}\right)\right)\Bigg|_0^h \\
&= C + \frac{1}{\pi}\sum_{j=1}^{2m} Z_j \sum_{k=1}^{\infty} \frac{1}{k^{r+1}}\left[\sin\left(k(x-(j-1)h) - \frac{\alpha\pi}{2}\right) - \sin\left(k(x-jh) - \frac{\alpha\pi}{2}\right)\right] \\
&= C + \frac{2}{\pi}\sum_{j=1}^{2m} Z_j \sum_{k=1}^{\infty} \frac{\sin\frac{kh}{2}\cos(k(x-(j-1/2)h) - \varphi)}{k^{r+1}}.
\end{aligned}$$

Since $S(\tau + (\nu + 1/2)h) = y_\nu$, $\nu \in \mathbf{Z}$, it follows that

$$C + \frac{2}{\pi}\sum_{j=1}^{2m} Z_j \sum_{k=1}^{\infty} \frac{\sin(kh/2)\cos(k(\tau + (\nu - j + 1)h) - \varphi)}{k^{r+1}} = y_\nu. \tag{18}$$

The condition $\Psi \perp C$ implies that

$$\sum_{j=1}^{2m} Z_j = 0. \tag{19}$$

We have obtained a system of $2m+1$ linear algebraic equations for the unknowns $Z_1, Z_2, \dots, Z_{2m}, C$. To prove Theorem 1 we need to establish that this system has a unique solution. To this end we first calculate the constant C.

Adding up all equations (18) and taking the equality

$$\sum_{j=0}^{2m-1} e^{ijsh} = \begin{cases} 2m, & \text{if } s \text{ is divisible by } 2m, \\ 0, & \text{if } s \text{ is not divisible by } 2m, \end{cases} \tag{20}$$

into account, we get

(21)

$$\begin{aligned} 2mC &+ \frac{2}{\pi}\sum_{j=1}^{2m} Z_j \sum_{k=1}^{\infty}\frac{1}{k^{r+1}}\sum_{\nu=0}^{2m-1}\sin\frac{kh}{2}\cos(k(\tau+(\nu-j+1)h)-\varphi) \\ &= 2mC + \frac{2m}{\pi}\sum_{j=1}^{2m} Z_j \sum_{s=1}^{\infty}\frac{\sin msh}{(2ms)^{r+1}}[e^{i(2ms(\tau+(-j+1)h)-\varphi)} + e^{-i(2ms(\tau+(-j+1)h)-\varphi)}] \\ &= \sum_{\nu=0}^{2m-1} y_\nu. \end{aligned}$$

In view of the fact that $\sin mhs = \sin \pi s = 0$ the double sum on the left in (21) is equal to zero. Consequently,

$$C = \frac{1}{2m}\sum_{\nu=0}^{2m-1} y_\nu. \tag{22}$$

Substituting this C in (18), we obtain, using (19), a system of $2m+1$ equations for the unknown numbers Z_j, $j = 1, \ldots, 2m$. In order to solve it we consider the system of $2m$ equations obtained by adding equation (19) to each equation of system (18):

$$\begin{aligned} &\sum_{j=1}^{2m} Z_j\left(1 + \frac{2}{\pi}\sum_{k=1}^{\infty}\frac{1}{k^{r+1}}\sin\frac{kh}{2}\cos(k(\tau+(\nu-j+1)h)-\varphi)\right) \\ &\qquad = y_\nu - \frac{1}{2m}\sum_{s=0}^{2m-1} y_s \qquad (\nu = 0, \ldots, 2m-1). \end{aligned} \tag{23}$$

The matrix T of the system (23) is a circulant with entries of the first row

$$\tau_j = 1 + \frac{2}{\pi}\sum_{k=1}^{\infty}\frac{1}{k^{r+1}}\sin\frac{kh}{2}\cos(k(\tau+(-j+1)h)-\varphi) \qquad (j = 1, \ldots, 2m).$$

Using the fact that for a circulant there are explicit expressions for the eigenvalues $\lambda_j = \lambda_j(\tau)$ (see, for example, [30], p. 133], we obtain

$$\begin{aligned} \lambda_j(\tau) = \sum_{s=0}^{2m-1} e^{ijsh}\tau_{s+1} = \sum_{s=0}^{2m-1} e^{ijsh}\left(1 + \frac{2}{\pi}\sum_{k=1}^{\infty}\frac{1}{k^{r+1}}\sin\frac{kh}{2}\cos(k(\tau - sh) - \varphi\right) \\ (j = 1, \ldots, 2m). \end{aligned} \tag{24}$$

Considering (20), for $j = 2m$ we have

$$\begin{aligned} \lambda_{2m}(\tau) &= 2m + \frac{2}{\pi}\sum_{s=0}^{2m-1}\sum_{k=1}^{\infty}\frac{\sin\frac{kh}{2}}{k^{r+1}}\cos(k(\tau - sh) - \varphi) \\ &= 2m + \frac{2}{\pi}\sum_{k=1}^{\infty}\frac{\sin\frac{kh}{2}}{k^{r+1}}\sum_{s=0}^{2m-1}\cos(k(\tau - sh) - \varphi) = 2m. \end{aligned}$$

For $j = 1, \ldots, 2m-1$ by (20) $\sum_{s=0}^{2m-1} e^{ijsh} = 0$, and so

(25)

$$\begin{aligned}
\lambda_j(\tau) &= \frac{2}{\pi} \sum_{s=0}^{2m-1} \sum_{k=1}^{\infty} \frac{1}{2k^{r+1}} \sin \frac{kh}{2} (e^{i(k(\tau - sh)-\varphi)} e^{ijsh} + e^{-i(k(\tau-sh)-\varphi)} e^{ijsh}) \\
&= \frac{1}{\pi} \sum_{k=1}^{\infty} \frac{\sin \frac{kh}{2}}{k^{r+1}} \left[e^{i(k\tau-\varphi)} \sum_{s=0}^{2m-1} e^{ish(j-k)} + e^{-i(k\tau-\varphi)} \sum_{s=0}^{2m-1} e^{ish(j+k)} \right] \\
&= \frac{2m}{\pi} \sum_{l=0}^{\infty} \frac{\sin \frac{(j+2ml)h}{2}}{(j+2ml)^{r+1}} e^{i((j+2ml)\tau-\varphi)} + \frac{2m}{\pi} \sum_{l=1}^{\infty} \frac{\sin \frac{(2ml-j)h}{2}}{(2ml-j)^{r+1}} e^{-i((2ml-j)\tau-\varphi)} \\
&= \frac{2m}{\pi} \sin \frac{jh}{2} \left(e^{-i\varphi} \sum_{l=0}^{\infty} \frac{(-1)^l e^{i(2ml+j)\tau}}{(2ml+j)^{r+1}} - e^{i\varphi} \sum_{l=1}^{\infty} \frac{(-1)^l e^{-i(2ml-j)\tau}}{(2ml-j)^{r+1}} \right).
\end{aligned}$$

Comparing (25) with (2), we conclude that

$$\lambda_j(\tau) = \frac{2m}{\pi} \sin \frac{jh}{2} a_j(\tau), \qquad j = 1, \ldots, 2m-1. \tag{26}$$

From this and Lemma 4 it follows that $\lambda_j(\tau) \neq 0$, $j = 1, \ldots, 2m-1$. Consequently,

$$\det T = \prod_{j=1}^{2m} \lambda_j(\tau) \neq 0, \tag{27}$$

and system (23) has a unique solution $\{Z_s^0\}_{s=1}^{2m}$. It is clear that for the proof of Theorem 1 we need to show that this solution satisfies (19) as well. We use the fact that the inverse matrix T^{-1} of a circulant matrix T has the form (see, for example, [30], p. 134)

$$T^{-1} = \begin{vmatrix} \eta_1 & \eta_2 & \ldots & \eta_{2m} \\ \eta_{2m} & \eta_1 & \ldots & \eta_{2m-1} \\ \ldots & \ldots & \ldots & \ldots \\ \eta_2 & \eta_3 & \ldots & \eta_1 \end{vmatrix},$$

where

$$\eta_j = \frac{1}{2m} \sum_{\nu=1}^{2m} \lambda_\nu^{-1}(\tau) e^{-i\nu(j-1)h}, \qquad j = 1, \ldots, 2m. \tag{28}$$

From this we obtain the following representations for the numbers Z_s^0:

$$Z_s^0 = \frac{1}{2m} \sum_{\nu=0}^{2m-1} \left(y_\nu - \frac{1}{2m} \sum_{l=0}^{2m-1} y_l \right) \eta_{\nu-s+2}, \qquad s = 1, \ldots, 2m.$$

Let us transform the last equalities using the definition of the numbers η_j and equality (20). Then

(29)

$$\begin{aligned}
Z_1^0 &= \frac{1}{2m} \left(y_0 \sum_{\nu=1}^{2m-1} \lambda_\nu^{-1} + y_1 \sum_{\nu=1}^{2m-1} \lambda_\nu^{-1} e^{-i\nu h} + \cdots + y_{2m-1} \sum_{\nu=1}^{2m-1} \lambda_\nu^{-1} e^{-i(2m-1)\nu h} \right), \\
Z_2^0 &= \frac{1}{2m} \left(y_0 \sum_{\nu=1}^{2m-1} \lambda_\nu^{-1} e^{-i(2m-1)\nu h} + y_1 \sum_{\nu=1}^{2m-1} \lambda_\nu^{-1} + \cdots + y_{2m-1} \sum_{\nu=1}^{2m-1} \lambda_\nu^{-1} e^{-i(2m-2)\nu h} \right), \\
&\cdots\cdots\cdots\cdots\cdots\cdots\cdots\cdots\cdots\cdots \\
Z_{2m}^0 &= \frac{1}{2m} \left(y_0 \sum_{\nu=1}^{2m-1} \lambda_\nu^{-1} e^{-i\nu h} + y_1 \sum_{\nu=1}^{2m-1} \lambda_\nu^{-1} e^{-2i\nu h} + \cdots + y_{2m-1} \sum_{\nu-1}^{2m-1} \lambda_\nu^{-1} \right).
\end{aligned}$$

Summing these relations, we get

$$\sum_{s=1}^{2m} Z_s^0 = \frac{1}{2m} \sum_{l=0}^{2m-1} y_l \sum_{\nu=1}^{2m-1} \lambda_\nu^{-1} \sum_{j=0}^{2m-1} e^{-ij\nu h} = 0$$

because the inner sum is equal to zero. Thus, the numbers $\{Z_s^0\}_{s=1}^{2m}$ are a solution of (19) and (23), and, therefore, satisfy (18). Theorem 1 is proved.

Remark. The method of proof of Theorem 1 involving the use of properties of a circulant matrix for proving existence and uniqueness of an interpolation polynomial spline was used earlier by Daugavet [31] for $\alpha = r \in \mathbf{N}$, and by the author [19], [32] and Kushpel′ [13] for other kernels.

3. Interpolational conjugate ω-splines

Let $\omega = \omega(t)$ $(0 \le t < \infty)$ be a given convex modulus of continuity and

$$\Omega(t) = \begin{cases} \frac{1}{2}\omega(2t), & 0 \le t \le h/2, \\ \frac{1}{2}\omega(2h - 2t), & h/2 < t \le h. \end{cases}$$

Definition. We call a function $S \in C_{2\pi}$ of the form

$$S = C + \pi^{-1} K_{r,r-1} * \Psi,$$

where

$$\begin{aligned} &\Psi \perp C, \qquad \Psi(t + (j-1)h) = Z_j \Omega(t), \\ &0 \le t \le h, \qquad Z_{j+2m} = Z_j, \qquad j \in \mathbf{Z}, \end{aligned}$$

and $Z_1, Z_2, \dots, Z_{2m}$ and C are arbitrary constants, a *periodic conjugate ω-spline.*

Conjugate ω-splines are functions trigonometrically conjugate to known ω-splines (for the definition and properties of ω-splines see, for example, [9] and [26]). Let $\theta_r = 0$ if r is even, and $\theta_r = h/2$ if r is odd.

Theorem 2. *For any sequence $y = \{y_\nu\}_{\nu=-\infty}^{\infty}$ satisfying the condition $y_{\nu+2m} = y_\nu$ there exists a unique conjugate ω-spline S such that*

$$S(\theta_r + \nu h) = y_\nu, \qquad \nu \in \mathbf{Z}.$$

Proof. We note that Subbotin [26] proved an analogous theorem for ω-splines (see also [19] and [20]). From the definition of a conjugate ω-spline and of the function $K_{r,\alpha}$, we have

$$\begin{aligned} S(x) &= C + \frac{1}{\pi}\int_0^{2\pi} K_{r,r-1}(x-t)\Psi(t)\,dt \\ &= C + \frac{1}{\pi}\sum_{j=1}^{2m} Z_j \int_0^h K_{r,r-1}(x - t - (j-1)h)\Omega(t)\,dt \\ &= C + \frac{1}{\pi}\sum_{j=1}^{2m} Z_j \int_0^h \sum_{k=1}^{\infty} \frac{1}{k^r}\cos\left(k(x - t - jh + h) - \frac{(r-1)\pi}{2}\right)\Omega(t)\,dt \\ &= C - \frac{1}{\pi}\sum_{j=1}^{2m} Z_j \sum_{k=1}^{\infty}\frac{1}{k^r}\int_0^h \Omega(t)\sin\left(k(x - t - jh + h) - \frac{r\pi}{2}\right)dt. \end{aligned}$$

By the conditions of interpolation

$$S(\theta_r+\nu h) = C-\frac{1}{\pi}\sum_{j=1}^{2m} Z_j\int_0^h \sum_{k=1}^{\infty}\frac{1}{k^r}\sin\left(k(\theta_r+\nu h-jh+h-t)-\frac{r\pi}{2}\right)\Omega(t)\,dt=y_\nu, \quad \nu=0,\dots,2m-1. \tag{30}$$

In addition, the condition $\Psi\perp C$ implies

$$\int_0^h \Omega(t)\,dt\sum_{j=1}^{2m} Z_j=0,$$

and, therefore,

$$\sum_{j=1}^{2m} Z_j=0. \tag{31}$$

As in the previous section, we have obtained a system of $2m+1$ linear algebraic equations for the unknowns $Z_1, Z_2, \dots, Z_{2m}, C$. We prove that this system has a unique solution. First we calculate the constant C. To this end we add up all the equations of system (30) using (20):

$$\begin{aligned}2mC&-\frac{1}{\pi}\sum_{j=1}^{2m} Z_j\int_0^h \Omega(t)\sum_{k=1}^{\infty}\frac{1}{k^r}\sum_{\nu=0}^{2m-1}\sin\left(k(\theta_r+\nu h-jh+h-t)-\frac{r\pi}{2}\right)dt\\ &=2mC-\frac{2m}{\pi}\sum_{s=1}^{\infty}\int_0^h\Omega(t)\frac{\sin(2ms(\theta_r-t)-r\pi/2)}{(2ms)^r}\sum_{j=1}^{2m}Z_j\,dt=\sum_{\nu=0}^{2m-1}y_\nu.\end{aligned}$$

From this and (31) we get

$$C=\frac{1}{2m}\sum_{\nu=0}^{2m-1}y_\nu.$$

Substituting this C in (30), we obtain the following system of $2m+1$ linear algebraic equations for the numbers Z_j, $j=1,\dots,2m$:

$$\begin{aligned}-\frac{1}{\pi}\sum_{j=1}^{2m} Z_j\int_0^h \sum_{k=1}^{\infty}\frac{1}{k^r}\sin\left(k(\theta_r+\nu h-jh+h-t)-\frac{r\pi}{2}\right)\Omega(t)\,dt&\\ =y_\nu-\frac{1}{2m}\sum_{s=0}^{2m-1}y_s,\qquad \nu=0,\dots,2m-1,&\\ \sum_{j=1}^{2m}Z_j=0.&\end{aligned}\tag{32}$$

The solution of this system depends on the parity of r.

1. r is even, $r=2l$. Then $\theta_r=0$. Adding each of the first $2m$ equations to the last in system (32), we obtain

$$\sum_{j=1}^{2m} Z_j d_{j-\nu}=y_\nu-\frac{1}{2m}\sum_{s=0}^{2m-1}y_s,\qquad \nu=0,\dots,2m-1, \tag{33}$$

where

$$d_j = 1 - \frac{1}{\pi}\int_0^h \sum_{k=1}^{\infty} \frac{\sin(k(-jh+h-t)-\pi l)}{k^{2l}}\Omega(t)\,dt.$$

The matrix D of the system of equations (33) is a circulant. Let us calculate its eigenvalues λ_j, $j = 1, \ldots, 2m$ (see [30], p. 133):

$$\begin{aligned}\lambda_j &= \sum_{s=0}^{2m-1} e^{ijsh} d_{s+1} \\ &= \sum_{s=0}^{2m-1} e^{ijsh}\left(1 + \frac{1}{\pi}\sum_{k=1}^{\infty}\frac{1}{k^{2l}}\int_0^h \Omega(t)\sin(k(t+sh)+\pi l)\,dt\right).\end{aligned} \tag{34}$$

For $j = 2m$, using (20), we get

$$\begin{aligned}\lambda_{2m} &= 2m + \frac{1}{\pi}\sum_{s=0}^{2m-1}\int_0^h \Omega(t)\sum_{k=1}^{\infty}\frac{1}{k^{2l}}\sin(k(sh+t)+\pi l)\,dt \\ &= 2m + \frac{1}{\pi}\int_0^h \Omega(t)\sum_{k=1}^{\infty}\frac{1}{k^{2l}}\sum_{s=0}^{2m-1}\sin(k(sh+t)+\pi l)\,dt \\ &= 2m + \frac{2m}{\pi}\int_0^h \Omega(t)\sum_{j=1}^{\infty}\frac{1}{(2mj)^{2l}}\sin(2mj(sh+t)+\pi l)\,dt \\ &= 2m + \frac{2m}{\pi}\int_0^h \Omega(t)\sum_{j=1}^{\infty}\frac{(-1)^l}{(2mj)^{2l}}\sin 2mjt\,dt.\end{aligned}$$

The function Ω is, by definition, even with respect to the middle point of the interval of integration, and the function $\sin 2mjt$ is odd with respect to that same point. Consequently

$$\int_0^h \Omega(t)\sin 2mjt\,dt = 0, \qquad j \in \mathbf{N},$$

and

$$\lambda_{2m} = 2m. \tag{35}$$

For $j = 1, \ldots, 2m-1$ from (34) we obtain

(36)

$$\begin{aligned}\lambda_j &= \frac{1}{\pi}\int_0^h \Omega(t)\sum_{k=1}^{\infty}\frac{1}{k^{2l}}\sum_{s=0}^{2m-1} e^{ijsh}\sin(k(t+sh)+\pi l)\,dt \\ &= \frac{1}{\pi}\int_0^h \Omega(t)\sum_{k=1}^{\infty}\frac{1}{k^{2l}}\frac{1}{2i}\left[e^{i(kt+\pi l)}\sum_{s=0}^{2m-1} e^{ish(j+k)} - e^{-i(kt+\pi l)}\sum_{s=0}^{2m-1} e^{ish(j-k)}\right]dt \\ &= \frac{m}{\pi i}\int_0^h \Omega(t)\left[\sum_{\nu=1}^{\infty}\frac{e^{i((2m\nu-j)t+\pi l)}}{(2m\nu-j)^{2l}} - \sum_{\nu=0}^{\infty}\frac{e^{-i((2m\nu+j)t+\pi l)}}{(2m\nu+j)^{2l}}\right]dt \\ &= \frac{(-1)^{l+1}}{\pi i}\int_0^h \Omega(t)\left[\sum_{\nu=0}^{\infty}\frac{e^{-i(2m\nu+j)t}}{(2m\nu+j)^{2l}} - \sum_{\nu=1}^{\infty}\frac{e^{i(2m\nu-j)t}}{(2m\nu-j)^{2l}}\right]dt \\ &= \frac{(-1)^l}{\pi i}e^{-ijh/2}\int_{-h/2}^{h/2}\Omega\left(\frac{h}{2}-t\right)\left[\sum_{\nu=0}^{\infty}\frac{(-1)^\nu e^{i(2m\nu+j)t}}{(2m\nu+j)^{2l}} - \sum_{\nu=1}^{\infty}\frac{(-1)^\nu e^{-i(2m\nu-j)t}}{(2m\nu-j)^{2l}}\right]dt.\end{aligned}$$

The expression in the square brackets coincides with the function $a_{j,r}(t)$ (see (2)) for $\varphi = 0$ and $r = 2l - 1$. By (11) and Lemma 3 we have $\operatorname{Re} a_{j,r}(t) > 0$ for $-h/2 < t < h/2$ and $j = 1, \dots, 2m-1$. Consequently, each λ_j $(j = 1, \dots, 2m-1)$ is different from zero. In addition, from (35) we have $\lambda_{2m} = 2m$. Therefore the determinant of the matrix D, equal to the product $\lambda_1 \cdots \lambda_{2m}$, is not zero and the system (33) has a unique solution. By a familiar property of circulants, this solution can be written in the form

$$Z_s^0 = \frac{1}{2m} \sum_{\nu=0}^{2m-1} \left[y_\nu - \frac{1}{2m} \sum_{s=0}^{2m-1} y_s \right] \eta_{\nu-s+2}, \qquad s = 1, \dots, 2m,$$

where

$$\eta_j = \frac{1}{2m} \sum_{\nu=1}^{2m} (\lambda_\nu e^{i\nu h(j-1)})^{-1}, \qquad j = 1, \dots, 2m.$$

Transforming the expressions for Z_s^0, we finally obtain

$$Z_s^0 = \frac{1}{2m} \sum_{l=0}^{2m-1} y_l \sum_{\nu=1}^{2m-1} (\lambda_\nu e^{i(l-s+1)\nu h})^{-1}, \qquad s = 1, \dots, 2m. \tag{37}$$

For the proof of Theorem 2 in case 1 it remains to show that the solution Z_s^0 satisfies the last equation of (32). In fact,

$$\begin{aligned} \sum_{s=1}^{2m} Z_s^0 &= \frac{1}{2m} \sum_{s=1}^{2m} \sum_{l=0}^{2m-1} y_l \sum_{\nu=1}^{2m-1} \lambda_\nu^{-1} e^{-i(l-s+1)\nu h} \\ &= \frac{1}{2m} \sum_{\nu=1}^{2m-1} \lambda_\nu^{-1} \sum_{l=0}^{2m-1} y_l e^{-i(l+1)\nu h} \sum_{s=1}^{2m} e^{is\nu h} = 0, \end{aligned}$$

since the inner sum is equal to zero by (20).

2. r is odd, $r = 2l - 1$. Then $\theta_r = h/2$. We prove that the system of equations consisting of the first $2m$ equations of (32) for the unknown $\{Z_s\}_{s=1}^{2m}$ has a unique solution; we find this solution and establish that it also satisfies the last equation of (32), i.e.

$$\sum_{s=1}^{2m} Z_s = 0.$$

We have

$$\sum_{j=1}^{2m} Z_j c_{j-\nu} = y_\nu - \frac{1}{2m} \sum_{s=0}^{2m-1} y_s, \qquad \nu = 0, \dots, 2m-1, \tag{38}$$

where

$$c_j = -\frac{1}{\pi} \int_0^h \sum_{k=1}^{\infty} \frac{1}{k^{2l-1}} \sin\left(k\left(\left(\frac{3}{2} - j\right)h - t\right) - \pi l + \frac{\pi}{2}\right) \Omega(t)\, dt.$$

The matrix C of system (38) is a circulant. We calculate its eigenvalues λ_j, $j =$

$1, \dots, 2m$:

$$\lambda_j = \sum_{s=0}^{2m-1} e^{ijsh} c_{s+1}$$

(39)
$$= \sum_{s=0}^{2m-1} e^{ijsh} \left(-\frac{1}{\pi} \int_0^h \sum_{k=1}^{\infty} \frac{1}{k^{2l-1}} \cos(k((s-\tfrac{1}{2})h+t)+\pi l) \cdot \Omega(t)\,dt \right)$$

$$= -\frac{1}{\pi} \int_0^h \Omega(t) \sum_{k=1}^{\infty} \frac{(-1)^l}{k^{2l-1}} \sum_{s=0}^{2m-1} e^{ijsk} \cos k((s-\tfrac{1}{2})h+t)\,dt.$$

For $j = 1, \dots, 2m-1$ we obtain
(40)

$$\lambda_j = \frac{(-1)^{l+1}}{\pi} \int_0^h \Omega(t) \sum_{k=1}^{\infty} \frac{1}{2k^{2l-1}} \left[e^{i\left(kt-\frac{kh}{2}\right)} \sum_{s=0}^{2m-1} e^{ish(j+k)} + e^{-i\left(kt-\frac{kh}{2}\right)} \sum_{s=0}^{2m-1} e^{ish(j-k)} \right] dt$$

$$= \frac{(-1)^{l+1} m}{\pi} \int_0^h \Omega(t) \left[\sum_{\nu=1}^{\infty} \frac{1}{(2m\nu - j)^{2l-1}} e^{i(2m\nu-j)(t-h/2)} \right.$$

$$\left. + \sum_{\nu=0}^{\infty} \frac{1}{(2m\nu+j)^{2l-1}} e^{-i(2m\nu+j)(t-h/2)} \right] dt$$

$$= \frac{(-1)^l m}{\pi} \int_{-h/2}^{h/2} \Omega\left(\frac{h}{2}-t\right) \left[\sum_{\nu=0}^{\infty} \frac{e^{i(2m\nu+j)t}}{(2m\nu+j)^{2l-1}} + \sum_{\nu=1}^{\infty} \frac{e^{-i(2m\nu-j)t}}{(2m\nu-j)^{2l-1}} \right] dt.$$

Let us denote the function in square brackets by $f(t)$. It can be represented in the form $f = f_1 + f_2$, where

$$f_1(t) = \tfrac{1}{2}[f(t) + f(-t)], \qquad f_2(t) = \tfrac{1}{2}[f(t) - f(-t)].$$

The function f_1 is even with respect to zero, and f_2 is odd. Consequently, by the fact that $\Omega(h/2 - t)$ is even with respect to zero we have

(41)
$$\int_{-h/2}^{h/2} \Omega\left(\frac{h}{2}-t\right) f(t)\,dt = \int_{-h/2}^{h/2} \Omega\left(\frac{h}{2}-t\right) f_1(t)\,dt$$
$$= 2\int_0^{h/2} \Omega\left(\frac{h}{2}-t\right) \left[\sum_{\nu=0}^{\infty} \frac{\cos(2m\nu+j)t}{(2m\nu+j)^{2l-1}} + \sum_{\nu=1}^{\infty} \frac{\cos(2m\nu-j)t}{(2m\nu-j)^{2l-1}} \right] dt.$$

It follows from Lemma 5 and the nonnegativity of Ω that the last integral is strictly positive. Thus,

(42)
$$\lambda_j \neq 0, \qquad j = 1, \dots, 2m-1.$$

For $j = 2m$ it follows from (39) that

$$\lambda_{2m} = \frac{(-1)^{l+1}}{\pi} \int_0^h \Omega(t) \sum_{k=1}^{\infty} \frac{1}{k^{2l-1}} \sum_{s=0}^{2m-1} \cos k((s-\tfrac{1}{2})h+t)\,dt$$

$$= \frac{(-1)^{l+1} m}{\pi} \int_0^h \Omega(t) \sum_{\nu=1}^{\infty} \frac{1}{(2m\nu)^{2l-1}} \cos 2m\nu \left(t - \frac{h}{2}\right) dt$$

$$= \frac{2m(-1)^{l+1}}{\pi} \int_0^{h/2} \Omega\left(t+\frac{h}{2}\right) \sum_{\nu=1}^{\infty} \frac{\cos 2m\nu t}{(2m\nu)^{2l-1}}\,dt.$$

We integrate the last equality by parts:

(43)
$$\lambda_{2m} = \frac{2m(-1)^{l+1}}{\pi}\left[\Omega\left(t+\frac{h}{2}\right)\sum_{\nu=1}^{\infty}\frac{\sin 2m\nu t}{(2m\nu)^{2l}}\Big|_0^{h/2} - \int_0^{h/2}\Omega'\left(t+\frac{h}{2}\right)\sum_{\nu=1}^{\infty}\frac{\sin 2m\nu t}{(2m\nu)^{2l}}\,dt\right]$$
$$= \frac{2m(-1)^{l}}{\pi}\int_0^{h/2}\Omega'\left(t+\frac{h}{2}\right)\sum_{\nu=1}^{\infty}\frac{\sin 2m\nu t}{(2m\nu)^{2l}}\,dt.$$

Let us investigate the function

$$g_{2l}(t) = \sum_{\nu=1}^{\infty}\frac{\sin 2m\nu t}{(2m\nu)^{2l}}$$

for $0 \le t \le h/2$. For $l = 1$ it is easy to verify that $g_2''(t) = -\frac{1}{2}\cos(mt)/\sin(mt)$. Hence $g_2''(t) < 0$ for $0 < t < h/2$, and so the function g_2 is convex. For $t = 0$ and $t = h/2$ we have $g_{2l}(0) = g_{2l}(h/2) = 0$. Therefore, g_2 is positive for $0 < t < h/2$. Using the equality $g_{2l+2}' = -g_{2l}$, we now easily deduce that $g_{2l}(t) > 0$ for any $l \in \mathbf{N}$ and $0 < t < h/2$.

By the properties of a modulus of continuity and the definition of the function Ω we have $\Omega'(t + h/2) < 0$ for $0 < t < h/2$. Therefore, the expression under the integral sign in (43) is negative for $0 < t < h/2$, and so $\lambda_{2m} \neq 0$. Taking (42) into account, we get $\det C = \lambda_1 \cdots \lambda_{2m} \neq 0$. Consequently, (38) has a unique solution Z_s^0. We prove that this solution satisfies the equality $\sum_{s=1}^{2m} Z_s = 0$. This could be done by writing Z_s^0 explicitly by the known properties of circulants. However, we proceed differently. We add up all the equations of system (38):

$$\sum_{\nu=0}^{2m-1}\sum_{j=1}^{2m} Z_j c_{j-\nu} = \sum_{\nu=0}^{2m-1}\left(y_\nu - \frac{1}{2m}\sum_{s=0}^{2m-1} y_s\right) = 0.$$

Since, by (39), $\lambda_{2m} = \sum_{s=0}^{2m-1} c_s$, it follows that

$$0 = \sum_{\nu=0}^{2m-1}\sum_{j=1}^{2m} Z_j c_{j-\nu} = \sum_{j=1}^{2m} Z_j \sum_{\nu=0}^{2m-1} c_{j-\nu} = \sum_{j=1}^{2m} Z_j \sum_{s=0}^{2m-1} c_s = \lambda_{2m}\sum_{j=1}^{2m} Z_j.$$

As already proved, $\lambda_{2m} \neq 0$. Therefore $\sum_{j=1}^{2m} Z_j = 0$, i.e. any solution of system (38) automatically satisfies the last equality.

It is easily verified that the solution of (38), and, consequently, also of (30)–(31), has the form (37) and, in particular, λ_{2m} does not appear in the formula for the solution.

Theorem 2 is completely proved.

4. Width estimates

In §1, for $\alpha \in \mathbf{R}$ and $r \in \mathbf{N}$ we proved (Lemma 1) that the function

$$A(t) = \sum_{\nu=0}^{\infty}\frac{(-1)^\nu \sin((2\nu+1)mt - \alpha\pi/2)}{(2\nu+1)^r}$$

has a unique zero τ on the half-open interval $[-h/2, h/2)$. Following Dzyadyk [4], [5], let us denote by β the number satisfying the condition

$$\sum_{\nu=0}^{\infty}\frac{\cos((2\nu+1)\beta - \alpha\pi/2)}{(2\nu+1)^r} = 0.$$

Comparing the two sums, we obtain $\beta = m(r + h/2) + \pi k$, $k \in \mathbf{Z}$.

Theorem 3 (Dzyadyk [4], [5]). *Let* $\alpha \in \mathbf{R}$. *Then*

$$E_{m-1}(W_\infty^{r,\alpha})_\infty = E_{m-1}(W_1^{r,\alpha})_1 = \frac{4}{\pi}\frac{M_{r,\alpha}}{m^r},$$

where

$$M_{r,\alpha} = \sin\frac{\alpha\pi}{2}\sum_{\nu=0}^{\infty}(2\nu+1)^{-r-1} \quad \textit{for } r \in (0,1] \textit{ and } \alpha \in [r, 2-r]$$

and

$$M_{r,\alpha} = \left|\sum_{\nu=0}^{\infty}\frac{\sin((2\nu+1)\beta - \alpha\pi/2)}{(2\nu+1)^{r+1}}\right|$$

for $r \in (0,1]$ *and* $\alpha \in [0,2]\backslash[r, 2-r]$, *and for* $r > 1$ *and any* $\alpha \in \mathbf{R}$.

Theorem 4. *Let* $\alpha \in \mathbf{R}$, $r \in \mathbf{N}$, $m \in \mathbf{N}$, *and let* p *be either* ∞ *or* 1. *Then*

(a) $d_{2m-1}(W_p^{r,\alpha})_p \geq 4M_{r,\alpha}/\pi m^r$,

(b) $d_{2m}(W_p^{r,\alpha})_p \geq 4M_{r,\alpha}/\pi m^r$.

Theorems 3 and 4 imply the equalities

$$d_{2m-1}(W_p^{r,\alpha})_p = d_{2m}(W_p^{r,\alpha})_p = \frac{4}{\pi}\frac{M_{r,\alpha}}{m^r} \tag{44}$$
$$(r \in \mathbf{N};\ \alpha \in \mathbf{R};\ m \in \mathbf{N};\ p = 1, \infty).$$

For $\alpha = r \in \mathbf{N}$ Theorem 4 was proved by Tikhomirov [7] ($p = \infty$), by Subbotin [26] ($p = 1$, part (a)) and by Ruban [10] ($p = 1$, part (b)). For $r = 1$ and any α the estimate for the odd width in Theorem 4 was obtained by Babenko [18]. For $\alpha + 1 = r \in \mathbf{N}$, $p = \infty$ and certain interrelations between r and m Theorem 4 was proved by the author in 1988 (see [19] and [38]). In the present paper we show how this theorem can be proved for $p = \infty$, any $\alpha \in \mathbf{R}$ and any $r, m \in \mathbf{N}$ satisfying the condition $m \leq C^{-1}(\alpha)(r+1)$, where $C(\alpha) = 4.000206$ for $\alpha \in \mathbf{N}$ and $C(\alpha) = 4.035314$ for $\alpha \notin \mathbf{N}$. Since Nguyễn Thị Thiêu Hoa [37] recently proved Theorem 4 by a different method in the general case, i.e., without restrictions on the numbers $r, m \in \mathbf{N}$, we present here only an outline of our proof. We note that the complete proof was presented in author's preprint [36].

Outline of the proof of Theorem 4 *for* $m \leq C^{-1}(\alpha)(r+1)$. Let $\eta_j = \eta_j(\alpha)$ be the sequence of numbers defined by (26), and let $r \in \mathbf{N}$. The following lemma plays the central role.

Lemma 6. *All numbers* η_j $(j = 1, \dots, 2m)$ *are real, and for* $r + 1 \geq C(\alpha)m$ *and for any* r *if* $m = 1$

$$\operatorname{sgn}\eta_j = (-1)^{j-1}\operatorname{sgn}\lambda_m, \qquad j = 1, \dots, 2m,$$

where

$$\lambda_m = \frac{4}{\pi m^r}\sum_{l=0}^{\infty}\frac{(-1)^l\cos\left(m\tau(1+2l) - \frac{\alpha\pi}{2}\right)}{(1+2l)^{r+1}}.$$

That the signs of the numbers η_j $(j = 1, \dots, 2m)$ in Lemma 6 alternate for $\alpha = r \in \mathbf{N}$ follows from Subbotin's paper [28] (another proof was proposed by Kushpel′ [13]). Calculations performed on a BÈSM-6 computer show that for $\alpha \neq r$ and $r+1 < C(\alpha)m$ the numbers η_j do not always alternate in sign. Example: $r = 1$, $m = 3$, $\alpha = r - 1$. In this case $\lambda_3 > 0$, $\eta_3 < 0$ and, consequently, the equality $\operatorname{sgn}\eta_j = (-1)^{j-1}\operatorname{sgn}\lambda_m$ does not hold for all $j = 1, \dots, 2m$.

From (29) for the parameters Z_s of the interpolation SK-spline S corresponding to the kernel $K_{r,\alpha}$ and satisfying the condition

$$S(\tau+\nu h+h/2)=y_\nu\,,\qquad \nu\in\mathbf{Z},$$

we have the relations

$$Z_s=\sum_{j=0}^{2m-1} y_j\eta_{j-s+2}\,,\qquad s=1\,,\,\dots\,,\,2m.$$

Hence, using Lemma 6, we easily obtain

$$|Z_s|\le|\lambda_m|^{-1}\max_j|S(\tau+(j+1/2)h)|. \tag{45}$$

Therefore, by Tikhomirov's theorem on the width of a sphere [33],

$$d_{2m-1}(W_\infty^{r,\alpha})_\infty\ge|\lambda_m|\,, \tag{46}$$

and in view of the equality $\beta=m(\tau+h/2)+\pi k$, $k\in\mathbf{Z}$, we deduce that

$$|\lambda_m|=4M_{r,\alpha}/\pi m^r.$$

A lower estimate for the even width d_{2m} is carried out by Tikhomirov's well-known method [7].

Theorem 5. *Let $\omega=\omega(\delta)$ be a convex modulus of continuity, $r, m\in\mathbf{N}$, and let $m\le(6/13\pi)(r+1)$ for r even and $m\le(1/2\pi)(r+1)$ for r odd. Then*

$$d_{2m}\left(\widetilde{W^rH^\omega}\right)_\infty\ge\frac{2}{\pi m^{r-1}}\int_0^{h/2}\omega(h-2t)\sum_{s=0}^{\infty}\frac{(-1)^{s(r-1)}\cos mt(1+2s)}{(1+2s)^r}\,dt.$$

Proof. Let $\eta_j=\eta_{j,\omega,r}$ be the sequence of numbers

$$\eta_j=\eta_{j,\omega,r}=\frac{1}{2m}\sum_{\nu=1}^{2m-1}\lambda_\nu^{-1}e^{-i\nu h(j-1)}\,,\qquad j=1\,,\,\dots\,,\,2m\,,$$

where

$$\begin{aligned}\lambda_\nu&=\lambda_{\nu,\omega,r}\\&=\frac{(-1)^l}{hi}e^{-i\nu h/2}\int_{-h/2}^{h/2}\Omega\left(\frac{h}{2}-t\right)\left[\sum_{s=0}^{\infty}\frac{(-1)^se^{i(2ms+\nu)t}}{(2ms+\nu)^{2l}}-\sum_{s=1}^{\infty}\frac{(-1)^se^{-i(2ms-\nu)t}}{(2ms-\nu)^{2l}}\right]dt\end{aligned}$$

for r even, $r=2l$, and

$$\lambda_\nu=\lambda_{\nu,\omega,r}=\frac{(-1)^l}{h}\int_{-h/2}^{h/2}\Omega\left(\frac{h}{2}-t\right)\left[\sum_{s=0}^{\infty}\frac{e^{i(2ms+\nu)t}}{(2ms+\nu)^{2l-1}}+\sum_{s=1}^{\infty}\frac{e^{-i(2ms-\nu)t}}{(2ms-\nu)^{2l-1}}\right]dt$$

for r odd, $r=2l-1$ (see the equalities (36) and (40)), and

$$\Omega(t)=\begin{cases}\frac12(\omega)(2t)\,, & 0\le t\le h/2\,,\\ \frac12\omega(2h-2t)\,, & h/2\le t\le h.\end{cases}$$

The next lemma is similar to Lemma 6.

Lemma 7. *Let $\omega=\omega(\delta)$ be a given convex modulus of continuity. Let also*

$$l\ge 1.000472\cdot\pi m-0.5\,,$$

for even $r=2l$, $m\ge2$, and

$$l\ge\pi m$$

for odd $r = 2l-1$, $m \geq 2$. *Then, and also for* $m = 1$ *and any* r,

$$\operatorname{sgn} \eta_j = (-1)^{j-1} \operatorname{sgn} \lambda_m, \qquad j = 1, \ldots, 2m.$$

Proof. We have

$$\eta_j = \frac{1}{2m}\left(\frac{(-1)^{j-1}}{\lambda_m} + G\right), \tag{47}$$

where

$$G = \sum_{\nu=1}^{m-1} [(\lambda_\nu e^{i\nu h(j-1)})^{-1} + (\lambda_{2m-\nu} e^{-i\nu h(j-1)})^{-1}].$$

For $m-1$ the lemma is obvious. Let us show that for $m \geq 2$ all η_j are real and that the sign of η_j in (47) is decided by the first term. To this end we estimate $|\lambda_m|^{-1}$ from below and $|G|$ from above, and compare the results.

First, we consider the case $r = 2l$. Then

$$\lambda_m = \frac{4(-1)^{l+1}}{\pi m^{2l-1}} \int_0^{h/2} \Omega\left(\frac{h}{2} - t\right) \sum_{s=0}^{\infty} \frac{(-1)^s \cos mt(1+2s)}{(1+2s)^{2l}} dt.$$

From this, using the fact that $2l + 1 \geq 7$, by the condition of Lemma 7 and by Lemma 3 we obtain

$$|\lambda_m|^{-1} \geq \frac{\pi m^{2l}}{4\Omega(h/2)} \left(\sum_{s=0}^{\infty} (2s+1)^{-2l-1}\right)^{-1} > \frac{\pi m^{2l}}{4\Omega(h/2) \cdot 1.000472}. \tag{48}$$

Using the evenness of the function $\Omega(h/2 - t)$, and the equalities (2) and (11) for $\varphi = 0$, we deduce that

$$G = 2(-1)^l \sum_{\nu=1}^{m-1} \sin \nu h \left(j - \frac{3}{2}\right) \left(\frac{1}{h} \int_{-h/2}^{h/2} \Omega\left(\frac{h}{2} - t\right) b_{\nu, 2l-1}(t)\, dt\right)^{-1}. \tag{49}$$

By the convexity of $\omega(\delta)$ we have $\Omega(h/2 - t) \geq \Omega(h/2)(1 - 2t/h)$ for $0 \leq t \leq h/2$. Integrating by parts, for $\nu = 1, \ldots, m-1$ we obtain

$$|\lambda_\nu| \geq \frac{1}{h} \int_{-h/2}^{h/2} \Omega\left(\frac{h}{2} - t\right) b_{\nu, 2l-1}(t)\, dt \geq \frac{2}{h} \int_0^{h/2} \Omega\left(\frac{h}{2}\right) \left(1 - \frac{2t}{h}\right) b_{\nu, 2l-1}(t)\, dt$$

$$= \frac{4}{h^2} \Omega\left(\frac{h}{2}\right) \left[\sum_{s=0}^{\infty} \frac{(-1)^s - \cos(\nu h/2)}{(2ms + \nu)^{2l+2}} + \sum_{s=1}^{\infty} \frac{(-1)^{s+1} + \cos(\nu h/2)}{(2ms - \nu)^{2l+2}}\right]$$

$$\geq \frac{8}{h^2} \Omega\left(\frac{h}{2}\right) \frac{\sin^2(\nu h/4)}{\nu^{2l+2}} > \frac{8((2\sqrt{2}/\pi)(\nu h/4))^2}{h^2 \nu^{2l+2}} \Omega\left(\frac{h}{2}\right) = \frac{4}{\pi^2 \nu^{2l}} \Omega\left(\frac{h}{2}\right).$$

In deriving the last inequality we have used the fact that $\sin x > (2\sqrt{2}/\pi)x$ for $0 < x < \pi/4$. Hence and from (49) we obtain

$$|G| \leq 2 \sum_{\nu=1}^{m-1} |\lambda_\nu|^{-1} < \frac{\pi^2}{2\Omega(h/2)} \sum_{\nu=1}^{m-1} \nu^{2l} < \frac{\pi^2 m^{2l+1}}{2\Omega(h/2)(2l+1)}. \tag{50}$$

Since

$$\frac{\pi m^{2l}}{1.000472 \cdot 4 \cdot \Omega(h/2)} \geq \frac{\pi^2 m^{2l+1}}{2(2l+1)\Omega(h/2)},$$

for $l \geq 1.000472\pi m - 0.5$, we obtain the assertion of Lemma 7 for even r from the inequalities (48) and (50) and the equality (47).

Now let $r = 2l - 1$. Then

$$\lambda_m = \frac{4(-1)^l}{\pi m^{2l-2}} \int_0^{h/2} \Omega\left(\frac{h}{2} - t\right) \sum_{s=0}^{\infty} \frac{\cos mt(1+2s)}{(1+2s)^{2l-1}}\, dt,$$

whence

$$(51) \qquad |\lambda_m|^{-1} \geq \frac{\pi m^{2l-1}}{4\Omega(h/2)} \left(\sum_{s=0}^{\infty} \frac{(-1)^s}{(1+2s)^{2l}}\right)^{-1} > \frac{\pi m^{2l-1}}{4\Omega(h/2)}.$$

It is easy to see that $\lambda_{2m-\nu} = \lambda_\nu$ $(\nu = 1, \dots, m-1)$. Consequently,

$$G = 2 \sum_{\nu=1}^{m-1} \frac{\cos \nu h(j-1)}{\lambda_\nu}.$$

As in the previous case, using the convexity of $\omega(\delta)$ and Lemma 5, integrating by parts we obtain

$$|\lambda_\nu| \geq \frac{2}{h} \int_0^{h/2} \Omega\left(\frac{h}{2} - t\right) \left[\sum_{s=0}^{\infty} \frac{\cos(2ms+\nu)t}{(2ms+\nu)^{2l-1}} + \sum_{s=1}^{\infty} \frac{\cos(2ms-\nu)t}{(2ms-\nu)^{2l-1}}\right] dt$$

$$\geq \frac{4}{h^2} \Omega\left(\frac{h}{2}\right) \left[\sum_{s=0}^{\infty} \frac{1-(-1)^s \cos(\nu h/2)}{(2ms+\nu)^{2l+1}} + \sum_{s=1}^{\infty} \frac{1-(-1)^s \cos(\nu h/2)}{(2ms-\nu)^{2l+1}}\right]$$

$$> \frac{8 \sin^2(\nu h/4)}{h^2 \nu^{2l+1}} \Omega\left(\frac{h}{2}\right) > \frac{4\Omega(h/2)}{\nu^{2l-1}\pi^2}.$$

Thus,

$$(52) \qquad |G| = \left|\sum_{\nu=1}^{m-1} \frac{2\cos \nu h(j-1)}{\lambda_\nu}\right| \leq 2 \sum_{\nu=1}^{m-1} |\lambda_\nu|^{-1} < \frac{\pi^2 m^{2l}}{4l\Omega(h/2)}.$$

Since

$$\frac{\pi^2 m^{2l}}{4l\Omega(h/2)} \leq \frac{\pi m^{2l-1}}{4\Omega(h/2)},$$

for $l \geq \pi m$, from (47), (51), and (52) we obtain the assertion of Lemma 7 for r odd. Lemma 7 is completely proved.

We continue the proof of Theorem 5. It follows from Theorem 2 that for any sequence $y = \{y_\nu\}_{\nu=-\infty}^{\infty}$ satisfying the condition $y_{\nu+2m} = y_\nu$ there exists a unique conjugate ω-spline S such that

$$S(\theta_r + \nu h) = y_\nu, \qquad \nu \in \mathbf{Z},$$

where $\theta_r = 0$ for r even, and $\theta_r = h/2$ for r odd. For the parameters Z_j of the interpolation spline S we have obtained the following relations:

$$Z_j = \frac{1}{2m} \sum_{l=0}^{2m-1} y_l \sum_{\nu=1}^{2m-1} (\lambda_\nu e^{i\nu h(l-j+1)})^{-1}, \qquad j = 1, \dots, 2m,$$

where $\lambda_\nu = \lambda_{\nu,\omega,r}$. Considering the definition of the numbers $\eta_j = \eta_{j,\omega,r}$, we get

$$Z_j = \sum_{l=0}^{2m-1} y_l \eta_{l-j+2}, \qquad j = 1, \dots, 2m.$$

Using Lemma 7, we obtain

$$\begin{aligned} |Z_j| &\le \left|\sum_{s=1}^{2m} (-1)^s \eta_s\right| \max_l |S(\theta_r + lh)| \\ &= \frac{1}{2m}\left|\sum_{\nu=1}^{2m-1} \lambda_{\nu,\omega,r}^{-1} \sum_{s=1}^{2m} (-1)^s e^{-i\nu h(s-1)}\right| \max_l |S(\theta_r + lh)| \\ &= |\lambda_{m,\omega,r}|^{-1} \max_l |S(\theta_r + lh)|. \end{aligned} \tag{53}$$

From Theorem 2, in view of the (unique) solvability of the interpolation problem, it follows that the dimension of the space of 2π-periodic conjugate ω-splines is equal to $2m$. From the definition of the conjugate ω-spline S we have

$$S = C + \pi^{-1} K_{r,r-1} * \Psi,$$

where $\Psi(t + (j-1)h) = Z_j \Omega(t)$, $0 \le t \le h$, $Z_{j+2m} = Z_j$. By the definition of Ω, we now see that $\Psi \in C_{2\pi}$. From the convexity of $\omega = \omega(\delta)$ we deduce that

$$\omega(\Psi, \delta) \le \max_j |Z_j| \omega(\Omega, \delta),$$

where the function Ω is extended as an odd function (i.e., $\Omega(-t) = -\Omega(t)$) for $-h \le t \le 0$, and further periodically with period $2h$. Since $\omega(\Omega, \delta) = \omega(\delta)$ $(0 \le \delta \le h)$, the last inequality implies that

$$\omega(\Psi, \delta) \le |\lambda_{m,\omega,r}|^{-1} \max_l |S(\theta_r + lh)| \omega(\delta) \le |\lambda_{m,\omega,r}|^{-1} \|S\|_\infty \omega(\delta). \tag{54}$$

From the representations for $\lambda_{m,\omega,r}$ obtained in Lemma 7, taking account of the definition of Ω, we derive that

$$|\lambda_{m,\omega,r}| = \frac{2}{\pi m^{r-1}} \int_0^{h/2} \omega(h - 2t) \sum_{s=0}^{\infty} \frac{(-1)^{s(r-1)} \cos mt(1 + 2s)}{(1 + 2s)^r}\, dt,$$

where the sum of the series under the integral sign is nonnegative for $0 \le t \le h/2$. Thus, for any function S from the $2m$-dimensional subspace of conjugate ω-splines such that $\|S\|_\infty = |\lambda_{m,\omega,r}|$, we obtain from the inequality (54) that $\omega(\Psi, \delta) \le \omega(\delta)$, i.e., that $S \in \widetilde{W^r H^\omega}$. From this, by Tikhomirov's theorem on the width of a sphere, we obtain the following lower estimate for the odd width:

$$d_{2m-1}(\widetilde{W^r H^\omega})_\infty \ge \frac{2}{\pi m^{r-1}} \int_0^{h/2} \omega(h - 2t) \sum_{s=0}^{\infty} \frac{(-1)^{s(r-1)} \cos mt(1 + 2s)}{(1 + 2s)^r}\, dt.$$

Such a lower estimate for the even width is derived by the well-known methods of Tikhomirov. The central role is played by (54) and by Lemma 10.4.3 from [34].

BIBLIOGRAPHY

1. J. Favard, *Application de la formule sommatoire d'Euler à la démonstration de quelques propriétés extrémales des intégrales des fonctions périodiques ou presque-périodiques*, Mat. Tidsskr. B **1936**, 81–94.

2. ———, *Sur les meilleurs procédés d'approximation de certaines classes de fonctions par des polynomes trigonométriques*, Bull. Sci. Math. (2) **61** (1937), 209–224, 243–256.

3. N. Achyeser [Akhiezer] and M. Kreĭn, *Sur la meilleure approximation des fonctions périodiques dérivables au moyen de sommes trigonométriques*, C. R. (Dokl.) Acad. Sci. URSS **15** (1937), 107–111.

4. V. K. Dzyadyk, *Best approximation on classes of periodic functions defined by kernels which are integrals of absolutely monotone functions*, Izv. Akad. Nauk SSSR Ser. Mat. **23** (1959), 933–950.

5. ——, *Best approximation on classes of periodic functions that are defined by integrals of a linear combination of absolutely monotone kernels*, Mat. Zametki **16** (1974), 691–701; English transl. in Math. Notes **16** (1974).

6. V. F. Babenko, *Inequalities for rearrangements of differentiable periodic functions, approximation problems, and approximate integration*, Dokl. Akad. Nauk SSSR **272** (1983), 1038–1041; English transl. in Soviet Math. Dokl. **28** (1983).

7. V. M. Tikhomirov, *Best methods of approximation and interpolation of differentiable functions in the space* $C_{[-1,1]}$, Mat. Sb. **80** (**122**) (1969), 290–304; English transl. in Math. USSR Sb. **9** (1969).

8. A. P. Buslaev and V. M. Tikhomirov, *Some questions of nonlinear analysis and approximation theory*, Dokl. Akad. Nauk SSSR **283** (1985), 13–18; English transl. in Soviet Math. Dokl. **32** (1985).

9. N. P. Korneĭchuk, *Splines in approximation theory*, "Nauka", Moscow, 1984. (Russian)

10. ——, *Exact constants in approximation theory*, "Nauka", Moscow, 1987; English transl., Cambridge Univ. Press, Cambridge, 1991.

11. Allan Pincus, *On n-widths of periodic functions*, J. Analyse Math. **35** (1979), 209–235.

12. ——, *n-widths of Sobolev spaces in* L^p, Constr. Approx. **1** (1985), 15–62.

13. A. K. Kushpel′, *SK-splines and sharp estimates for widths of function classes in the space* $C_{2\pi}$, Preprint No. 85-51, Inst. Mat. Akad. Nauk Ukrain. SSR, Kiev, 1985. (Russian) MR **87g**:41061.

14. Nguen Thi Th′eu Khoa [Nguyễn Thị Thiêu Hoa], *Rolle's theorem for differential operators, and some extremal problems of approximation theory*, Dokl. Akad. Nauk SSSR **295** (1987), 1313–1318; English transl. in Soviet Math. Dokl. **36** (1988).

15. V. T. Shevaldin, *L-splines and widths*, Mat. Zametki **33** (1983), 735–744; English transl. in Math. Notes **33** (1983).

16. S. I. Novikov, *Widths of a class of periodic functions that is defined by a differential operator*, Mat. Zametki **42** (1987), 194–206; English transl. in Math. Notes **42** (1987).

17. A. A. Zhensykbaev, *Approximation of certain classes of differentiable periodic functions by interpolation splines of a uniform subdivision*, Mat. Zametki **15** (1974), 955–966; English transl. in Math. Notes **15** (1974).

18. V. F. Babenko, *On the widths of certain classes of convolutions*, Ukrain. Mat. Zh. **35** (1983), 603–607; English transl. in Ukrainian Math. J. **35** (1983).

19. V. T. Shevaldin, *Lower estimates for the widths of classes of sourcewise representable functions*, Trudy Mat. Inst. Steklov. **189** (1989), 185–201; English transl. in Proc. Steklov Inst. Math. **1990**, no. 4 (189).

20. ——, *Some problems of extremal interpolation in the mean for linear differential operators*, Trudy Mat. Inst. Steklov. **164** (1983), 203–240; English transl. in Proc. Steklov Inst. Math. **1985**, no. 2 (164).

21. S. I. Novikov and V. T. Shevaldin, *Lower bound for the even width of a class of periodic functions that is defined by a differential operator and a modulus of continuity*, Approximation in Concrete and Abstract Banach Spaces (Yu. N. Subbotin, editor), Ural. Nauchn. Tsentr Akad. Nauk SSSR, Sverdlovsk, 1987, pp. 107–112. (Russian)

22. V. F. Babenko, *Approximation of function classes defined by means of a modulus of continuity*, Dokl. Akad. Nauk SSSR **298** (1988), 1296–1299; English transl. in Soviet Math. Dokl. **37** (1988).

23. I. Feschiev and M. Khemeamin, *Upper bounds for the best approximations by trigonometric polynomials of the classes* $\widetilde{W^rH^\omega}$ *in the space* C, Plovdiv. Univ. Nauchn. Trud. **22** (1984), 45–67 (1985). (Russian)

24. ——, *Upper bounds for the best approximations by trigonometric polynomials of the classes* $\widetilde{W^rH^\omega}$ *in the space* L, Plovdiv. Univ. Nauchn. Trud. **22** (1984), 105–123 (1985). (Russian)

25. L. V. Taĭkov, *Best approximation in the mean of certain classes of analytic functions*, Mat. Zametki **1** (1967), 155–162; English transl. in Math. Notes **1** (1967).

26. Yu. N. Subbotin, *Approximation by spline functions and estimates of widths*, Trudy Mat. Inst. Steklov. **109** (1971), 35–60; English transl. in Proc. Steklov Inst. Math. **109** (1971).

27. S. B. Stechkin and Yu. N. Subbotin, *Splines in computational mathematics*, "Nauka", Moscow, 1976. (Russian)

28. Yu. N. Subbotin, *On the relations between finite differences and the corresponding derivatives*, Trudy Mat. Inst. Steklov. **78** (1965), 24–42; English transl. in Proc. Steklov Inst. Math. **78** (1965).

29. J. H. Ahlberg, E. N. Nilson, and J. L. Walsh, *Best approximation and convergence properties of higher-order spline approximations*, J. Math. Mech. **14** (1965), 231–243.

30. ———, *The theory of splines and their applications*, Academic Press, New York, 1967.

31. I. K. Daugavet, *Introduction to the theory of approximation of functions*, Izdat. Leningrad. Gos. Univ., Leningrad, 1977. (Russian)

32. V. T. Shevaldin, *Interpolation periodic L-splines with uniform nodes*, Approximation of Functions by Polynomials and Splines, (Yu. N. Subbotin, editor), Ural. Nauchn. Tsentr Akad. Nauk SSSR, Sverdlovsk, 1985, pp. 140–147. (Russian)

33. V. M. Tikhomirov, *Widths of sets in function spaces and the theory of best approximation*, Uspekhi Mat. Nauk **15** (1960), no. 3 (93), 81–120; English transl. in Russian Math. Surveys **15** (1960).

34. N. P. Korneĭchuk, *Extremal problems of approximation theory*, "Nauka", Moscow, 1976. (Russian)

35. A. P. Prudnikov, Yu. A. Brychkov, and O. I. Marichev, *Integrals and series*. Vol. I: *Elementary functions*, "Nauka", Moscow, 1981; English transl., Gordon and Breach, New York, 1986.

36. V. T. Shevaldin, *Estimates from above for the widths of certain classes of periodic functions*, preprint, Ural. Nauchn. Tsentr Akad. Nauk SSSR, Sverdlovsk, 1989. (Russian)

37. Nguen Tkhi Tkh′en Khoa [Nguyễn Thị Thiêu Hoa], *Widths and quadrature formulas on periodic classes of functions that are conjugate to Sobolev classes*, Mat. Zametki **46** (1989), no. 2, 83–93; English transl. in Math. Notes **46** (1989).

38. V. T. Shevaldin, *On widths and classes of conjugate functions*, Extremal Problems, Functional Analysis, and Their Applications (P. L. Ul′yanov and V. M. Tikhomirov, editors), Izdat. Moskov. Gos. Univ., Moscow, 1988, 8–10. (Russian)

Translated by S. IVANOV